# Elastomers: From Theory to Applications

# Elastomers: From Theory to Applications

Editors

**Gert Heinrich**
**Michael Lang**

MDPI • Basel • Beijing • Wuhan • Barcelona • Belgrade • Manchester • Tokyo • Cluj • Tianjin

*Editors*
Gert Heinrich
Technische Universität Dresden
Germany

Michael Lang
Leibniz-Institut für Polymerforschung Dresden e.V.
Germany

*Editorial Office*
MDPI
St. Alban-Anlage 66
4052 Basel, Switzerland

This is a reprint of articles from the Special Issue published online in the open access journal *Polymers* (ISSN 2073-4360) (available at: https://www.mdpi.com/journal/polymers/special_issues/Elastomers_Theory_Applications).

For citation purposes, cite each article independently as indicated on the article page online and as indicated below:

LastName, A.A.; LastName, B.B.; LastName, C.C. Article Title. *Journal Name* **Year**, *Volume Number*, Page Range.

**ISBN 978-3-0365-6930-7 (Hbk)**
**ISBN 978-3-0365-6931-4 (PDF)**

© 2023 by the authors. Articles in this book are Open Access and distributed under the Creative Commons Attribution (CC BY) license, which allows users to download, copy and build upon published articles, as long as the author and publisher are properly credited, which ensures maximum dissemination and a wider impact of our publications.

The book as a whole is distributed by MDPI under the terms and conditions of the Creative Commons license CC BY-NC-ND.

# Contents

**About the Editors** .................................................................. vii

**Preface to "Elastomers: From Theory to Applications"** .......................... ix

**Sergey Panyukov**
Theory of Flexible Polymer Networks: Elasticity and Heterogeneities
Reprinted from: *Polymers* **2020**, *12*, 767, doi:10.3390/polym12040767 .................. 1

**Thridsawan Prasopdee and Wirasak Smitthipong**
Effect of Fillers on the Recovery of Rubber Foam: From Theory to Applications
Reprinted from: *Polymers* **2020**, *12*, 2745, doi:10.3390/polym12112745 ................ 29

**Umut D. Çakmak, Michael Fischlschweiger, Ingrid Graz and Zoltán Major**
Adherence Kinetics of a PDMS Gripper with Inherent Surface Tackiness
Reprinted from: *Polymers* **2020**, *12*, 2440, doi:10.3390/polym12112440 ................ 47

**Rami Bouaziz, Laurianne Truffault, Rouslan Borisov, Cristian Ovalle,**
**Lucien Laiarinandrasana, Guillaume Miquelard-Garnier and Bruno Fayolle**
Elastic Properties of Polychloroprene Rubbers in Tension and Compression during Ageing
Reprinted from: *Polymers* **2020**, *12*, 2354, doi:10.3390/polym12102354 ................ 59

**Alexander V. Agafonov, Anton S. Kraev, Alexander E. Baranchikov and Vladimir K. Ivanov**
Electrorheological Properties of Polydimethylsiloxane/ $TiO_2$-Based Composite Elastomers
Reprinted from: *Polymers* **2020**, *12*, 2137, doi:10.3390/polym12092137 ................ 73

**Jan Plagge and Manfred Klüppel**
Micromechanics of Stress-Softening and Hysteresis of Filler Reinforced Elastomers with
Applications to Thermo-Oxidative Aging
Reprinted from: *Polymers* **2020**, *12*, 1350, doi:10.3390/polym12061350 ................ 87

**Gea Prioglio, Silvia Agnelli, Lucia Conzatti, Winoj Balasooriya, Bernd Schrittesser**
**and Maurizio Galimberti**
Graphene Layers Functionalized with A *Janus* Pyrrole-Based Compound in Natural Rubber
Nanocomposites with Improved Ultimate and Fracture Properties
Reprinted from: *Polymers* **2020**, *12*, 944, doi:10.3390/polym12040944 ................ 107

**Wenbo Luo, Youjian Huang, Boyuan Yin, Xia Jiang and Xiaoling Hu**
Fatigue Life Assessment of Filled Rubber by Hysteresis Induced Self-Heating Temperature
Reprinted from: *Polymers* **2020**, *12*, 846, doi:10.3390/polym12040846 ................ 131

**Khwanchat Promhuad and Wirasak Smitthipong**
Effect of Stabilizer States (Solid Vs Liquid) on Properties of Stabilized Natural Rubber
Reprinted from: *Polymers* **2020**, *12*, 741, doi:10.3390/polym12040741 ................ 141

**Mariapaola Staropoli, Dominik Gerstner, Aurel Radulescu, Michael Sztucki, Benoit Duez,**
**Stephan Westermann, Damien Lenoble and Wim Pyckhout-Hintzen**
Decoupling the Contributions of ZnO and Silica in the Characterization of Industrially-Mixed
Filled Rubbers by Combining Small Angle Neutron and X-Ray Scattering
Reprinted from: *Polymers* **2020**, *12*, 502, doi:10.3390/polym12030502 ................ 151

**Shota Akama, Yusuke Kobayashi, Mika Kawai and Tetsu Mitsumata**
Efficient Chain Formation of Magnetic Particles in Elastomers with Limited Space
Reprinted from: *Polymers* **2020**, *12*, 290, doi:10.3390/polym12020290 ................ 167

**Zhifei Chen, Shuxin Li, Yuwei Shang, Shan Huang, Kangda Wu, Wenli Guo and Yibo Wu**
Cationic Copolymerization of Isobutylene with 4-Vinylbenzenecyclobutylene: Characteristics and Mechanisms
Reprinted from: *Polymers* **2020**, *12*, 201, doi:10.3390/polym12010201 . . . . . . . . . . . . . . . . . **177**

**Christopher G. Robertson, Sankar Raman Vaikuntam and Gert Heinrich**
A Nonequilibrium Model for Particle Networking/Jamming and Time-Dependent Dynamic Rheology of Filled Polymers
Reprinted from: *Polymers* **2020**, *12*, 190, doi:10.3390/polym12010190 . . . . . . . . . . . . . . . . . **193**

# About the Editors

**Gert Heinrich**

Gert Heinrich graduated at the University in Jena (G) in Quantum Physics in 1973. At the University of Technology (TH) Leuna-Merseburg, he finished his doctorate in 1978 in polymer network physics and his Habilitation in 1986 about the theory of polymer networks and topological constraints. In 1990, he received a position at the tire manufacturer Continental in Hanover (G) as a senior research scientist and the head of Materials Research. Heinrich continued his academic activities as lecturer at Universities of Hanover (G) and Halle/Wittenberg (G). In 2002, he was appointed as a full professor for "Polymer Materials and Rubber Technology" at the University of Technology Dresden and as director of the Institute of Polymer Materials at the Leibniz Institute of Polymer Research Dresden e.V. (IPF). Since 2017, he has been a Senior Professor. His work has been recognized by several grants and awards, e.g., the George Stafford Whitby Award for distinguished teaching and research from the Rubber Division of the American Chemical Society (ACS); the Colwyn Medal in UK for outstanding services to the rubber industry; the Carl Dietrich Harries Medal from the German Rubber Society; and the Lifetime Achievement Award from the Tire Technology International Magazine.

**Michael Lang**

Michael Lang graduated at the University in Regensburg (G) in Physics in 2001 and finished his doctorate about the formation and structure of polymer networks at the same place in 2004. After a post-doc in the lab of Michael Rubinstein (UNC North Carolina), he joined the Leibniz Institute of Polymer Research in Dresden in 2007 where he became deputy head of the theory department in 2015.

# Preface to "Elastomers: From Theory to Applications"

This Special Issue "Elastomers: From Theory to Applications" focuses on the current state of the art of elastomers, both in modern applications and from a theoretical perspective. The main characteristic of elastomer materials is the high elongation and (entropy) elasticity of these materials, and the ability to swell multiple times in a suitable solvent. The use of filled elastomers, especially of new kinds of elastomer nanocomposites, is of high interest for rubber technologies. Nanostructured adaptable gels allow for the tailoring of responsive materials or filtering systems. These materials enable widespread applications in engineering fields, ranging from modern tire technologies to medical applications and consumer goods. Elastomers are also useful in various biomaterial applications. Bio-elastomers are widely available in nature and have been shown to have specific properties that are often far superior to their synthetic counterparts. Furthermore, nowadays, additional functions play a major role in elastomeric composites, e.g., in the use of dielectric, electrorheological, or magnetorheological elastomers for smart applications in soft robotics, etc.

All elastomers share typical features, such as entropy-driven elasticity, the presence of entanglements, and topological constraints of network chain conformations. These features still offer fascinating scientific challenges in the synthesis, characterization and application, as well as for the theory of polymer networks and the cross-scale modeling of elastomeric materials, their polymeric constituents and, finally, the corresponding solids under real-life conditions.

The series of papers within this Special Issue offer the latest research in several specific sub-areas of elastomer research or review selected fields. The scope of the Special Issue encompasses the leading scientific contributions in the synthesis, characterization, and theory of elastomers. Of particular interest are new structures and functionalities incorporated into elastomers, leading to enhanced properties of crosslinked elastomeric materials, and/or to a better understanding of the structure–property relationships and application behaviour.

Thus, the present volume provides a comprehensive overview of the recent developments in the field of elastomers and will be of interest to both academic researchers and industrial professionals.

G. H. acknowledges the DFG (German Research Foundation) Project 380321452/GRK2430 for financial support.

**Gert Heinrich and Michael Lang**
*Editors*

Review

# Theory of Flexible Polymer Networks: Elasticity and Heterogeneities

Sergey Panyukov

P. N. Lebedev Physics Institute, Russian Academy of Sciences, Moscow 117924, Russia; panyukov@lpi.ru

Received: 28 February 2020; Accepted: 21 March 2020; Published: 1 April 2020

**Abstract:** A review of the main elasticity models of flexible polymer networks is presented. Classical models of phantom networks suggest that the networks have a tree-like structure. The conformations of their strands are described by the model of a combined chain, which consists of the network strand and two virtual chains attached to its ends. The distribution of lengths of virtual chains in real polydisperse networks is calculated using the results of the presented replica model of polymer networks. This model describes actual networks having strongly overlapping and interconnected loops of finite sizes. The conformations of their strands are characterized by the generalized combined chain model. The model of a sliding tube is represented, which describes the general anisotropic deformations of an entangled network in the melt. I propose a generalization of this model to describe the crossover between the entangled and phantom regimes of a swollen network. The obtained dependence of the Mooney-Rivlin parameters $C_1$ and $C_2$ on the polymer volume fraction is in agreement with experiments. The main results of the theory of heterogeneities in polymer networks are also discussed.

**Keywords:** elastomers; polymer networks; elastic modulus; loops; heterogeneities

## 1. Introduction

Polymer networks and gels belong to a unique class of materials that have the properties of both solids and liquids. Such "soft solids" are elastically deformed at macroscopic scales, while at short distances and short times the network strands experience liquid fluctuations, which are responsible for the exceptional properties of polymer networks, such as the reversibility of huge deformations (when stretched up to 3000%, see [1]).

In this work, I review the main microscopic (molecular) approaches to the description of the elasticity of flexible polymer networks and establish relationships between these approaches. Phantom networks are considered in Section 2, and networks with topological entanglements are studied in Section 3. Spatial heterogeneities developed in swollen and deformed polymer networks are investigated in Section 4.

The classical theory proposed by Flory more than half a century ago, Ref. [2] suggests that polymer networks have a tree-like structure. Although this theory takes into account the presence of loops in the network, it is assumed that they all are infinite in size. Such an approach is insufficient to describe the thermodynamics of polymer networks with loops of finite size, and explicit consideration of these loops is required. The mathematical model of such networks is based on the replica method. This approach takes into consideration that the properties of the network depend not only on the "current" conditions in which the network is placed but also on the conditions of its preparation since it is these conditions that determine the molecular structure of the formed network. The replica approach takes into account the excluded volume interaction of linear chains both at preparation and experiment conditions by generalizing the theory of the "$n \to 0$-component" field $\varphi^4$ proposed by de Gennes.

The main properties of the order parameter $\varphi$ introduced for the description of such "soft solids" are discussed in Section 2.

The conformations of the chains in the network are described by the combined chain model [3]. This model is based on the concept of virtual chains, which determine the effective elasticity of the tree-like structures in the network. In polydisperse networks, the distribution of virtual chains is also polydisperse. I show that the order parameter in the replica network model stores information about this distribution. The found solution of the replica model is used to calculate the distribution function of the lengths of virtual chains in polydisperse networks.

The structure of real networks differs significantly from the classical picture of ideal trees. Typical loops of such networks have finite dimensions and strongly overlap with each other. The impact of such loops on the conformations of the network chain is described by the extended combined chain model [4]. I discuss the connection of this model with the predictions of the replica theory, which offers an analytical description of such mutually overlapping loops. Along with typical loops, the network may contain topological defects, which are also taken into account by the replica method. The condition for the loss of elasticity by polymer networks, both due to the presence of topological defects and near the gel point is discussed.

With an increase in the length of polymer chains, the effects of topological entanglements become significant. The properties of entangled polymer liquids are significantly different from the properties of entangled soft solids—polymer networks. While non-concatenated rings in the melt are compacted into fractal loopy globules [5], internal stresses in stretched polymer networks prevent such a collapse. Entangled networks are described by the non-affine tube model that generalizes the concept of virtual chains to the case of entangled polymer systems [6]. I discuss the physics of deformation of network strands in this model in Section 3.

Anisotropic deformations of the entangled network lead to slippage of chains along the contour of the entangled tube, which is described by the slip tube model [3]. This model suggests the additivity of the phantom and entangled contributions to the free energy of the network. Although this additivity approximation describes well the uniaxial deformations of the network in the melt, it cannot be used to describe the crossover of the Mooney Rivlin dependence between entangled and phantom deformation regimes observed during swelling of the polymer network. I propose the generalization of the slip tube model to account for this crossover and show that the obtained concentration dependences of the Mooney-Rivlin parameters $C_1$ and $C_2$ are in agreement with the experimental data.

The irregularity of the molecular structure of polymer networks leads to the presence of spatial inhomogeneities. The main results of the theory of heterogeneities in polymer networks are discussed in Section 4.

## 2. Phantom Networks

The polymer network is obtained by crosslinking linear chains. In models of phantom polymer networks, it is assumed that in the process of thermodynamic fluctuations, polymer chains can freely "pass" through each other. The phantom approximation works well only for not too long network strands; otherwise, topological restrictions related to the mutual impermeability of polymer chains should be taken into account. In this section, I discuss the main approaches to the theoretical description of phantom networks. The elastic free energy of a network of Gaussian chains is quadratic in the deformation ratios $\lambda_\alpha = L_\alpha / L_{0\alpha}$ of the network dimensions along the principal axes of deformation $\alpha = x, y, z$ in the deformed state ($L_\alpha$) and under conditions of its preparation ($L_{0\alpha}$):

$$\mathcal{F}_{ph} = G \sum_\alpha \frac{\lambda_\alpha^2 - 1}{2} \tag{1}$$

The interaction of strand monomers is usually taken into account by imposing the incompressibility condition

$$\lambda_x \lambda_y \lambda_z = 1/\phi. \tag{2}$$

where $\phi$ is the polymer volume fraction. In the case of uniaxial deformation we have

$$\lambda_x = \phi^{-1/3}\lambda, \quad \lambda_y = \lambda_z = \phi^{-1/3}\lambda^{-1/2}. \tag{3}$$

In the affine network model, it is assumed that the ends of each network strand are pinned to a "non-fluctuating elastic background" which deforms affinely with the network surface [3]. The elastic modulus of such a network is proportional to the density $\nu$ of its strands,

$$G = kT\nu \tag{4}$$

and $kT$ is the thermal energy. Equation (4) is attractive for its simplicity. However, in real networks, the strand ends are not fixed in space—they are connected to neighboring strands through cross-links. The elastic modulus $G$ depends on the molecular structure of the network and the polydispersity of its strands. At first glance, the "bookkeeping" of the order of connection of all network strands with each other and the numbers of monomers of each of these strands seems impossible for networks of macroscopic sizes. However, this is precisely what the replica method does.

### 2.1. Replica Method

The replica method was proposed a long time ago [7] and was used later to calculate correlation functions of density fluctuations in polymer networks [8]. The main goal of this approach is to build a mathematical model that takes into account most of effects inherent to real polymer networks. Unlike widespread numerical simulations, which can also take these effects into account, the replica method is an analytical theory, which is more amenable for understanding network elasticity as a function of many different molecular features.

When the polymer network is deformed, its structure remains unchanged, formed under the conditions of its preparation. The main idea of the replica approach is to consider the polymer network in an expanded space with coordinates $\hat{\mathbf{x}} = (\mathbf{x}^{(0)}, \mathbf{x}^{(1)}, \ldots, \mathbf{x}^{(m)})$ of dimension $3(1+m)$. The replica space consists of an "initial system" under conditions of the network preparation (with coordinates $\mathbf{x}^{(0)}$) and $m$ replicas of the "final system" under experimental conditions (with coordinates $\mathbf{x}^{(k)}, k = 1, \ldots, m$), see Figure 1.

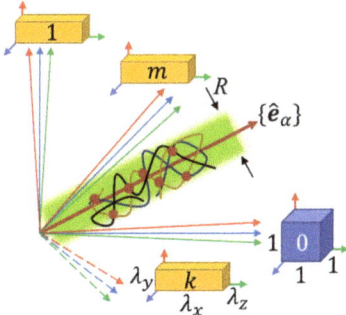

**Figure 1.** The replica system consists of the initial system ($k = 0$)—the polymer network at preparation conditions (blue cube) and $m$ identical replicas of the final system ($k = 1, \ldots, m$)—the network, deformed by factors $\lambda_\alpha$ along the principal axes $\alpha = x, y, z$ of deformation (yellow cuboids). The polymer network in the replica space of dimension $3(1+m)$ is mainly localized inside the (green) cylinder with the diameter $R \simeq a\tilde{N}^{1/2}$ directed along unit vectors $\{\hat{\mathbf{e}}_\alpha\}$, corresponding to affine deformation of the polymer. The networks (and all their monomers and strands) in the initial system and any of the $m$ replicas of the final system are the projections of the network in the replica space onto the corresponding subspaces $k = 0, \ldots, m$.

In each of these systems, the polymer network with all of its monomers and bonds is the projection of the network in the replica space onto the corresponding subspaces. Therefore, in all these systems, the network structure is exactly the same, while the conformations of network strands can vary significantly. The free energy $F$ of the final system (a deformed network) is expressed through the analytic continuation to $m = 0$ of the free energy $F_m$ of the replica system [8,9]:

$$F = \lim_{m \to 0} \frac{dF_m}{dm}. \tag{5}$$

## 2.2. Liquid-Solid Order Parameter

According to de Gennes, polymer networks are "soft solids", which emphasizes the presence of liquid and solid-state degrees of freedom interacting with each other. In the Landau approach, solids are described by an order parameter—a number (in general case, a tensor), equal to zero for a liquid and nonzero only in the solid phase. Unlike ordinary low molecular weight solids, the properties of polymer networks depend on the characteristics of both the initial and final systems. Therefore, the order parameter for polymer networks is not a number, but a function of the difference of coordinates $x_\alpha^{(k)} - \lambda_\alpha x_\alpha^{(0)}$ of final and initial systems. This dependence reflects the dual nature of polymer networks: At scales larger than the characteristic size of the network cycles, the network deforms affinely as an ordinary solid, see Figure 1. On a smaller spatial scale, the network chains fluctuate in space and have conformations similar to the chains in a polymer liquid. The unique properties of polymer networks are determined by the interaction of their solid-state and liquid degrees of freedom [10].

In the liquid phase, the order parameter is independent of coordinates. In the case of Gaussian polymer networks, the order parameter $\varphi(\varsigma)$ is a function of a single variable

$$\varsigma = \frac{1}{2}\left[\hat{x}^2 - \sum_\alpha (\hat{e}_\alpha \hat{x})^2\right], \tag{6}$$

where $\hat{e}_\alpha$ is a unit vector along the direction $x_\alpha^{(k)} = \lambda_\alpha x_\alpha^{(0)}$ of the affine deformation in the replica space. At $\varsigma = 0$ (that is, at $x_\alpha^{(k)} = \lambda_\alpha x_\alpha^{(0)}$), the order parameter $\varphi(0)$ is determined only by the conditions of the network preparation and does not depend on the experimental conditions. This last dependence is encrypted in the dimensionless order parameter $\chi(t) = \varphi(\varsigma)/\varphi(0)$ which is the function of the dimensionless variable $t = \varsigma/R$. In the case of well-developed networks obtained by crosslinking polymer chains with $f$-functional monomers far beyond the gel point, the characteristic radius is $R \simeq b\tilde{N}^{1/2}$, see Figure 1. Here $b$ is the monomer size and $\tilde{N}$ is the average number of network strand monomers.

In the case of monodisperse networks, the function $\chi(t)$ was calculated by Edwards (see Equation (3.19) of Ref. [11]):

$$\chi(t) = e^{-\frac{f-2}{f-1}t} \tag{7}$$

In the case of polydisperse networks, this function $\chi(t)$ is determined by the differential equation (see Appendix A)

$$t\chi''(t) = \chi(t) - \chi^{f-1}(t) \tag{8}$$

For large $t \gg 1$, the function $\chi(t)$ decreases more slowly compared to the monodisperse case (7), as a stretched exponent

$$\chi(t) \sim t^{1/4} e^{-2t^{1/2}} \tag{9}$$

For small $t$, the solution of Equation (8) can be approximated by a power function

$$\chi(t) \simeq (1 + ct)^{-k} \tag{10}$$

Substituting it in Equation (8) and equating to zero the first two expansion coefficients ($t$ and $t^2$) in powers of $t$, we find

$$k = \frac{3}{f-2}, \quad c = \frac{(f-2)^2}{f+1} \tag{11}$$

The replica method is also called "replica trick" because of the hidden meaning of its mathematical constructions. As will be shown below, Equation (8) describes the relationship between distribution functions of the molecular trees that characterize the structure of the polymer network. Note that Equation (8) is obtained by estimating the Hamiltonian (A2) in the Appendix A using the steepest descent method. Therefore, it corresponds to the mean-field approximation, which works due to the presence of a large number of overlapping network strands.

### 2.3. Overlap Parameter

Like in liquid polymer systems, in a typical network there are many overlapping network strands. An important characteristic of the network is the overlap parameter, which is defined as the number of network strands within the volume $R^3$ pervaded by one network strand. In the case of a solution of chains with monomer density $\rho^{(0)}$, $P^{(0)} \simeq \rho^{(0)} R^3 / N$. In Gaussian networks, the size of the network strand with Kuhn length $b$ is $R \simeq bN^{1/2}$, and

$$P^{(0)} \simeq \rho^{(0)} b^3 N^{1/2} \tag{12}$$

with a large parameter $P^{(0)}$, polymer networks obtained far from the gel point can be considered as $P^{(0)}$ overlapping yet independent "elementary" networks, the strands of which consist of $N$ monomers. There is on the order of one network strand per volume $R^3$ in the "elementary" network, and its elastic modulus is estimated as $G_{elem} \simeq kT/R^3$, $kT$ is the thermal energy. Since the real network consists of $P^{(0)}$ overlapping elementary networks, its elastic modulus is $P^{(0)}$-times larger [12]:

$$G = P^{(0)} G_{elem} \simeq kT\rho^{(0)}/N \tag{13}$$

The polymer network can also be represented as $P^{(0)}$ polymer trees (or "layers") overlapping with each other. The perfect network model assumes an infinite number of layers, and also that these trees have an infinite number of generations. In real networks $P^{(0)} < P_e$ is finite and not too large (the entanglement overlap parameter $P_e \sim 20$–30, see Section 3 below), and the different root trees are interconnected by loops. The number of network strands forming a closed loop of minimum length, which binds together different layers of the network, depends logarithmically on the overlap parameter [4],

$$l \approx \frac{1}{f-1} \ln P^{(0)} \tag{14}$$

The case $P^{(0)} \approx 1$ with the loop size $l \approx 1$ describes the "elementary networks" in Equation (13). In accordance with the results of numerical simulations [13], the loop length $l$ decreases with dilution and with increasing functionality $f$ of cross-links. Note that the minimal size loops do not directly determine the network elasticity since each elastically effective chain is simultaneously part of a large number of loops. The cumulative effect of these typical loops of the network is self-averaged, and can be described in the effective mean-field approximation, see Equation (37) below.

The overlap parameter also determines the density fluctuation of the network obtained in the case of incomplete conversion above the gel point, at $p > p_c$. In the mean-field approximation, the density of gel monomers is

$$\rho_g^{(0)} \simeq (p - p_c) \rho^{(0)}, \tag{15}$$

and its connectivity radius is estimated as $\xi \simeq b\left[N/\left(p-p_c\right)\right]^{1/2}$. Fluctuations in the number of monomers of the polymer network on the scale of the connectivity radius occur by attaching or detaching sol molecules with a characteristic number of monomers

$$L \simeq N/\left(p-p_c\right)^2 \tag{16}$$

Therefore, relative fluctuations in the gel density are estimated as

$$\frac{\delta \rho_g^{(0)}}{\rho_g^{(0)}} \simeq \frac{L}{R_{bb}^3 \rho_g^{(0)}} \simeq \frac{1}{\left(p-p_c\right)^{3/2} P^{(0)}}, \tag{17}$$

In the case of a large overlap parameter, these fluctuations are small outside the narrow Ginzburg region,

$$p - p_c \gg \left(P^{(0)}\right)^{-2/3} \tag{18}$$

### 2.4. Combined Chain Model: Monodisperse Networks

The elasticity of the polymer network has an entropic origin and is determined by the change in strand conformations during the network deformation. The conformations of a strand depend on its interaction with the elastic environment, which can be described by virtual chains attached to the ends of this strand. The network strand together with the two virtual chains is called a combined chain. The ends of this combined chain are attached to an affine deformable non-fluctuating background, see Figure 2b.

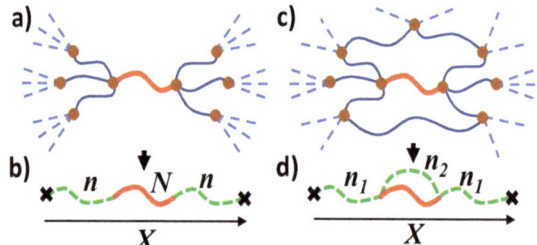

**Figure 2.** (**a**) The model of perfect network with tree-like connection of network strands (solid curves) and infinite size loops (dotted lines). (**b**) A network strand with $N$ monomers can be considered as a part of the combined chain, connected through two effective chains with $n$ monomers (dashed green curves) to the non-fluctuating background. (**c**) Actual network with finite size loops. (**d**) The combined chain consists of three effective chains. The filled circles and crosses indicate cross-links and nonfluctuating background, respectively. The end-to-end vector of the combined chain **X** deforms affinely with network deformation.

In the perfect network model, it is assumed that the network has a tree structure, whereas all its cycles are of infinite size, see Figure 2a. Each of the trees attached to the ends of the network strand is modeled by a virtual chain. The effective number $n$ of its monomers is related to the numbers $n_i$ of virtual chain monomers on the next generation of the tree $i = 1, \ldots, f-1$ and the corresponding numbers of strand monomers $N_i$ as

$$\frac{1}{n} = \sum_{i=1}^{f-1} \frac{1}{m_i}, \qquad m_i = N_i + n_i \tag{19}$$

In the case of a monodisperse network with a fixed number $N_i = N$ of monomers of its strands, all $n_i = n$ and the solution of Equation (19) determines the number of monomers of the virtual chain [14]

$$n = N/(f-2) \qquad (20)$$

The average vector **R** between the ends of the network strand is determined from the force balance condition. In the case of a monodisperse network,

$$\mathbf{R} = \frac{\mathbf{X}}{1 + 2n/N} \qquad (21)$$

where $n$ is the number of monomers of the virtual chain, Equation (20). The mean square fluctuation of the vector **R** is determined by the parallel connection of the real chain and two serially-connected virtual chains

$$\left\langle (\Delta \mathbf{R})^2 \right\rangle = \frac{b^2}{1/N + 1/(2n)} = \frac{2}{f} b^2 N \qquad (22)$$

The vector **X** between the ends of the combined chain is deformed affinely with macroscopic network deformation. The elastic modulus of the perfect network with monomer density $\rho^{(0)}$ and chain concentration $\rho^{(0)}/N$ is

$$G = \frac{kT\rho^{(0)}/N}{1 + 2n_1/N} = kT\nu \left(1 - \frac{2}{f}\right) \qquad (23)$$

Here $\nu = \rho^{(0)}/N$ is the concentration of strands and $(1 + 2n_1/N)^{-1}$ is the strand fraction of the combined chain. A comparison of expressions (23) and (4) shows that the elastic modulus of the perfect network is less than the modulus of the affine network by a factor of $1 - 2/f$. It is generally accepted that the $2/f$ factor is associated with the presence of fluctuations in the strand size, which are also proportional to this factor (see Equation (22)). In fact, fluctuations have nothing to do with it [15]: an exact elastic modulus of Gaussian phantom network is different from that of the perfect network model and coincides with the elastic modulus of the classical non-fluctuating grid, in which each strand is replaced by a corresponding elastic thread with the same elastic stiffness coefficient [16]. This coincidence is due to affine deformation of the average distances between the cross-links in such networks.

Equation (23) can be recast in more general form with the network modulus proportional to the difference of the number densities of network strands $\nu$ and cross-links $\mu = 2\nu/f$, since there are $f/2$ strands per crosslink [14]:

$$G = kT(\mu - \nu) \qquad (24)$$

This expression is quite universal; with general densities $\nu$ and $\mu$, it describes polydisperse networks, as well as networks with incomplete conversion, prepared above the gel point far from the Ginzburg region defined by Equation (18). In general, $\mu - \nu$ makes a sense of the cyclic rank of the network of unit volume. To define it, mentally cut one of the strands so that the network does not break into two disconnected parts. The cyclic rank is equal to the maximum number of such cuts, and it makes sense the number of independent network loops. An important limitation of expression (24) is the assumption of a model of perfect networks about the infinite sizes of all of the network cycles.

In real networks, not all loops transmit stress in the network. For example, primary loops, connected to the network at only one crosslink, are not capable of permanently supporting a stress. To exclude the contribution of such elastically ineffective loops and dangling chain ends, it was proposed to account in Equation (24) only for the elastically effective strands and crosslinks. Elastically effective strands deform and store elastic energy upon network deformation [17]. Elastically effective crosslinks connect at least two elastically effective strands [18]. The elastic modulus of real networks can significantly differ from such a modified expression (24), since each of finite size loops is characterized by its "elastic effectiveness", which depends on the type of the loop. The replica method provides a universal tool for accounting for such effects, see Equation (37) below.

## 2.5. Combined Chain Model: Polydisperse Networks

Real networks are polydisperse, and therefore, the numbers of monomers $n$ of virtual chains are random variables. Information on the distribution of the lengths of virtual chains is important for studying the distribution of tension in real chains of the network. Below I calculate the distribution function $p(n)$ of virtual chain lengths. The distribution of inverse variables $s = 1/n$ is described by the function

$$q(s) = \frac{1}{s^2} p\left(\frac{1}{s}\right), \quad s = \frac{1}{n} \tag{25}$$

Appendix B provides an algorithm for calculating this distribution for an arbitrary distribution $P(N)$ of the number of monomers $N$ of the real network strands. The problem is reduced to solving a system of nonlinear integral equations. In the most interesting case of the exponential distribution $P(N) = e^{-N/\tilde{N}}/\tilde{N}$, the asymptotic behavior $n \gg \tilde{N}$ of the solution of these equations can be found

$$p(n) \sim e^{-(f-1)^2 n/\tilde{N}} \tag{26}$$

Of course, knowledge of only asymptotic is not enough to describe conformations of a chain in the polydisperse network. The replica method offers a much more constructive approach, allowing us to calculate this function over the entire range of $n$ values. One can show, generalyzing calculations of Ref. [9] that the distribution function $q(s)$ in Equation (25) is determined by inverse Laplace transform of the function $\chi^{f-1}(t)$ with the dimensionless order parameter $\chi(t) = \varphi(\varsigma)/\varphi(0)$,

$$\int q(s) e^{-s\tilde{N}t} ds = \chi^{f-1}(t) \tag{27}$$

Due to the boundary condition $\chi(0) = 1$ for the function $\chi(t)$, the distribution function $q(s)$ is normalized to unity. To better understand the meaning of Equation (27), we substitute the exponential function $\chi(t)$ for the monodisperse networks, Equation (7), into this equation. The result is a $\delta$-functional distribution,

$$q(s) = \delta[s - (f-2)/N], \tag{28}$$

in accordance with expression (20) for the number of monomers $n = 1/s$ of virtual chains.

In the case of polydisperse networks, this correspondence, Equation (27), establishes a connection between the standard method of distribution functions and the replica approach. Since $\chi^{f-1}(t)$ is the Laplace transform of the convolution function $\tilde{\chi}^{(f-1)}(t)$ (see Appendix B), the function $\chi(t)$ is the Laplace transform of the distribution function $\tilde{\chi}(s')$ defined by Equation (A7). Equation (8) for this function corresponds to Equation (A8) of the replica approach for the exponential distribution of the numbers of monomers of real strands.

Substituting the asymptotic expression (9) for the function $\chi(t)$, into Equation (27), we find a more accurate asymptotic expression for the distribution function

$$p(n) \simeq \frac{1}{\tilde{N}} \left(\frac{n}{\tilde{N}}\right)^{f/2-1} e^{-(f-1)^2 n/\tilde{N}}, \quad n \gg \frac{\tilde{N}}{(f-2)^2} \tag{29}$$

Using the function $\chi(t)$ from Equation (10), we get

$$p(n) \simeq \frac{c}{\tilde{N}\Gamma[(f-1)k]} \left(\frac{\tilde{N}}{cn}\right)^{k(f-1)+1} e^{-\frac{\tilde{N}}{cn}} \tag{30}$$

Thus, the distribution $p(n)$ of virtual chains vanishes with all its derivatives as $n \to 0$ and decreases exponentially on the scale $\tilde{N}/(f-2)^2$ for large $n$. Calculating the average number of monomers of the virtual chain with the distribution (30), we find

$$\bar{n} \simeq \frac{f+1}{2f-1} \frac{\tilde{N}}{f-2} \qquad (31)$$

Note that the obtained distribution of the lengths of the virtual chains is narrower than the initial distribution of the lengths of the network strands.

### 2.6. Generalized Combined Chain Model

A perfect network assumes a tree-like structure on all spatial scales. In real networks, due to the excluded volume effect, trees cannot "grow" to infinity (the Malthus effect), since too large trees cannot fit in real 3D space. Therefore, in real networks, the size of a typical tree is finite, and the network consists of a large number of loops of finite length, see Figure 2c. They are strongly overlapped and interconnected with each other. The impact of such typical loops, which are responsible for the solid-state elasticity of the network, can be described by the generalized combined chain model.

In this model, in addition to two virtual chains of $n_1$ monomers, the combined chain includes an additional virtual chain of $n_2$ monomers, see Figure 2d. This virtual chain represents an effective elastic of all loops of finite size in the network, shunting the real strand of $N$ monomers. The average end-to-end vector $R$ of the network strand is related to the end-to-end vector $X$ of the combined chain via the force balance condition, generalizing Equation (21):

$$\mathbf{R} = \frac{\mathbf{X}}{1 + 2n_1(1/N + 1/n_2)}. \qquad (32)$$

In this model, the mean square fluctuation of the vector $\mathbf{R}$ is

$$\left\langle (\Delta \mathbf{R})^2 \right\rangle = \frac{b^2}{1/N + 1/(2n_1) + 1/n_2} = \frac{b^2}{1/N + 1/(2n)} \qquad (33)$$

The second equality can be treated as the contribution of two effective virtual chains connecting the network strand to the non-fluctuating background. They have renormalized number of monomers

$$2n = \frac{2n_1 n_2}{2n_1 + n_2}, \qquad (34)$$

corresponding to parallel connection of two effective chains: with $2n_1$ and $n_2$ monomers.

The elastic modulus of this model is

$$G = \frac{kT\rho^{(0)}/N}{1 + 2n(1/N + 1/n_2)} \frac{1}{1 + N/n_2} \qquad (35)$$

The factor $(1 + N/n_2)^{-1}$ is the fraction of energy related to the network strand when it is connected in parallel with the effective chain of $n_2$ monomers, see Figure 2d. This result can be rewritten in the form

$$G = \frac{kT\rho^{(0)}}{N + 2n} - \frac{kT\rho^{(0)}}{N + n_2}. \qquad (36)$$

where $2n$ is the number of monomers of the effective virtual chains, Equation (34). The first term in this expression has the same meaning as for the perfect network in Equation (23). The negative second contribution in Equation (36) is due to the finite size of typical loops in the network. The number of monomers of the shunt virtual chain, $n_2$, can be calculated using the replica method.

## 2.7. Finite-Size Loops of Real Networks

In real networks, there are always cyclic defects of finite dimensions. Sparse structural defects can be taken into account within the framework of the perfect network model, assuming that they are connected with the affinely deformed non-fluctuating background through the root trees. These trees can be modeled by the virtual chains [3]. Since part of the network strands is "spent" on creating the cyclic fragment itself, the length of the effective chain linking this fragment to the elastic non-fluctuating background is different from Equation (20). In polydisperse networks, the distribution function of such virtual chains is determined by Equation (27), in which $f - 1$ has the meaning of the number of branches on the first generation of this tree. In the ideal defect gas approximation, the contribution of the structural defects to the elastic modulus of a perfect network was calculated in works [19–22]. Note that typical network loops are ignored in this approximation, which takes into account only explicitly treated loops of small concentration.

In real networks, there are both loops, binding different network layers (see Equation (14)) and topological defects, which are not sparsely distributed. The replica method allows calculating the effect of both typical loops and cyclic defects of arbitrary concentration on the elasticity of the network. The elastic modulus of the network obtained by random end-linking polydisperse chains by cross-links with functionality $f$ and concentration $\rho_f^{(0)}$ and cyclic fragments with functionality $\{f_i\}$ and arbitrary concentrations $\left\{x_i \rho_f^{(0)}\right\}$ is [4]

$$G = kT\rho_f^{(0)} \frac{f/2 - 1 + \sum_i (f_i/2 - 1) x_i}{1 + \sum_i i x_i} - kT v^{(0)} \rho^{(0)} \frac{Q(0)}{2} \qquad (37)$$

The functionality $f_i$ of the cyclic defect defined as the number of "external" network strands joined to the cyclic fragment. For the ring fragment with $i$ "internal" strands and cross-links, $f_i = i(f - 2)$, since two of the functional groups of each cross-link are involved in creating the ring. In the case of incomplete conversion, $p < 1$ (but still far from the gelation threshold, $|1 - p| \ll 1$), in expression (37), instead of functionalities $f$ and $f_i$, their average values should be substituted, $pf$ and $pf_i$. Such a dependence of the network elastic modulus on the conversion is in agreement with numerical simulations of Gaussian networks, performed in Ref. [16].

The fractions $x_i$ of cyclic defects (small fragments of the network with $i \geqslant 1$ "internal" cross-links) depend on the type of reactions leading to the formation of the network [23,24]. All $x_i$ universally depend on one dimensionless parameter $x_1$ characterizing the conditions for network preparation [25]. The primary loops with concentration $x_1$ are tied to the network at only one crosslink. The denominator of the first term in expression (37) describes an increase in the effective number of monomers between cross-links due to primary loops, $\tilde{N} \to \tilde{N}(1 + x_1)$. Therefore, although primary loops cannot bear shear stress in the network, they renormalize the length of the elastically effective chains, which deform and store elastic energy upon network deformation [7,9].

For general $f > 2$, the first term in Equation (37) can be interpreted as the contribution of elastically effective network strands with renormalized number of monomers due to the presence of topological defects in the network—primary loops and cyclic fragments of finite size. Only this contribution is predicted in the classical model of phantom networks. Its expansion in a series in the parameters $x_i$ with accuracy up to first order terms reproduces the result of the so-called "network theory with strand pre-strain" [21]. At network preparation conditions, the strands of a cyclic fragment are contracted and the surrounding strands are stretched (pre-strained) with respect to Gaussian sizes of these chains [22]. These effects are "automatically" taken into account in the framework of the replica approach in Equation (37).

The last term in Equation (37) is always negative and gives an impact of a large number of typical loops of the network. The factor $Q(0)$ in the second term of expression (37) determines the probability of the formation of typical closed loops. These loops shunt elastically effective chains of the network, an effect that is taken into account by an additional effective chain in the model of the generalized combined chain in Figure 2d. Unlike perfect networks, the monomers of real networks mutually repel each other, "crowding out" the network loops on small length scales, see Figure 2. Therefore, the elasticity of a polymer network, Equation (37), explicitly depends on excluded volume parameter $v^{(0)}$, characterizing monomer interaction at the preparation conditions. This effect is ignored by the classical theory of polymer networks.

Note that the elastic modulus of the polymer network $G$ in Equation (37) vanishes at a finite fraction $x_1$ of cyclic fragments. Such a network with overlap parameter $P^{(0)} \gg 1$ has tree-like connections of overlapped cyclic fragments with a finite monomer density. To better understand this result, we compare it with the more familiar case of a disordered low-molecular-weight solid. In this case, the network fragments cannot overlap since the overlap parameter $P^{(0)} = 1$. Therefore, the elastic modulus turns to zero only at the point of percolation transition, at which the network density also vanishes.

A similar effect was observed in numerical simulations of networks obtained near the gel point, which is an analog of the percolation transition in low-molecular-weight solids. It is shown that the elastic modulus of such networks vanishes at a finite density of the network monomers [26]. One can estimate the corresponding shift in critical conversion $p_{c\mu}$ at which $G = 0$, compared with the gel point at the conversion $p_{c\gamma}$, at which the average degree of polymerization of soluble molecules diverges. For end-linked networks, the looping probability is $Q(0) \simeq (bN^{1/2})^{-3} \sim 1/(NP^{(0)})$. Dropping all the numerical factors, we find from Equation (37) the mean-field estimate for the elastic modulus

$$G \simeq kT\frac{\rho_g^{(0)}}{L} - kT\frac{\rho_g^{(0)}}{NP^{(0)}} \qquad (38)$$

Here $L \simeq N/(p - p_{c\gamma})^2$ and $\rho_g^{(0)} \simeq (p - p_{c\gamma})\rho^{(0)}$ is the density of gel monomers, see Equations (16) and (15). Therefore, $G = 0$ at $p_{c\mu} - p_{c\gamma} \sim 1/\sqrt{P^{(0)}}$, in accordance with the result of numerical simulations [26]. A more rigorous description of this effect requires the inclusion of fluctuation corrections. Inside the strongly fluctuating Ginzburg region, Equation (18), the overlap parameter of elastically effective chains is small (which is why strong density fluctuations develop in this region), and they can be described by critical exponents of percolation theory, see Reference [26].

## 3. Entangled Networks

The physics of topological entanglements is perhaps one of the most "entangled" sections of physics of polymer networks since it is impossible to give a sufficiently rigorous and constructive formalism that adequately describes entanglements for networks of macroscopic size. Therefore, the consideration of topological entanglements is usually carried out using some uncontrolled assumptions, allowing to switch from substantially "multi-chain" problem to considering the conformations of one network strand. The transfer of elastic shear stresses from individual strands of a network to its solid-state degrees of freedom (represented by the affinely deformed non-fluctuating background) is described by virtual chains. In phantom networks, such a transfer is performed only through cross-links at the strand ends. In entangled networks, the stress is carried by all entangled segments of the strand due to its topological interaction with the loops formed by adjacent network strands.

Entangled segments are characterized by the large overlap parameter

$$P_e \simeq \rho^{(0)} b^3 N_e^{1/2} \qquad (39)$$

where $N_e$ is the number of monomers of the entangled strand. In the case of flexible polymers, the parameter $P_e \simeq 20$–$30$. The number of monomers in an entanglement strand $N_e$ and the overlap parameter $P_e$ depend on the polymer chemistry [27].

As shown in [5], in the melt of non-concatenated rings the condition $P_e = const$ is satisfied at all scales down to the distance $a_0 \simeq bN_e^{1/2}$ between adjacent entanglements. Entangled loops of different sizes overlap with similar size loops at the same overlap parameter $P_e$. As a result, ring molecules are packaged into the self-similar structures called fractal loopy globules. Unlike ordinary globules, in which chains with Gaussian statistics experience random reflections from the surface of the globule, the ring chains in the fractal loopy globules are not Gaussian and form loops at all scales, starting from a size

$$a_0 \simeq bN_e^{1/2} \tag{40}$$

of the entangled segment. Ring sections with the number of monomers $g < N_e$ smaller than entanglement strands are still Gaussian with size $r \simeq bg^{1/2}$. Larger subsections of a ring are compressed with fractal dimension $d_f = 3$ and size

$$r \simeq a_0 \left(g/N_e\right)^{1/3}, \quad a_0 < r < R \tag{41}$$

up to the size $R$ of a loopy globule in a melt

$$R \simeq bN^{1/3}N_e^{1/6} \tag{42}$$

Large typical values of the parameter $N_e \gg 1$ allow developing the mean-field theory of entanglements in polymer networks, since the total effect of large number $P_e \gg 1$ of loops interacting with the entanglement segment is effectively averaged. The strands of the entangled polymer networks are confined in an effective entangled tube with a diameter $a$, which, under the conditions of the network preparation, is equal to the size $a_0$ of the entangled segment. The tube rotates by a random angle of about 90° on the scale $a_0$, see Figure 3a.

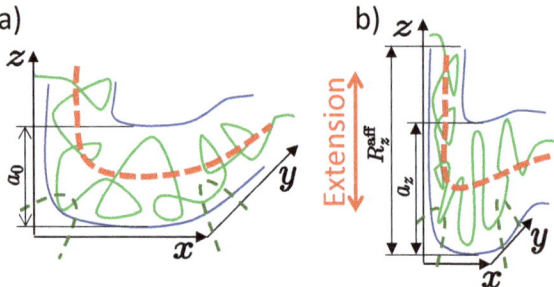

**Figure 3.** Nonaffine tube model. (**a**) In an undeformed state, the chain fluctuates in entangled tube of diameter $a_0$ (**b**) The tube diameter $a_z$ in the stretching direction is less than the affine length $R_z^{aff}$. The dashed lines show the trajectory of the primitive path, which is obtained by smoothing the affine deformed coordinates of the original Gaussian chain on the tube diameter scale $a$.

In the case of networks obtained by crosslinking a melt of linear chains, the presence of a large number of overlapping entangled chains allows using the mean-field approximation. In this approximation, topological interactions are described by virtual chains attached to all monomers ($s = 1, \ldots, N$) of the chain with coordinates $\mathbf{x}(s)$. The elastic energy of Gaussian virtual chains is

$$\sum_s \frac{3\left(\mathbf{r}(s) - \mathbf{X}(s)\right)^2}{2b^2 N_e^2} \tag{43}$$

Here $r_\alpha(s) = x_\alpha(s)/\lambda_\alpha$ and $X_\alpha(s)$ ($\alpha = x,y,z$) are undeformed coordinates of two ends of the $s$-th virtual chain. As shown in [3], with this choice of the potential of virtual chains, they do not contribute to the elastic stress of the polymer network, which is determined only by the contribution of all real polymer chains.

## 3.1. Physics of Entangled Network Deformation

Consider the case of a highly entangled network, $N/N_e \gg 1$, with many entanglements per network chain. As shown in [28], the network deforms affinely like a solid only at spatial scales exceeding the affine length $R_\alpha^{aff}$, see Figure 3. On shorter scales, chains take on liquid-like conformations. In the case of anisotropic deformation of the gel by the ratios $\lambda_\alpha$ along the axes $\alpha = x, y, z$, the affine length is also anisotropic and depends on the direction $\alpha$:

$$R_\alpha^{aff} \simeq \lambda_\alpha a_0 \qquad (44)$$

We define an affine strand of the size $R_\alpha^{aff}$ directed along the axis $\alpha$, which consists of $N_\alpha^{aff}$ monomers and deforms by the stretching ratio $\lambda_\alpha$. The number of chain entanglements does not change upon network deformation; therefore, the diameter of the deformed tube $a_\alpha$ is equal to the size of the entangled segment of $N_e$ monomers. In the stretched network this segment is a section of a longer stretched affine strand of size $R_\alpha^{aff} > a_\alpha$:

$$a_\alpha \simeq R_\alpha^{aff} \left( N_e / N_\alpha^{aff} \right) \qquad (45)$$

Fluctuations of the affine strand $b \left( N_\alpha^{aff} \right)^{1/2}$ are limited by entanglements and therefore equal to the tube diameter

$$b \left( N_\alpha^{aff} \right)^{1/2} \simeq a_\alpha \qquad (46)$$

The solution of Equations (45) and (46) is

$$N_\alpha^{aff} = \lambda_\alpha N_e \qquad (47)$$

Thus, the entanglement efficiency decreases with a network stretching, and the entanglements are not effective at $N_\alpha^{aff} > N$; such an entangled network behaves like a phantom one. The free energy of the entangled network is estimated as the elastic energy of all its stretched entangled segments,

$$\frac{\mathcal{F}_e}{V} \simeq \frac{\rho k T}{N_e} \sum_\alpha \frac{\left( R_\alpha^{aff} \right)^2}{a_\alpha^2} \simeq \frac{\rho k T}{N_e} \sum_\alpha \lambda_\alpha \qquad (48)$$

A more rigorous estimate for the free energy of the non-affine tube model with the number of monomers of the chain $N \gg N_e$ gives [29]

$$\frac{\mathcal{F}_e}{V} = \frac{\rho k T}{2 N_e} \sum_\alpha \left( \lambda_\alpha + \frac{1}{\lambda_\alpha} \right), \qquad (49)$$

This expression includes an additional to Equation (48) contribution $\sim 1/\lambda_\alpha$ describing the compression of the chain in the tube.

## 3.2. Slip-Tube Model

This model generalizes the nonaffine tube model by taking into account the slippage of chains along the tube contour. In the case of anisotropic deformation of the network, its chains are more stretched in the direction of maximum extension relative to other directions. The increased chain tension in this direction draws the chain from the tube segments along these directions. The redistribution of the chain length between different sections of the tube is taken into account by renormalization of parameters of the non-affine tube model for each direction of deformation

$$\lambda_\alpha \to \lambda_\alpha/g_\alpha^{1/2}, \qquad N \to N/g_\alpha \qquad (50)$$

The parameters $g_\alpha$ are normalized by the condition of preserving the full length of the chain,

$$\sum_\alpha g_\alpha = 3 \qquad (51)$$

and in the case of uniaxial deformation, Equation (3), there is only one independent redistribution parameter $g_z$:

$$g_x = (3 - g_z)/2 \qquad (52)$$

In addition to the above renormalization of Equation (49), the free energy of this model has an entropic contribution taking into account the chain slippage along the tube,

$$\frac{\mathcal{F}_e}{V} = \frac{\rho kT}{2N_e} \sum_\alpha \left( \frac{\lambda_\alpha}{g_\alpha^{1/2}} + \frac{g_\alpha^{1/2}}{\lambda_\alpha} \right) - \frac{\rho kT}{3N_e} \sum_\alpha \ln g_\alpha, \qquad (53)$$

The parameters $g_\alpha$ are found by minimizing the free energy (62), and the minimization equation reads as

$$h_z(g_z) = h_x(g_x), \quad h_\alpha(g_\alpha) = \frac{\partial}{\partial g_\alpha} \frac{\mathcal{F}_e}{\nu kT} \qquad (54)$$

The elastic stress of the slip-tube model is

$$\frac{\sigma_{\alpha\alpha}^e}{\nu kT} = \frac{\lambda_\alpha}{\nu kT} \frac{d\mathcal{F}_e}{d\lambda_\alpha}. \qquad (55)$$

The effect of entanglements substantially depends on the degree of network stretching, and the chains of strongly stretched networks become effectively phantom. Such a crossover from entangled to affine behavior is usually described using the assumption of additivity of the phantom and affine contributions to the free energy:

$$\mathcal{F} = \mathcal{F}_{ph} + \mathcal{F}_e \qquad (56)$$

This assumption also leads to the additivity of the corresponding contributions to the elastic tensor,

$$\sigma_{\alpha\alpha} = \sigma_{\alpha\alpha}^{ph} + \sigma_{\alpha\alpha}^e \qquad (57)$$

In the case of uniaxial network deformation, Equation (3), a convenient representation of stress is the Mooney ratio of the total stress $\sigma_{zz} - \sigma_{xx}$ to the functional dependence $\lambda^2 - \lambda^{-1}$ predicted by phantom network models:

$$f^*\left(\lambda^{-1}\right) = \frac{\sigma_{zz} - \sigma_{xx}}{\lambda^2 - \lambda^{-1}} \qquad (58)$$

As was shown in Ref. [3], the Mooney plot of the experimental data is well described by the slip tube model.

*3.3. Crossover from Unentangled to Entangled Regimes*

The Mooney-Rivlin dependence (58) is usually fitted by the phenomenological equation

$$f^*\left(\lambda^{-1}\right) \simeq 2C_1 + 2C_2/\lambda \qquad (59)$$

It is usually assumed that the parameter $2C_1$ is only due to chemical cross-links, while the parameter $2C_2$ includes all of the entanglement contributions. According to the non-affine tube model, part of the entanglement contribution is also included in the parameter $2C_1$ [6].

In this section, we study the concentration dependence of the Mooney-Rivlin coefficients. Although the free energy additivity approximation (56) can be used to describe the polymer network in the melt, a refined theory should be developed to find the Mooney dependence in the entire concentration range. The importance of such a study is emphasized by the fact that the networks prepared near the crossover from unentangled to entangled regimes are a fairly typical case.

Consider a network, consisting of $\nu$ Gaussian chains with $N$ monomers whose endpoints are displaced affinely with the global deformation of the network. We assume that each chain of this network is confined in an effective "non-affinely deformed" tube, described by the virtual chain's potential (43). The free energy of this model is calculated as the sum of contributions of all its strands and is found in Appendix C:

$$\frac{\mathcal{F}}{\nu kT} = \sum_\alpha \left( \frac{\lambda_\alpha^2 - 1}{2} \frac{k_\alpha}{\tanh k_\alpha} + \ln \frac{\sinh k_\alpha}{k_\alpha} \right), \tag{60}$$

where $k_\alpha = N/(N_e \lambda_\alpha)$. In the "phantom" limit $N \ll N_\alpha^{aff} = \lambda_\alpha N_e$ Equation (60) turns into the free energy of affine network model, Equation (4). In the limit of strongly entangled networks, $N \gg N_\alpha^{aff}$, this expression reproduces the free energy of the non-affine tube model, Equation (49). The first term in Equation (60) can be rewritten as

$$\left( \frac{\lambda_\alpha R_0}{R_{fl,\alpha}} \right)^2 \tag{61}$$

where $R_0 = bN^{1/2}$ is the chain size under network preparation conditions and $R_{fl,\alpha}$ is the fluctuation size in the $\alpha$-direction, equal to $R_0$ in the phantom regime and the tube diameter $a_\alpha$ in the entangled regime. The logarithmic term in this expression describes the entropy of the chain compression in the tube.

By renormalizing the parameters of this model (50) to take into account the effect of chain redistribution in the entangled tube, and adding the contribution of entropy of slippage along the tube (53), we arrive at the expression for free energy

$$\frac{\mathcal{F}}{\nu kT} = \sum_\alpha \left( \frac{\lambda_\alpha^2 - g_\alpha}{2} \frac{k_\alpha}{\tanh k_\alpha} + g_\alpha \ln \frac{\sinh k_\alpha}{k_\alpha} - \frac{N}{3N_e} \ln g_\alpha \right), \tag{62}$$

Solving equations (54) for this model, we find the stress (55) in the polymer network, which determines the Mooney-Rivlin dependence (58). The parameter $C_2(\phi)$ of this dependence is determined by the expression

$$2C_2(\phi) = \frac{df^*(\lambda^{-1})}{d\lambda^{-1}}. \tag{63}$$

We plot the dependences $C_1(\phi)$ and $C_2(\phi)$ (63) for $\lambda^{-1} = 0.7$ and $N/N_e = 2.4$ in Figure 4. The parameter $C_1$ of the Mooney-Rivlin dependence (59) weakly depends on the polymer volume fraction $\phi$. In accordance with the experimental data [30], the dependence of the parameter $C_2$ on $\phi$ is almost linear, and $C_2$ vanishes at $\phi \simeq 0.2$. The vanishing of $C_2$ at finite $\phi$ is related to the transition between phantom and entangled regimes. When the polymer swells, its chains move away from each other, which leads to a weakening of the effective topological potential that holds the chains in the tube. Chain fluctuations in this potential increase with the swelling and can reach the fluctuations of the phantom chain. This concentration corresponds to the disappearance of the contribution of entanglements to the network elasticity and the vanishing of the Mooney-Rivlin coefficient, $C_2 = 0$.

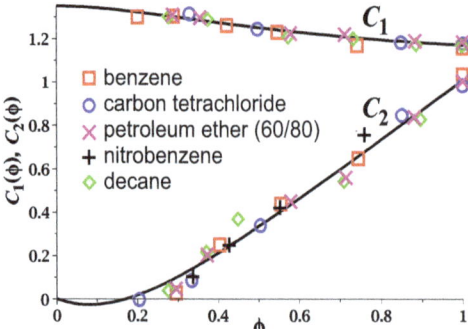

**Figure 4.** Dependence of the Mooney-Rivlin coefficients $C_1(\phi)$ (upper curve) and $C_2(\phi)$ (lower curve) on the volume fraction of polymer $\phi$: theory (solid curves) and experimental data (points, Kg/cm$^2$) for swollen rubber [30].

## 4. Heterogeneities in Polymer Networks

Network inhomogeneity is a common feature of polymer networks and gels due to the randomness of the crosslinking process and the presence of additional topological defects such as dangling chain ends, cross-linker shortcuts, and chains forming loops. The origin of nanostructural inhomogeneities and their characterization by light, neutron, and X-ray scattering as well as by NMR spectroscopy and optical, electron, and X-ray microscopies is reviewed in Ref. [31], and the main methods of their study are outlined in the review [32]. The first attempts to take such heterogeneities into account were made by using the model of randomly cross-linked networks containing fractal regions, such as percolation clusters [33]. Networks with fractal heterogeneities can be swollen and deformed by unfolding the fractal regions without significant elastic entropy penalty [34]. It is shown that strong heterogeneities lower the shear modulus of the network if a part of strands is so short to be considered rigid and not able to deform. Networks with a large number of such very short strands have higher breaking energies.

The effect of entanglements on heterogeneities in polymer networks is studied in Ref. [35] using molecular dynamics simulations of polymer networks made by either end-linking or randomly crosslinking a melt of linear precursor chains. The end-linking leads to nearly ideal monodisperse networks, while random cross-linking produces strongly polydisperse networks. It is shown that the microscopic strain response, the diameter of the entanglement tube, and stress–strain relation weakly depend on the linking process by which the networks were made.

The replica method was used in Ref. [28] to calculate the characteristic size and amplitude of the spatial nonuniformities of the network due to defects of its structure and topological restrictions. Using this method it is shown that inhomogeneities can arise as consequences of a stretching of polymer networks [36]. Although the replica theory [8] provides a complete solution of the statistical mechanics of polymer gels, it uses replica trick which is unfamiliar to the majority of people in the polymer community. A more intuitive phenomenological approach capturing all the main physical ingredients of the complete theory is developed in Ref. [37].

### 4.1. Theory of Heterogeneities in Polymer Networks

In this section, we review the main results of the theory of random spatial heterogeneities developed in swollen and deformed polymer networks [38]. The non-triviality of this problem stems from the fact that information about network structure is "encrypted" in the pattern of cross-links joining polymer chains, which represent a very small fraction of the network volume. The initial crosslinking pattern is reproduced only partially due to thermal fluctuations arising in the new equilibrium state after crosslinking. Conformations of polymer strands in such networks with fixed topological structure can be varied in a wide range depending on experimantal conditions.

The density profile of monomers in the polymer network can be recovered from the Fourier component of the deviation of the density from its average value $\rho$:

$$\rho(\mathbf{x}) = \rho + \int \tilde{\rho}(\mathbf{q}) e^{i\mathbf{q}\mathbf{x}} \frac{d\mathbf{q}}{(2\pi)^3} = \rho_{eq}(\mathbf{x}) + \delta\rho(\mathbf{x}) \tag{64}$$

Here, $\delta\rho(\mathbf{x})$ are random deviations (due to thermal fluctuations) of the density from the equilibrium density profile $\rho_{eq}(\mathbf{x})$ describing spatial inhomogeneities in polymer networks. The fluctuation contribution to the free energy of the network with a given distribution of monomer density can be represented as the sum of the entropy contribution and the (osmotic) contribution of volume interactions $\sim v$:

$$\mathcal{F}[\tilde{\rho}] = \frac{kT}{2} \int \left[ \frac{|\tilde{\rho}(\mathbf{q}) - \tilde{n}(\mathbf{q})|^2}{\tilde{g}(\mathbf{q})} + v|\tilde{\rho}(\mathbf{q})|^2 \right] \frac{d\mathbf{q}}{(2\pi)^3} \tag{65}$$

Here $\tilde{n}(\mathbf{q})$ is the density profile in the "elastic reference state" [37], maximizing the entropy of the polymer network. The density $\tilde{n}(\mathbf{q})$ is determined by the molecular structure of the network and it vanishes in the short-wavelength limit $q \gg R^{-1}$, since solid-state degrees of freedom are determined only at length scales exceeding the monomer fluctuation radius $R$. On a smaller scale, the gel has liquid degrees of freedom, contributing to the temperature structural factor

$$\tilde{g}(\mathbf{q}) = \rho N \left[ \frac{1}{Q^2/2 + (4Q^2)^{-1} + 1} + \frac{2Q^2}{(1+Q^2)^2(\lambda \cdot \mathbf{Q})^2} \right] \tag{66}$$

The first term in square brackets determines the contribution of the liquid degrees of freedom of the polymer network. The dimensionless wavevector $\mathbf{Q} = R\mathbf{q}$ is normalized by the monomer fluctuating radius, and the vector $\lambda \cdot \mathbf{Q}$ has components $(\lambda_x Q_x, \lambda_y Q_y, \lambda_z Q_z)$. At large $Q \gg 1$ the term $Q^2/2$ in the denominator of the first term of the right hand side of Equation (66) gives the usual Lifshitz entropy of polymer solutions, $\tilde{g}(\mathbf{q}) = 2\rho N/Q^2$ [39]. The term $(4Q^2)^{-1}$, first introduced by de Gennes for heteropolymer networks [40], describes the suppression of density fluctuations on length scales larger than the monomer fluctuation radius $R$. The second term in square brackets in Equation (66) determines the contribution of solid-state degrees of freedom of the polymer network. It remains finite in the long-wavelength limit and retains its angular dependence on the anisotropic deformation even in the limit $q \to 0$.

Equilibrium monomer density profile is found by minimizing the free energy (65)

$$\tilde{\rho}_{eq}(\mathbf{q}) = \frac{\tilde{n}(\mathbf{q})}{1 + v\tilde{g}(\mathbf{q})} \tag{67}$$

As can be seen from Equation (67), in good solvent with positive excluded volume $v > 0$, the equilibrium density profile is more homogeneous than that of the corresponding elastic reference state. The excluded volume interaction suppresses not only thermal fluctuations in the polymer network, but also frozen-in spatial inhomogeneities. Although the static spatial fluctuations are permanent inhomogeneities, they can reversibly increase and diverge at the spinodal line, at which $v\tilde{g}(0) = -1$. This observation is experimentally confirmed in Ref. [41].

In (ergodic) heterogeneous systems, there are two types of averages, i.e., the *thermal* or *time* *averages* and *ensemble* or *space averages*, denoted by $\langle X \rangle$ and $\overline{X}$, respectively. The Fourier component of the thermal correlator of density fluctuations $\delta\rho(\mathbf{x}) = \rho(\mathbf{x}) - \rho_{eq}(\mathbf{x})$ is found by averaging with the Gibbs weight $e^{-\mathcal{F}[\tilde{\rho}]/kT}$, Equation (65):

$$\tilde{D}(\mathbf{q}) = \left\langle |\delta\tilde{\rho}(\mathbf{q})|^2 \right\rangle = \frac{\tilde{g}(\mathbf{q})}{1 + v\tilde{g}(\mathbf{q})} \tag{68}$$

Therefore, in addition to permanent spatial heterogeneities, a polymer network undergoes thermal dynamical density fluctuations which diverge at the spinodal line.

All information about the heterogeneity of the molecular structure of the network is contained in the monomer density $\tilde{n}(\mathbf{q})$ in the "elastic reference state", which can be considered as a random variable characterized by a correlator

$$\nu(\mathbf{q}) = \overline{\tilde{n}(\mathbf{q})\tilde{n}(-\mathbf{q})} = \frac{1}{(1+Q^2)^2}\left[6\rho N + \frac{9}{\lambda_x \lambda_y \lambda_z}\tilde{S}^{(0)}(\lambda \cdot \mathbf{q})\right] \tag{69}$$

Therefore, it is important to understand the physical meaning of the different terms of this expression:

Each of two factors $(1+Q^2)^{-1}$ describes the thermal "smearing" of the density response to random stresses in the polymer network. The first term in square brackets comes from the correlator of frozen random stresses, which exist even in a homogeneous polymer network. Such stresses appear due to heterogeneities in the distribution of cross-links. The second term in brackets in Equation (69), which is proportional to the correlator of the affine deformed density pattern in the reference state $S^{(0)}(\lambda \cdot \mathbf{q})$, strongly depends on network preparation conditions.

(1) In the case of cross-linking in a melt, density fluctuations are suppressed and $\tilde{S}^{(0)}(\mathbf{q}) = 0$. The fact that $\nu(\mathbf{q})$ does not vanish upon swelling of such a network, despite the absence of density fluctuations at the preparation conditions, means that there are still inhomogeneities in the crosslink density, which manifest themselves upon swelling.

(2) In the case of instant cross-linking of a semi-dilute polymer solution, $\tilde{S}^{(0)}(\mathbf{q})$ is given by the Ornstein-Zernicke expression

$$\tilde{S}^{(0)}(\mathbf{q}) = \left\langle \left|\delta\tilde{\rho}^{(0)}(\mathbf{q})\right|^2 \right\rangle = \frac{\rho^{(0)}N}{v^{(0)}\rho^{(0)}N + Q^2/2}, \tag{70}$$

which is finite for any second virial coefficient $v_0$ at network preparation conditions.

(3) In case of equilibtium chemical cross-linking, the total structure factor of the gel in the state of preparation is determined by the expression for the polymer liquid, Equation (70), with renormalized excluded volume parameter, $v^{(0)} \to v_{ren}^{(0)} = v^{(0)} - (\rho^{(0)}N)^{-1}$. The decrease in $v^{(0)}$ is due to the fact that the monomers forming cross-links give an additional attractive contribution to the effective second virial coefficient $v_{ren}^{(0)}$. The amplitude of heterogeneities increases when approaching the *cross-link saturation threshold* at $v^{(0)}\rho^{(0)}N = 1$ at which both the structure factor $S^{(0)}(\mathbf{q} \to 0)$ and the characteristic size of heterogeneities at preparation conditions

$$\xi \cong R/\sqrt{v^{(0)}\rho^{(0)}N - 1}$$

diverge.

The theory also predicts the appearance of a maximum degree of spatial gel inhomogeneity at a critical polymer network concentration, as confirmed by experiments on PAAm gels [42,43].

### 4.2. Scattering Intensity

The scattering intensity on wavevector $\mathbf{q}$ is proportional to the structure factor, which is given by the sum

$$\tilde{S}(\mathbf{q}) = \left\langle |\tilde{\rho}(\mathbf{q})|^2 \right\rangle = \tilde{D}(\mathbf{q}) + \tilde{C}(\mathbf{q}) \tag{71}$$

of contributions of the thermal fluctuations, Equation (68), and the inhomogeneous equilibrium density variations due to defects of the topological structure of the network,

$$\tilde{C}(\mathbf{q}) = \overline{|\tilde{\rho}_{eq}(\mathbf{q})|^2} = \frac{\nu(\mathbf{q})}{[1 + v\tilde{g}(\mathbf{q})]^2} \qquad (72)$$

where bar means ensemble average.

The experimental data can be visualized using contour plots of neutron scattering from random inhomogeneities of network structure in anisotropic-deformed swollen gels [44]. Under uniaxial gel stretching, an increase in the scattered intensity on small wave vectors $q$ in the stretching direction is observed, which is enveloped by elliptical patterns at larger $q$ values with a maximum oriented normal to this axis. This behavior is opposite of what is expected in theories assuming only thermal fluctuations and is called "abnormal butterfly patterns". The elliptical patterns at large wavevector $q$ originate from the correlator of static inhomogeneities, Equations (72) and (69), which contains the term $S_0(\lambda \cdot \mathbf{q})$ that "remembers" the affinely deformed inhomogeneous structure of the network. The butterfly patterns along the stretching axis in the small $q$ range is due to strong angular dependence of the thermal structure factor, (68). This function comes into numerator of the thermal correlator in Equation (68), describing "normal butterfly patterns". At high cross-link concentrations the correlator of static inhomogeneities gives the main contribution to the scattering intensity. Since the function $\tilde{g}(\mathbf{q})$ is included in the denominator of Equation (72), this expression describes the so named "abnormal butterfly patterns". As a result, under uniaxial extension, a crossover from normal to abnormal butterfly scattering patterns occurs with an increase of the strength of inhomogeneity or the swelling ratio [45]. In addition to such butterfly scattering patterns, the theory also predicted "Lozenge" patterns if only part of all network chains, i.e., their deuterated fragments, could scatter neutrons, in accordance with scattering experiments [46].

The static inhomogeneity in poly (N-isopropyl acrylamide) gel (PNIPA) has been investigated in Ref. [47] by the methods of small-angle neutron scattering (SANS) and neutron spin echo. The obtained SANS scattering amplitude $\tilde{S}(q)$ was successfully decomposed into the thermal and static components, respectively, $\tilde{D}(q)$ and $\tilde{C}(q)$ in Equation (71). It was revealed that $\tilde{C}(q)$ becomes dominant in the $q$-region, where the abnormal butterfly scattering under stretching is observed. As the temperature increases toward the temperature for volume phase transition, $\tilde{C}(q)$ approximated by the square of the Lorentzian shape increases more drastically than $\tilde{D}(q)$ of the Lorentzian shape. These experimental findings are also well described in the theoretical framework of this section.

In Ref. [48] gels prepared by two cross-linking methods were studied using SANS technique. One is chemical cross-linking with BIS and the other is gamma-ray cross-linking of a PNIPA solution. It is shown that the degree of the inhomogeneity is much larger in chemically cross-linked gels than in the gamma-ray cross-linked gels. The experimental data are in quantitative agreement with predictions of the theory of scattering intensity $\tilde{S}(\mathbf{q})$ on chemically and instantaneously cross-linked gels, respectively.

*4.3. Heterogeneities in Charged Gels*

A study of the structure factor of weakly charged polyelectrolyte gels under uniaxial stretching was performed by Mendes et al. [49], who observed after introducing ions to the gel, the disappearance of the "butterfly pattern" in the small angle scattering intensity, as well as an increase in the scattering intensity in the direction perpendicular to the gel stretching. The origin of this maximum has been elucidated in SANS experiment by Shibayama et al. who studied the deformed state of weakly charged polymer gels PNIPA/AAc [50]. In this experiment, an anisotropic scattering maximum is observed, which indicates that the spatial distribution of the charged groups changes as a result of gel deformation and therefore is strongly coupled with the static inhomogeneities.

All the observed patterns of SANS intensity were well reproduced using a generalization of the above theory to the case of charged polymer networks [51]. The only effect of electrostatic interactions is to replace the excluded volume parameters $v^{(0)}$ and $v$ with effective virial coefficients for both the final state and for the state of preparation:

$$v \to v(\mathbf{q}) \equiv v + 1/s_{DH}(\mathbf{q}),$$
$$v^{(0)} \to v^{(0)}(\mathbf{q}) \equiv v^{(0)} + 1/s_{DH}^{(0)}(\mathbf{q})$$

where

$$\frac{1}{s_{DH}(\mathbf{q})} = \frac{4\pi l_B \alpha^2}{\kappa^2 + q^2}$$

is the inverse of the Debye–Hückel structure factor, $\kappa^{-1}$ is the Debye screening length, $\alpha$ is the degree of ionization (fraction of charged monomers) and $l_B$ is the Bjerrum length. $s_{DH}^{(0)}(\mathbf{q})$ is obtained by substituting $\kappa = \kappa^{(0)}$ and $\alpha = \alpha^{(0)}$ into the above equation. As shown in Ref. [52], the above theoretical prediction for $S(\mathbf{q})$ well reproduces the observed scattering intensity functions of wekly charged PNIPA/AAc gels.

In general, an introduction of cross-links to a polymer solution leads to an increase in the scattering intensity due to static inhomogeneities. However, a reverse phenomenon, called the "inflection" in scattering intensity, was predicted by the above theory [53] and observed in weakly charged gels and polymer solutions [54]. While the gel becomes more inhomogeneous with increasing the degree of cross-linking in a good solvent, the inhomogeneities can be suppressed in a poor solvent, although in a relatively small regions of cross-link concentration. However, this phenomenon is interesting due to its physical significance.

### 4.4. Good Solvent: Scaling Approach

Earlier in this paper, we considered the networks with Gaussian chains obtained in the melt or the $\theta$-solution of linear chains. In general, polymer networks can also be obtained by crosslinking semi-diluted polymer solutions. The networks can be placed in a good solvent, in which the polymer chains swell compared to the case of theta solvent. This section explains the basic ideas of the scaling approach to describing networks prepared/swollen in a good solvent.

The mean-field approach can be adapted to descripbe the elasticity of such gels [55] using the well-known de Gennes blob picture of semi-dilute solutions [56]. The key idea is the spatial scale separation: while static density inhomogeneities exist only on scales comparable to or larger than the monomer fluctuation radius $R$, thermal density fluctuations are dominated by smaller scales and are quite similar to those in semi-dilute polymer solutions.

Consider a gel prepared by random cross-linking of chains in a semi-dilute polymer solution in a good solvent at the monomer density $\rho^{(0)}$ that is swollen to density $\rho < \rho^{(0)}$. The monomer density fluctuations should be taken into account both in the initial (where the gel was prepared) and in the final (where it is being studied) states of the gel. On length scales shorter than the correlation lengths in these states,

$$\xi^{(0)} = b^{-5/4}\left(\rho^{(0)}\right)^{-3/4} \quad \text{and} \quad \xi = b^{-5/4}\rho^{-3/4}, \tag{73}$$

density fluctuations are large, and the gel behaves like a polymer solution ("liquid-like" regime). On scales exceeding the blob sizes (the corresponding wave vectors $q^{(0)} = 1/\xi^{(0)}$ and $q = 1/\xi$), density fluctuations are small, and the mean-field description with appropriately renormalized parameters can be used [8]

$$b^{(0)} \to b_{ren}^{(0)} = b\left(\rho^{(0)}b^3\right)^{-1/8},$$
$$b \to b_{ren} = b\left(\rho b^3\right)^{-1/8}, \quad \lambda_\alpha \to \lambda_\alpha b_{ren}^{(0)}/b_{ren}$$

where the subscript $_{ren}$ stands for renormalized values that differ from the "bare" ones. The renormalized second virial coefficients $v_{ren}^{(0)}$ and $v_{ren}$ are:

$$v^{(0)} \to v_{ren}^{(0)} = b^3 \left(\rho^{(0)} b^3\right)^{1/4}, \qquad v \to v_{ren} = b^3 \left(\rho b^3\right)^{1/4} \qquad (74)$$

Equations (73) and (74) complete the renormalization of the mean-field theory: in order to describe a gel in a good solvent on length scales larger than the thermal correlation length, one only have to replace the bare parameters in the previously derived expressions for the free energy, correlation functions, etc., by their renormalized values.

### 4.5. Amplification of Cross-Linking Density Pattern

Is it possible to "write" some "useful" information in a polymer network cross-link pattern, and under what conditions it can be "read" back? The answer to this question is given in Ref. [57]. After the formation of a homogeneous (on length scales large compared to its "mesh" size R) network, large-scale patterns can be generated in it by further cross-linking followed by swelling (and possibly stretching) of the network, which leads to a nonuniformly swollen gel. This additional cross-linking can be done by adding light-sensitive cross-links to the transparent network. By focusing the laser beam in the regions inside the gel, it is possible to "write" information into the gel structure in the form of 2D or 3D patterns of cross-link density.

It was shown that although such information is hidden at the preparation conditions, it can be reversibly "developed" by gel swelling, since unobservable variations of the cross-link density in the melt are transformed into observable variations of monomer density in the swollen gel. The gel regions with increased cross-link concentration can be considered as inclusions with enhanced elastic modulus. It has been shown that in swollen gels that stretch isotropically upon absorption of the solvent, the observed monomer density pattern is an affinity stretched initial cross-linking pattern. Such gels can serve as a magnifying glass that enlarges the initially written pattern without distorting its shape. The corresponding magnification factor can be very large in the case of super-elastic networks.

A strong enhancement of the contrast between the high and the low monomer density regions can be obtained by placing the gel in a poor solvent with a negative second virial coefficient $v < 0$. The observable image becomes distorted, especially near the edges and corners of the pattern. Gel boundaries should be fixed in order to avoid its contraction. Gel contraction can also be prevented by focusing a laser beam only on a part of the localized pattern and heating it, resulting in a local change of the quality of solvent.

## 5. Discussion

Despite its more than half-century history, even now the theory of polymer network elasticity asks more questions than it gives answers. In this work, we tried to bridge the gap between the main developed approaches, which allowed us to give answers to some of these questions.

In real networks, there are both topological defects and typical loops, which are not sparsely distributed. Using the replica method, the cumulative effect of a large number of strongly overlapping typical loops of finite size can be described in the effective mean-field approximation. This approximation works for polymer networks due to large overlap parameter of network strands, $P^{(0)} \gg 1$, see Equation (12).

In the replica approach, the properties of polymer networks are described by the liquid-solid (sol-gel) order parameter $\varphi_1(\varsigma)$, which is actually not a parameter, but a function of the variable $\varsigma$, and is determined by a complex integro-differential equation. The only, but very important exception is the elastic modulus, which is expressed through the value of the order parameter at $\varsigma = 0$, and the corresponding equations for $\varphi_1(0)$ become algebraic. This greatly simplifies the calculations of the elastic modulus of the network with an arbitrary number of cyclic fragments of finite size. In contrast to the classical theory of phantom networks, the mean field of loops, $\varphi_1(0)$ explicitly depends on the excluded volume parameter $v^{(0)}$ and the density of monomers $\rho^{(0)}$ at network preparation conditions. This dependence takes into account the limitations imposed by the packing of highly overlapping typical loops in real $3D$ space on the molecular structure of the network being formed.

The resulting elastic modulus of real networks has two main contributions, see Equation (37). The contribution of elastically effective network strands is taken into account in the classical model of phantom networks. The classical expression for this contribution is renormalized due to the presence of topological defects in the network—primary loops and cyclic fragments of finite size. The second contribution to the elastic modulus is always negative and gives an impact of a large number of typical loops of the network. It describes the effect of shunting of the elastically effective chains by finite size polymer loops and depends on the interaction of monomers at the preparation conditions of the network.

Both contributions to the network elastic modulus can be represented by the generalized combined chain model with an additional effective chain (see Figure 2b) that accounts the cumulative effect of finite size loops in real networks. The virtual chains transmit local stresses from the network fragments to the solid-state degrees of freedom of such soft solids. We established the connection of the distribution function of the lengths of virtual chains with the order parameter of the replica network model and calculated this distribution function for the exponential distribution of the lengths of the network strands.

We also proposed a generalization of the slip tube model of entangled polymer networks, which allows us to describe the crossover between the entangled and phantom regimes of network swelling. The dependence of the Mooney-Rivlin parameters $C_1$ and $C_2$ on the polymer concentration calculated for this model is in agreement with the experiments.

Although from the very beginning the heterogeneities were recognized as one of the most essential features of gels [58], it took long time to formulate their theoretical description due to complexity of this problem. Up to today, the understanding of the gel structure has greatly improved owing to both theoretical developments and a large number of experimental studies. Effect of cross-links on heterogeneous structure of polymer gels, abnormal butterfly patterns, microphase separation, and so on, are well understood with the aid of the theretical approaches discussed in Section 4.

The list of remaining questions is much longer:

Polymer networks can be obtained by crosslinking semidiluted polymer solutions. How does a strong nonlinear deformation of such networks occur in the phantom regime? How do such networks fracture when they are strongly stretched? The molecular theory (Lake-Thomas model) of the fracture of dry networks is constructed in Ref. [59]. What happens when compressing such networks? What is deformation mechanism of heterogeneous super-tough networks, kinetics of crack growth and phase transitions in such networks? Stress-induced microphase separation in multicomponent networks was studied in [60]. What is the effect of microphase separation on the elasticity of such networks?

These and many other questions are still awaiting answers.

**Funding:** This research received no external funding.

**Conflicts of Interest:** The authors declare no conflict of interest.

### Appendix A. Replica Model of the Network

Consider a network prepared by random end-linking a solution of polydisperse linear chains by $f$-functional cross-links. Following the idea of $\varphi^4$ formulation of the excluded volume problem [56], we introduce $n \to 0$ component field, $\varphi(\hat{x})$ in the replica space with components $\varphi_i$ ($i = 1, \ldots, n$), which is related to the monomer density in this space as

$$\rho(\hat{x}) = \frac{1}{2}\varphi^2(\hat{x}) = \frac{1}{2}\sum_{i=1}^{n}\varphi_i^2(\hat{x}). \tag{A1}$$

Unlike the main text, in this Appendix the symbol $n$ is used only to indicate the number of components of the field $\varphi$. The free energy $F_m$ of the network in the replica system is described by Hamiltonian

$$H[\varphi] = \int d\hat{x} \left[ \frac{1}{2}\mu\varphi^2 + \frac{b^2}{2}\left(\hat{\nabla}\varphi\right)^2 - \frac{z_f}{f!}\varphi_1^f \right]$$
$$+ \sum_k \int dx^{(k)} \frac{v^{(k)}}{8} \left[ \prod_{l \neq k} \int dx^{(l)} \varphi^2 \right]^2. \tag{A2}$$

Here, $\hat{\nabla}$ is the gradient of the variables $\hat{x}$, $\mu$ is monomer chemical potential and $z_f$ is the fugacity of $f$-functional cross-links. The last sum in Equation (A2) describes the two-body monomer interaction in the initial system characterized by the excluded volume parameter $v^{(0)}$ and in final systems ($k = 1, \ldots, m$) with the excluded volume parameters $v^{(k)} = v$.

In the mean-field approximation, the steepest descent value of the field $\varphi_1$ is found from the extremum of Hamiltonian in Equation (A13). The field $\varphi_1(\hat{x})$ has the meaning of the order parameter of the liquid-solid (sol-gel) phase transition. The steepest descent solution $\varphi_1(\varsigma) > 0$ depends on a single variable $\varsigma$, Equation (6). At $\varsigma = 0$, the differential Equation (8) for this function becomes algebraic, an explicit solution of the minimum conditions is found

$$\varphi_1(0) = \left[ \frac{(f-1)!}{\tilde{N}z_f} \right]^{1/(f-2)}, \tag{A3}$$

In the limit $m \to 0$, the dimensionless order parameter is determined by differential Equation (8). The free energy of the network depends only on $\varphi_1(0)$. Substituting Equation (A2) into Equation (5), we find the free energy of the network[8]:

$$F = V^{(0)} G \left[ \sum_\alpha \frac{\lambda_\alpha^2}{2} + \ln\left(aN^{1/2}\right) \right] \tag{A4}$$

with the elastic modulus

$$\frac{G}{kT} = \rho_f^{(0)} \left( \frac{f}{2} - 1 \right), \tag{A5}$$

where $\rho_f^{(0)}$ is the cross-link concentration.

**Appendix B. Distribution of Virtual Chains**

According to Equation (19) the distribution function $q(s) = \tilde{\chi}^{(f-1)}(s)$ of the inverse number of monomers $s = 1/n$ can be presented as a convolution of $f-1$ distribution functions $\tilde{\chi}(s')$ of random variables $s' = 1/m$. The multiple convolution is defined recurrently as

$$\tilde{\chi}^{(k)}(s) = \left( \tilde{\chi}^{(k-1)} * \tilde{\chi} \right)(s) \equiv \int_0^s \tilde{\chi}^{(k-1)}(s-s') \tilde{\chi}(s') ds' \tag{A6}$$

The distribution function $\tilde{\chi}(s')$ of an individual branch of the tree is related to the distribution function $\pi(m)$ of the inverse values $m = 1/s'$ by an equation similar to Equation (25):

$$\tilde{\chi}(s') = \frac{1}{(s')^2} \pi\left(\frac{1}{s'}\right) \tag{A7}$$

Finally, the distribution function $\pi(m)$ of the monomer numbers $m = N + n$ of one branch is a convolution of distribution functions of the monomer numbers of the network strands $P(N)$ and virtual chains $p(n)$,

$$\pi(m) = (P * p)(m) = \int_0^m P(m-n) p(n) dn \tag{A8}$$

Equations (25) and (A6)–(A8) completely determine the distribution function $p(n)$ of the number of monomers $n$ of virtual chains for a given distribution function $P(N)$ of the number of monomers $N$ of real network strands. The only problem is solving this system of nonlinear integral equations. Here the replica method comes to the rescue, see Appendix A.

**Appendix C. Free Energy of Entangled Networks**

The network free energy for given positions of attachment points $\mathbf{X}(s)$ of virtual chains to the affinely deformed background is

$$F[\mathbf{X}] = -\nu kT \sum_\alpha \ln \int D x_\alpha(s) e^{-H_\alpha[x_\alpha]/kT}, \tag{A9}$$

$$\frac{H_\alpha[x_\alpha]}{kT} = \int_0^N ds \left\{ \frac{3\dot{x}_\alpha^2}{2b^2} + \frac{3[x_\alpha(s)/\lambda_\alpha - X_\alpha(s)]^2}{2b^2 N_e^2} \right\}. \tag{A10}$$

The simplest way to calculate the above integral is to use the normal mode expansion of the chain trajectories, which satisfy the boundary conditions $x_\alpha(N) = \lambda_\alpha X_\alpha(N)$ and $x_\alpha(0) = \lambda_\alpha X_\alpha(0)$ ($\mathbf{X}(s)$ are the positions of attachment points of virtual chains in the undeformed network):

$$x_\alpha(s) = \lambda_\alpha R_\alpha \frac{s}{N} + \sum_\omega \tilde{x}_\alpha(\omega) e^{i\omega s}, \qquad \omega = \frac{2\pi k}{N}, \quad k = \pm 1, \cdots \tag{A11}$$

where $R_\alpha = X_\alpha(N) - X_\alpha(0)$ and

$$X_\alpha(s) = \lambda_\alpha R_\alpha \frac{s}{N} + \sum_\omega \tilde{X}_\alpha(\omega) e^{i\omega s}, \qquad \tilde{x}_\alpha(-\omega) = \tilde{x}_\alpha^*(\omega), \quad \tilde{X}_\alpha(-\omega) = \tilde{X}_\alpha^*(\omega). \tag{A12}$$

The mode with $\omega = 0$ is excluded from the summation because it corresponds to the translational displacement of the chain, which is prohibited in the network. Substituting the above expressions into Equation (A10) we get

$$\frac{H_\alpha[x_\alpha]}{kT} = \frac{3R_\alpha^2}{2b^2 N}\lambda_\alpha^2 + \frac{3}{2b^2}\sum_{\omega \neq 0}\left[\omega^2|\tilde{x}_\alpha(\omega)|^2 + \frac{|\tilde{x}_\alpha(\omega) - \lambda_\alpha \tilde{X}_\alpha(\omega)|^2}{N_e^2 \lambda_\alpha^2}\right] \tag{A13}$$

Changing the integration variables, $\tilde{x}_i(\omega) = \tilde{u}(\omega) + i\tilde{v}(\omega)$, we find

$$\frac{F[\mathbf{X}]}{\nu kT} = \sum_\alpha \left[ \frac{3R_\alpha^2}{2b^2 N}\lambda_\alpha^2 + \frac{3\lambda_\alpha^2}{2b^2}\sum_{\omega>0} \frac{\omega^2 |\tilde{X}_\alpha(\omega)|^2}{1+\omega^2 N_e^2 \lambda_\alpha^2} \right.$$
$$\left. - \sum_{\omega>0} \ln \frac{2b^2 N_e^2 \lambda_\alpha^2}{3N(1+\omega^2 N_e^2 \lambda_\alpha^2)} \right]. \tag{A14}$$

Here sums are taken only over positive $\omega$ since the only positive frequency components $\tilde{u}(\omega)$ and $\tilde{v}(\omega)$ can be taken as independent variables: the condition for the reality of the function $x(s)$, Equation (A12), imposes restrictions $\tilde{u}(-\omega) = \tilde{u}(\omega)$ and $\tilde{v}(-\omega) = -\tilde{v}(\omega)$ on the real and imaginary parts of $\tilde{x}(\omega)$.

This free energy should be averaged over positions of attachment points, $\mathbf{X}(s)$, with correlation functions

$$\overline{\tilde{X}_\alpha(\omega) \tilde{X}_\beta(-\omega)} = \frac{b^2}{3}\left(\frac{1}{\omega^2} + N_e^2\right)\delta_{\alpha\beta} \tag{A15}$$

which corresponds to randomly placed brash of chain with two virtual chains of $N_e^2$ monomers

$$\overline{[\mathbf{X}(s) - \mathbf{X}(s')]^2} = b^2|s-s'| + 2N_e^2 \tag{A16}$$

Averaging the free energy (A14) with the correlation function, (A15) and changing the integration variables $\tilde{X}(\omega) \to (R, \tilde{p}(\omega), \tilde{q}(\omega))$ when integrating over $\mathbf{X}(\omega)$ we find

$$\frac{\mathcal{F}}{\nu kT} = \sum_\alpha \left[ \frac{\lambda_\alpha^2}{2} + \sum_{\omega>0} \frac{\lambda_\alpha^2 - 1}{1 + \omega^2 N_e^2 \lambda_\alpha^2} \right.$$
$$\left. + \sum_{\omega>0} \ln\left(1 + \frac{1}{N_e^2 \lambda_\alpha^2 \omega^2}\right) \right] + Const. \tag{A17}$$

Here all the terms which do not depend on the extension coefficients $\lambda_\alpha$ are absorbed in *Const*. The sum and the product can be expressed in terms of hyperbolic functions, see Equation (60).

## References

1. Urayama, K.; Kohjiya, S. Extensive stretch of polysiloxane network chains with random- and super-coiled conformations. *Eur. Phys. J. B* **1998**, *2*, 75–78. [CrossRef]
2. Flory, P.J. *Principles of Polymer Chemistry*; Cornell University Press: Ithaca, NY, USA, 1953.
3. Rubinstein, M.; Panyukov, S. Elasticity of Polymer Networks. *Macromolecules* **2002**, *35*, 6670–6686. [CrossRef]
4. Panyukov, S. Loops in Polymer Networks. *Macromolecules* **2019**, *52*, 4145–4153. [CrossRef]
5. Ge, T.; Panyukov, S.; Rubinstein, M. Self-Similar Conformations and Dynamics in Entangled Melts and Solutions of Nonconcatenated Ring Polymers. *Macromolecules* **2016**, *49*, 708–722. [CrossRef]
6. Rubinstein, M.; Panyukov, S. Nonaffine Deformation and Elasticity of Polymer Networks. *Macromolecules* **1997**, *30*, 8036–8044. [CrossRef]
7. Deam, R.T.; Edwards, S.F. The Theory of Rubber Elasticity. *Philos. Trans. R. Sot. Lond. Ser. A* **1976**, *280*, 317. [CrossRef]
8. Panyukov, S.; Rabin, Y. Statistical physics of polymer gels. *Phys. Rep.* **1996**, *269*, 1–131. [CrossRef]
9. Panyukov, S.V.; Rabin, Y. Replica Field Theory Methods in Physics of Polymer Networks. In *Theoretical and Mathematical Models in Polymer Research*; Modern Methods in Polymer Research and Technology; Grosberg, A.Y., Ed.; Academic Press: Boston, MA, USA, 1989; pp. 83–185.
10. Panyukov, S.; Rabin, Y.; Fiegel, A. Solid Elasticity and Liquid-Like Behaviour in Randomly Crosslinked Polymer Networks. *Europhys. Lett.* **1994**, *28*, 149–154. [CrossRef]
11. Edwards, S.F. A field theory formulation of polymer networks. *J. Phys.* **1988**, *49*, 1673–1682. [CrossRef]
12. Cai, L.-H.; Panyukov, S.; Rubinstein, M. Hopping Diffusion of Nanoparticles in Polymer Matrices. *Macromolecules* **2015**, *48*, 847–862. [CrossRef]
13. Lang, M.; Kreitmeier, S.; Göritz, D. Trapped Entanglements in Polymer Networks. *Rubber Chem. Technol.* **2007**, *80*, 873–894. [CrossRef]
14. Rubinstein, M.; Colby, R.H. *Polymer Physics*; Oxford University Press: Oxford, UK; New York, NY, USA, 2003.
15. Schieber, J.D.; Horio, K. Fluctuation in entanglement positions via elastic slip-links. *J. Chem. Phys.* **2010**, *132*, 074905. [CrossRef] [PubMed]
16. Gusev, A.A. Numerical Estimates of the Topological Effects in the Elasticity of Gaussian Polymer Networks and Their Exact Theoretical Description. *Macromolecules* **2019**, *52*, 3244–3251. [CrossRef]
17. Helfand, E.; Tonelli, A.E. Elastically Ineffective Polymer Chains in Rubbers. *Macromolecules* **1974**, *7*, 832–834. [CrossRef]
18. Tonelli, A.E.; Laboratories, B.; Hill, M. Elastically Ineffective Cross-Links in Rubbers. *Macromolecules* **1974**, *7*, 59–63. [CrossRef]
19. Zhong, M.; Wang, R.; Kawamoto, K.; Olsen, B.D.; Johnson, J.A. Quantifying the impact of molecular defects on polymer network elasticity. *Science* **2016**, *353*, 1264–1268. [CrossRef]
20. Lin, T.-S.; Wang, R.; Johnson, J.A.; Olsen, B.D. Topological Structure of Networks Formed from Symmetric Four-Arm Precursors. *Macromolecules* **2018**, *51*, 1224–1231. [CrossRef]
21. Lin, T.-S.; Wang, R.; Johnson, J.A.; Olsen, B.D. Revisiting the Elasticity Theory for Real Gaussian Phantom Networks. *Macromolecules* **2019**, *52*, 1685–1694. [CrossRef]
22. Lang, M. Elasticity of Phantom Model Networks with Cyclic Defects. *ACS Macro Lett.* **2018**, *7*, 536–539. [CrossRef]

23. Lange, F.; Schwenke, K.; Kurakazu, M.; Akagi, Y.; Chung, U.I.; Lang, M.; Sommer, J.-U.; Sakai, T.; Saalächter, K. Connectivity and structural defects in model hydrogels: A combined proton NMR and Monte Carlo simulation study. *Macromolecules* **2011**, *44*, 9666–9674. [CrossRef]
24. Zhou, H.; Woo, J.; Cok, A.M.; Wang, M.; Olsen, B.D.; Johnson, J.A. Counting primary loops in polymer gels. *PNAS* **2012**, *109*, 19119–19124. [CrossRef] [PubMed]
25. Wang, R.; Alexander-Katz, A.; Johnson, J.A.; Olsen, B.D. Universal Cyclic Topology in Polymer Networks. *Phys. Rev. Lett.* **2016**, *116*, 188302. [CrossRef] [PubMed]
26. Lang, M.; Miller, T. Analysis of the Gel Point of Polymer Model Networks by Computer Simulations. *Macromolecules* **2020**, *53*, 498–512. [CrossRef]
27. Panagiotou, E.; Kruger, M.; Millett, K.C. Writhe and mutual entanglement combine to give the entanglement length. *Phys. Rev. E* **2013**, *88*, 062604. [CrossRef]
28. Panyukov, S.V. Topology fluctuations in polymer networks. *Sov. Phys. JETP* **1989**, *69*, 342–353.
29. Panyukov, S.V. Topological interactions in the statistical theory of polymers. *Sov. Phys. JETP* **1988**, *67*, 2274–2284.
30. Gumbrell, S.M.; Mullins, L.; Rivlin, R.S. Departures of the elastic behaviour of rubbers in simple extension from the kinetic theory. *Rubber Chem. Technol.* **1955**, *28*, 24–35. [CrossRef]
31. Di Lorenzo, F.; Seiffert, S. Nanostructural heterogeneity in polymer networks and gels. *Polym. Chem.* **2015**, *6*, 5515–5528. [CrossRef]
32. Seiffert, S. Origin of Nanostructural Inhomogeneity in Polymer-Network Gels. *Polym. Chem.* **2016**, *7*, 36–43. [CrossRef]
33. Vilgis, T.A.; Heinrich, G. The Essential Role of Network Topology in Rubber Elasticity. *Angew. Makromol. Chem.* **1992**, *202*, 243–259. [CrossRef]
34. Vilgis, T.A.; Sommer, J.-U.; Heinrich, G. Swelling and fractal heterogeneities in networks. *Macromol. Symp.* **1995**, *93*, 205–212. [CrossRef]
35. Svaneborg, C.; Grest, G.S.; Everaers, R. Disorder effects on the strain response of model polymer networks. *Polymer* **2005**, *46*, 4283–4295. [CrossRef]
36. Panyukov, S.V. Inhomogeneities as consequences of a stretching of polymer networks. *JETP Lett.* **1993**, *58*, 118–122.
37. Panyukov, S.; Rabin, Y. Polymer Gels: Frozen Inhomogeneities and Density Fluctuations. *Macromolecules* **1996**, *29*, 7960–7975. [CrossRef]
38. Panyukov, S.V. Theory of heterogeneities in polymer networks. *Polym. Sci. Ser. A* **2016**, *58*, 582–594. [CrossRef]
39. Lifshitz, I.M.; Grosberg, A.Y.; Khokhlov, A.R. Some problems of the statistical physics of polymer chains with volume interaction. *Rev. Mod. Phys.* **1978**, *50*, 683. [CrossRef]
40. de Gennes, P.G. Effect of cross-links on a mixture of polymers. *J. Phys. Lett.* **1979**, *40*, 69–72. [CrossRef]
41. Matsuo, E.S.; Orkisz, M.; Sun, S.-T.; Li, Y.; Tanaka, T. Origin of Structural Inhomogeneities in Polymer Gels. *Macromolecules* **1994**, *27*, 6791–6796. [CrossRef]
42. Kizilay, M.Y.; Okay, O. Effect of Initial Monomer Concentration on Spatial Inhomogeneity in Poly(acrylamide) Gels. *Macromolecules* **2003**, *36*, 6856–6862. [CrossRef]
43. Kizilay, M.Y.; Okay, O. Effect of swelling on spatial inhomogeneity in poly(acrylamide) gels formed at various monomer concentrations. *Polymer* **2004**, *45*, 2567–2576. [CrossRef]
44. Bastide, J.; Leibler, L.; Prost, J. Scattering by deformed swollen gels: butterfly isointensity patterns. *Macromolecules* **1990**, *23*, 1821–1825. [CrossRef]
45. Onuki, A. Scattering from deformed swollen gels with heterogeneities. *J. Phys. II* **1992**, *2*, 45–61. [CrossRef]
46. Panyukov, S.V. Microscopic theory of anisotropic scattering by deformed polymer networks. *Sov. Phys. JETP* **1992**, *75*, 347–352.
47. Koizumi, S.; Monkenbusch, M.; Richter, D.; Schwahn, D.; Farago, B. Concentration fluctuations in polymer gel investigated by neutron scattering: Static inhomogeneity in swollen gel. *J. Chem.* **2004**, *121*, 12721. [CrossRef] [PubMed]
48. Norisuye, T.; Masui, N.; Kida, Y.; Ikuta, D.; Kokufuta, E.; Ito, K.; Panyukov, S.; Shibayama, M. Small angle neutron scattering studies on structural inhomogeneities in polymer gels: Irradiation cross-linked gels vs chemically cross-linked gels. *Polymer* **2002**, *43*, 5289–5297. [CrossRef]

49. Mendes, E.; Schosseler, F.; Isel, F.; Boué, F.; Bastide, J.; Candau, S.J. A SANS Study of Uniaxially Elongated Polyelectrolyte Gels. *Europhys. Lett. (EPL)* **1995**, *32*, 273–278. [CrossRef]
50. Shibayama, M.; Kawakubo, K.; Ikkai, F.; Imai, M. Small-Angle Neutron Scattering Study on Charged Gels in Deformed State. *Macromolecules* **1998**, *31*, 2586–2592. [CrossRef]
51. Rabin, Y.; Panyukov, S. Scattering Profiles of Charged Gels: Frozen Inhomogeneities, Thermal Fluctuations, and Microphase Separation. *Macromolecules* **1997**, *30*, 301–312. [CrossRef]
52. Shibayama, M.; Kawakubo, K.; Norisuye, T. Comparison of the Experimental and Theoretical Structure Factors of Temperature Sensitive Polymer Gels. *Macromolecules* **1998**, *31*, 1608–1614. [CrossRef]
53. Shibayama, M.; Ikkai, F.; Shiwa, Y.; Rabin, Y.. Effect of Degree of Cross-linking on Spatial Inhomogeneity in Charged Gels. I. Theoretical Predictions and Light Scattering Study. *J. Chem. Phys.* **1997**, *107*, 5227–5235. [CrossRef]
54. Ikkai, F.; Shibayama, M.; Han, C.C. Effect of Degree of Cross-Linking on Spatial Inhomogeneity in Charged Gels. 2. Small-Angle Neutron Scattering Study. *Macromolecules* **1998**, *31*, 3275–3281. [CrossRef]
55. Panyukov, S.V. Scaling theory of high elasticity. *Sov. Phys. JETP* **1990**, *71*, 372–379.
56. de Gennes, P.-G. *Scaling Concepts in Polymer Physics*; Cornell University Press: Ithaca, NY, USA, 1979.
57. Panyukov, S.; Rabin, Y. Cross-Linking Patterns and Their Images in Swollen and Deformed Gels. *Macromolecules* **2015**, *48*, 7378–7381. [CrossRef]
58. Bastide, J.; Leibler, L. Large-scale heterogeneities in randomly cross-linked networks. *Macromolecules* **1988**, *21*, 2647–2649. [CrossRef]
59. Wang, S.; Panyukov, S.; Rubinstein, M.; Craig, S.L. Quantitative Adjustment to the Molecular Energy Parameter in the Lake–Thomas Theory of Polymer Fracture Energy. *Macromolecules* **2019**, *52*, 2772–2777. [CrossRef]
60. Panyukov, S.; Rubinstein, M. Stress-Induced Ordering in Microphase-Separated Multicomponent Networks. *Macromolecules* **1996**, *29*, 8220–8230. [CrossRef]

© 2020 by the author. Licensee MDPI, Basel, Switzerland. This article is an open access article distributed under the terms and conditions of the Creative Commons Attribution (CC BY) license (http://creativecommons.org/licenses/by/4.0/).

*Article*

# Effect of Fillers on the Recovery of Rubber Foam: From Theory to Applications

Thridsawan Prasopdee [1] and Wirasak Smitthipong [1,2,3,*]

[1] Specialized Center of Rubber and Polymer Materials in Agriculture and Industry (RPM), Department of Materials Science, Faculty of Science, Kasetsart University, Chatuchak, Bangkok 10900, Thailand; thridsawan@gmail.com

[2] Office of Research Integration on Target-Based Natural Rubber, National Research Council of Thailand (NRCT), Chatuchak, Bangkok 10900, Thailand

[3] Office of Natural Rubber Research Program, Thailand Science Research and Innovation (TSRI), Chatuchak, Bangkok 10900, Thailand

* Correspondence: fsciwssm@ku.ac.th

Received: 24 October 2020; Accepted: 17 November 2020; Published: 19 November 2020

**Abstract:** Natural rubber foam (NRF) can be prepared from concentrated natural latex, providing specific characteristics such as density, compression strength, compression set, and so on, suitable for making shape-memory products. However, many customers require NRF products with a low compression set. This study aims to develop and prepare NRF to investigate its recoverability and other related characteristics by the addition of charcoal and silica fillers. The results showed that increasing filler loading increases physical and mechanical properties. The recoverability of NRF improves as silica increases, contrary to charcoal loading, due to the higher specific surface area of silica. Thermodynamic aspects showed that increasing filler loading increases the compression force ($F$) as well as the proportion of internal energy to the compression force ($F_u/F$). The entropy ($S$) also increases with increasing filler loading, which is favorable for thermodynamic systems. The activation enthalpy ($\Delta H_a$) of the NRF with silica is higher than the control NRF, which is due to rubber–filler interactions created within the NRF. A thermodynamic concept of crosslinked rubber foam with filler is proposed. From theory to application, in this study, the NRF has better recoverability with silica loading.

**Keywords:** rubber foam; filler; charcoal; silica; compression set; recovery; thermodynamics

---

## 1. Introduction

Natural rubber (NR) is a biobased polymer usually applied in the rubber industry to manufacture products such as gloves, pillows, mattresses, tires, and so on. It is derived from the *Hevea brasiliensis* tree as colloidal latex, which presents other substances or nonrubber components, such as proteins, fatty acids, inorganic matters, and so on. This natural latex can be stabilized by the base chemical to maintain prolonged storage [1–4].

Many products, such as pillows and mattresses, are made from rubber foam prepared from concentrated natural latex, which provides specific characteristics. Rubber foams are generally porous and elastic with ventilated surfaces. These foams are made into lightweight products for comfort applications such as pillows and mattresses [5,6]. From the perspective of mechanical properties, rubber foam can either be soft or firm depending on the formula of the compounded latex. Concentrated natural latex is mixed with chemical agents, consisting of a blowing agent, vulcanizing agent, accelerators, an activator, antioxidants, a gelling agent, and so on, to form compounded latex [1,7]. All the chemical agents have to be approximately ground into the micron scale because they can be easily mixed with the concentrated natural latex since the rubber particle is also present in the micron scale [8].

Rubber foam typically presents shape-memory polymer characteristics, which can be altered by useful properties from the external stimuli [9]. This ability enables responsive materials with form-fitting, actuation, and sensing characteristics in industries such as furniture, biomedicine, aerospace, and so on [10,11]. The viscoelastic properties of rubber foam play an important role in the shape-memory effect. However, many customers require natural rubber foam products that return to their initial form in minimal time or have better recoverability. One can enhance the elasticity of rubber foam to improve the recovery performance of rubber foam products.

In 2015, Bashir et al. [12] used eggshell powder (ESP) as a filler in NRF. The contribution of ESP to an increase in mass increases the density of NRF with increasing ESP loading. At low ESP loading, the tensile strength of the ESP-filled NRF initially drops because the filler is not enough to reinforce the NRF. The tensile strength increases after adding more filler loadings due to the reinforcement effect of ESP. The increment of ESP loading causes the ESP-filled NRF to lose recoverability, caused by nonelastic deformation, indicating the deformation of the hard phase from the increase of ESP loading. Another filler in NRF is rice husk powder (RHP). Ramasamy et al. [13] found that the recovery percentage decreases with the increase of RHP loading, whereas the compression strength is increased with increasing RHP loading.

Many previous works have shown that the recoverability of NRF is decreased with increasing filler loading. The main objective of this study is to enhance the recoverability of NRF; thus, we investigated a crosslinked NRF filled with commercial filler consisting of activated charcoal and silica, which are widely used in the rubber industry. This is a useful channel to better understand the relationship of the mechanical properties and thermodynamic theory of NRF, which can be applied for the recovery of rubber foam products.

## 2. Materials and Methods

### 2.1. Materials

The following materials were used: high-ammonia-concentrated natural latex (Num Rubber and Latex Co., Ltd., Trang, Thailand), potassium oleate (KO: 20% aqueous dispersion prepared by Thanodom Technology Co., Ltd., Bangkok, Thailand), sulfur (S: 50% aqueous dispersion prepared by Thanodom Technology Co., Ltd., Bangkok, Thailand), zinc diethyldithiocarbamate (ZDEC: 50% aqueous dispersion prepared by Thanodom Technology Co., Ltd., Bangkok, Thailand), zinc-2-mercaptobenzothiazole (ZMBT: 50% aqueous dispersion prepared by Thanodom Technology Co., Ltd., Bangkok, Thailand), Wingstay L (WingL: 50% aqueous dispersion prepared by Thanodom Technology Co., Ltd., Bangkok, Thailand), zinc oxide (ZnO: 28% aqueous dispersion prepared by Thanodom Technology Co., Ltd., Bangkok, Thailand), diphenylguanidine (DPG: 28% aqueous dispersion prepared by Thanodom Technology Co., Ltd., Bangkok, Thailand), sodium silicofluoride (SSF: 23% aqueous dispersion prepared by Thanodom Technology Co., Ltd., Bangkok, Thailand), toluene (AR grade, RCI Labscan Co., Ltd., Bangkok, Thailand), bamboo charcoal (BET 2–4 m$^2$/g for a 40 μm charcoal particle size, CharcoalHome Co., Ltd., Bangkok, Thailand), precipitated silica (BET 170–190 m$^2$/g for a 10–20 nm silica particle size, Extol Technology Co., Ltd., Dongguan, China).

### 2.2. Preparation of Rubber Foams

Rubber foams were prepared in nine types, as presented in Table 1. For each sample, high-ammonia-concentrated natural latex was weighed, and the ammonia was eliminated using a blender at 80 rpm for 1 min. Filler was later added and mixed for 1 min. The speed of the blender was increased to 160 rpm, and KO was subsequently added and mixed for 10 min until the volume of foam increased by approximately three times. The blending speed was reduced to 80 rpm. A group of chemicals (S, ZDEC, and ZMBT as vulcanizing chemicals and WingL) was added, and mixing continued for 1 min. After that, another group of chemicals (ZnO and DPG) was added, and mixing continued for 1 min. SSF was added at last and mixed for 3 min. During the mixing of SSF, gel-forming was repeatedly tested until the gel point was almost reached. The foam was then poured into molds, and the lids were closed afterward. The samples were left at room temperature for 45 min. Finally, the samples were vulcanized in a hot air oven at 90 °C for 2 h, removed from the molds, washed, and dried at 70 °C.

Table 1. Composition of raw materials for the natural rubber foam (NRF) used in this study.

| Sample Name | Control NRF | NRF/ 2 Ch | NRF/ 4 Ch | NRF/ 6 Ch | NRF/ 8 Ch | NRF/ 2 Si | NRF/ 4 Si | NRF/ 6 Si | NRF/ 8 Si |
|---|---|---|---|---|---|---|---|---|---|
| Chemicals | | | | | (phr [1]) | | | | |
| NRL | 100 | 100 | 100 | 100 | 100 | 100 | 100 | 100 | 100 |
| KO | 3.63 | 3.63 | 3.63 | 3.63 | 3.63 | 3.63 | 3.63 | 3.63 | 3.63 |
| Vulcanizing chemicals | 4.00 | 4.00 | 4.00 | 4.00 | 4.00 | 4.00 | 4.00 | 4.00 | 4.00 |
| WingL | 1.00 | 1.00 | 1.00 | 1.00 | 1.00 | 1.00 | 1.00 | 1.00 | 1.00 |
| ZnO | 2.80 | 2.80 | 2.80 | 2.80 | 2.80 | 2.80 | 2.80 | 2.80 | 2.80 |
| DPG | 0.67 | 0.67 | 0.67 | 0.67 | 0.67 | 0.67 | 0.67 | 0.67 | 0.67 |
| SSF | 1.66 | 1.66 | 1.66 | 1.66 | 1.66 | 1.66 | 1.66 | 1.66 | 1.66 |
| Charcoal | - | 2.00 | 4.00 | 6.00 | 8.00 | - | - | - | - |
| Silica | - | - | - | - | - | 2.00 | 4.00 | 6.00 | 8.00 |

[1] Parts per hundred of rubber.

## 2.3. Characterization

Density was measured from the weight of the rubber foam sample and the measured volume of the rubber foam sample, calculated as follows:

$$\text{Density} = \frac{M}{V} \quad (1)$$

where $M$ is the weight of the rubber foam sample (kg), and $V$ is the volume of the rubber foam sample (m$^3$).

The crosslinking density of the NRF was determined by the swelling method. Small pieces of the NRF were immersed into toluene to reach the equilibrium of swelling. The crosslinking density of rubber ($v$) can be calculated according to the Flory–Rehner equation [14–16] as follows:

$$v = -\frac{1}{V_s}\left[\frac{\ln(1-V_r) + V_r + \chi V_r^2}{V_r^{\frac{1}{3}} - \frac{V_r}{2}}\right] \quad (2)$$

where $V_s$ is the molar volume of toluene (106.9 cm$^3$/mol), $V_r$ is the volume fraction of rubber in the swollen network, and $\chi$ is the parameter between the rubber and solvent interaction (0.43 + 0.05 $V_r$). The NRF sample in toluene was kept in the dark at an ambient temperature. It was removed from the toluene and weighed every day until the equilibrium was reached (approximately 7 days). Finally, the sample was dried in an oven at 60 °C for 24 h and weighed again.

The functional group of the NRF sample was determined by attenuated total reflection–Fourier transform infrared spectroscopy (ATR–FTIR VERTEX 70, Bruker, Billerica, MA, USA). The NRF sample was put on a Ge crystal probe at 500–4000 cm$^{-1}$.

Compression stress of the NRF was determined in 3 replications by a texture analyzer (TA.XT Plus, Stable Micro Systems, Godalming, Surrey, UK) using a platen probe with a diameter of 100 mm and a speed of 0.1 mm/s, adapted from ISO 3386. The NRF sample was 45 mm width × 45 mm length × 21.5 mm thickness, compressed at 75% from the foam surface at room temperature.

The compression set was conducted according to ISO 1856 Method C by measuring the height of the sample ($d_o$). The sample was then compressed at 75 ± 4%height for 72 h at an ambient temperature. Finally, the sample was released from the compression for 30 min, the height of the sample was measured again ($d_r$), and %compression set was calculated as follows:

$$\%\text{compression set} = \left[\frac{d_o - d_r}{d_o} \times 100\right] \quad (3)$$

Furthermore, %recovery of rubber foam can be calculated as follows:

$$\%\text{recovery} = 100 - \%\text{compression set} \quad (4)$$

The morphological properties of the NRF sample were characterized by scanning electron microscopy (SEM) analysis (FEI, Quanta 450 FEI, Eindhoven, Netherlands). The NRF sample was cut into small pieces and coated with gold in a sputter coater (Polaron Range SC7620, Quorum Technologies Ltd., Kent, UK).

For the thermodynamic aspects, the NRF was determined in 3 replications by a texture analyzer (TA.XT Plus, Stable Micro Systems, Godalming, Surrey, UK) using a platen probe with a diameter of 100 mm and speed of 0.1 mm/s. The sample was 45 mm width × 45 mm length × 21.5 mm thickness and prepared at the test temperature 10 min before starting the experiment. The sample was then compressed from 20% strain to 70% strain from the original foam shape at various temperatures (25, 35, 45, 55, and 65 °C). After that, the force–temperature relationship graph was plotted, and the obtained variables were used to calculate the compression force ($F$) due to the changes of internal energy ($F_u$) and entropy ($F_s$) associated with the deformation process.

Other thermodynamic aspects were studied by a temperature sweep of the NRF sample using dynamic mechanical analysis (DMA1, Mettler Toledo, Columbus, OH, USA). The NRF sample was cut into a sample of 7 mm width × 7 mm length × 7 mm thickness and tested at temperatures from 80 to 80 °C. The changes of Gibbs free energy ($\Delta G$) and entropy ($\Delta S$) were calculated. The activation enthalpy of the transition process (($\Delta H_a$)$_{avg}$) for the relaxation of the backbone motion of rubber chains is related to the area under the tan δ peak, which can be calculated by [17]:

$$(\Delta H_a)_{avg} = \frac{\left(\ln E_g - \ln E_r\right)\pi R T_g^2}{t_A} \quad (5)$$

where $t_A$ is the area under the tan δ peak, $E_g$ is the storage modulus at the glassy state, $E_r$ is the storage modulus at the rubbery state, $R$ is the gas constant (8.3145 J/mol·K), and $T_g$ is the glass transition temperature (K) of the NRF.

## 3. Results and Discussion

### 3.1. Physical and Chemical Properties

The density of the NRF samples increases with increasing filler loading, as presented in Figure 1. This is due to increasing filler loading, which causes an increase in the mass of the NRF with filler. Kudori and Ismail [18] found that foam density increases as the filler size decreases. The smaller filler size hinders pore formation and increases the continuous matrix amount. In the present study, silica filler (nanometer) is smaller than charcoal (micrometer). Therefore, NRF with silica loading exhibits a higher density than NRF with charcoal loading at a given filler concentration. Increasing charcoal loading barely affects the density, which may be because charcoal acts as a nucleating agent during the process of foam growth [19–21].

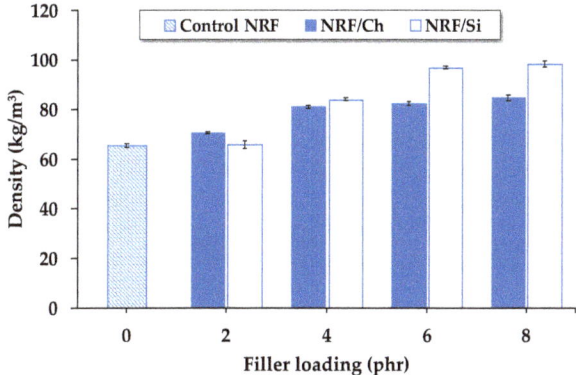

**Figure 1.** Effect of filler loading on the density of the NRF samples.

As presented in Figure 2, the crosslinking density of the NRF samples increases with the presence of filler, which may be due to the additional carbon–sulfur linkages formed by the chemical reaction between the rubber and filler [22]. Another reason is that the amplification of the deformation of rubber chains in the NRF with filler loading is more than the control NRF. Fillers in NRF extend rubber chains due to the interaction of rubber chains at the filler surface, i.e., some rubber chains may be occluded in the voids of the filler, causing the extension of rubber chains and leading to an increased crosslinking density. However, at the same loading of vulcanizing chemicals, increasing filler loading causes fewer differences in the crosslinking density.

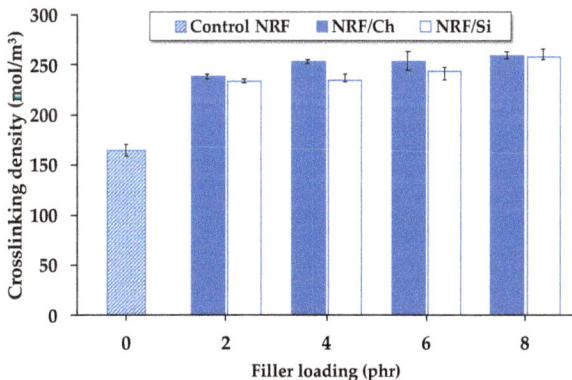

**Figure 2.** Effect of filler loading on the crosslinking density of the NRF samples.

The chemical compositions of the control NRF and NRF with fillers were analyzed by ATR–FTIR (Figure 3). There is no significant difference in the functional groups of NRF [7,23]. There is almost no difference for the NRF with charcoal loading due to carbonization at high temperatures, which causes the charcoal powder to exhibit a hydrophobic nature [24]. However, there is a band growing at 1100 cm$^{-1}$ for the NRF filled with silica. This band corresponds to the vibration absorption of the silane group (Si–O–C) [25], present in the rubber network, which usually exhibits within the ranges 800–850 and 1100–1200 cm$^{-1}$.

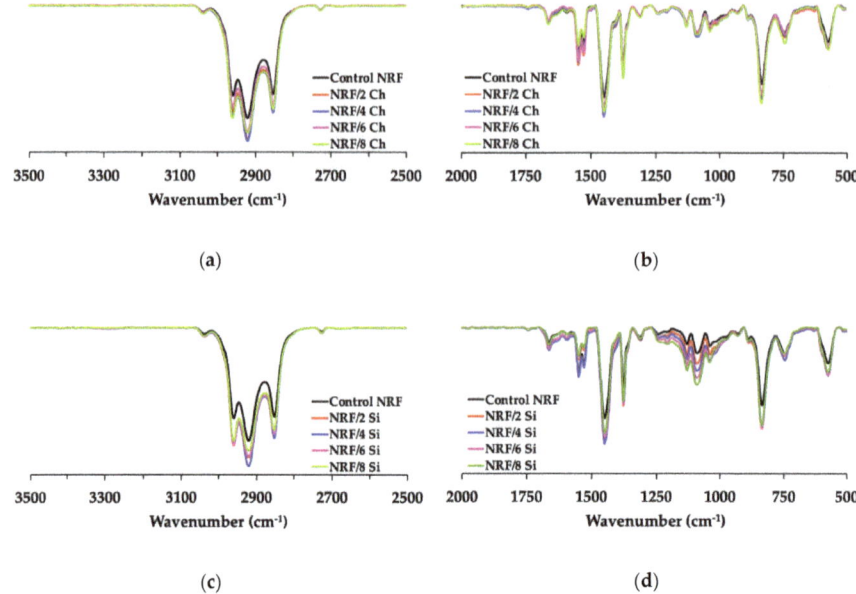

**Figure 3.** ATR–FTIR spectra of the NRF samples: (**a,b**) NRF with charcoal loading; (**c,d**) NRF with silica loading.

### 3.2. Mechanical and Morphological Properties

The control NRF and NRF with fillers were compressed up to 75% (Figure 4). The compression strength at maximum compression shows that the NRF with silica loading has higher compression strength than the NRF with charcoal loading at a given filler concentration. There are two different regions in the compression stress–strain curves of foam materials: elasticity at the low-strain region and solidity at the high-strain region [7]. Increasing filler loading increases the solidity or stiffness of the NRF at high strain, where the foam cells with each other, leading to the immediate increase of compression stress. There is also the stress-induced crystallization of rubber chains that affects the increase of compression stress at high strain. The addition of more than 2 phr of charcoal causes the foam to be sticky, explaining the unfavorable interaction within the foam structure. Although the crosslinking density of the NRF with various silica loadings is almost identical, the compression strength is significantly different. The better interaction within the foam structure is due to the smaller silica filler size, which has more specific surface areas compared to the charcoal filler. We found a linear relationship between compression strength and filler loading in both types of fillers. This relationship depends on the rubber–filler interaction, presented as:

$$C_{Ch} = 0.285 \cdot [Ch] + 5.304 \tag{6}$$

$$C_{Si} = 1.721 \cdot [Si] + 4.649 \tag{7}$$

where $C_{Ch}$ is the compression strength of the NRF with charcoal loading (kPa), $C_{Si}$ is the compression strength of the NRF with silica loading (kPa), [Ch] is the concentration of charcoal filler (phr), and [Si] is the concentration of the silica filler (phr).

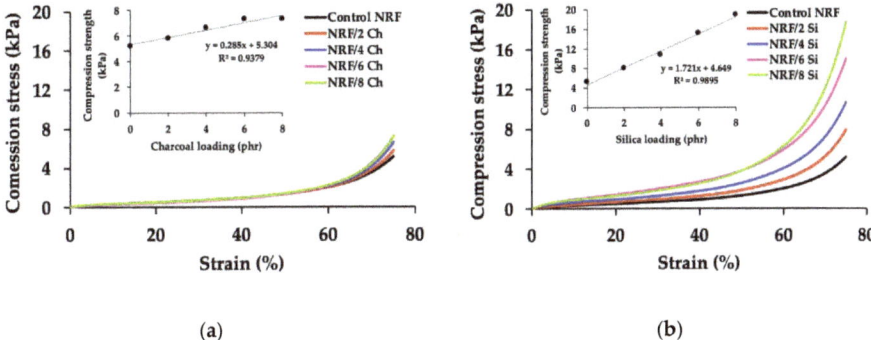

**Figure 4.** Compression stress of the NRFs: (**a**) NRF with charcoal loading; (**b**) NRF with silica loading.

The compression set describes the elastic behavior of the NRF, which relates to the material's recovery percentage. Figure 5 shows that increasing charcoal loading increases the compression set percentage and decreases the recovery percentage. As mentioned above, the addition of more than 2 phr of charcoal causes the foam to be sticky. Thus, when the NRF with more than 2 phr of charcoal loading is compressed at 75%height of its thickness for a long period (72 h), the ability to return to its original shape is decreased. On the contrary, increasing silica loading decreases the compression set percentage and increases the recovery percentage. Decreasing the compression set percentage indicates higher elasticity. Hence, the NRF with silica loading possesses higher elasticity than the NRF with charcoal loading. Microsized charcoal has been shown to behave like eggshell powder and rice husk powder in previous works [12,13], which decreased the percentage of NRF recovery when filler loading is increased and vice versa with NRF-filled nanosized silica. Therefore, we can propose the relationship between recoverability of NRF and filler concentration as the following polynomial equation:

$$\%R_{Ch} = -0.3057 \cdot [Ch]^2 + 1.5206 \cdot [Ch] + 94.697 \tag{8}$$

$$\%R_{Si} = -0.1165 \cdot [Si]^2 + 1.5602 \cdot [Si] + 94.789 \tag{9}$$

where $\%R_{Ch}$ is the recoverability of the NRF with charcoal loading (%), $\%R_{Si}$ is the recoverability of the NRF with silica loading (%), [Ch] is the concentration of charcoal filler (phr), and [Si] is the concentration of silica filler (phr).

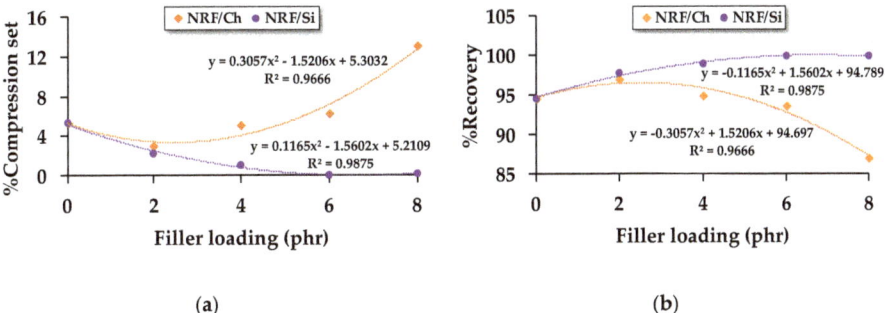

**Figure 5.** Compression recovery properties of the NRFs: (**a**) %compression set of the NRF with different filler loadings; (**b**) %recovery of the NRF with different filler loadings.

The morphological properties of NRF were investigated by SEM. The micrographs indicated that all types of NRF contain a cellular structure that exhibits an interconnected network of open cells, as presented in Figure 6. Porosity analysis was determined by the ImageJ software by adjusting the threshold of the images. The white region corresponds to the pore shape, whereas the dark region corresponds to the open holes or pores (Figure 6). The interconnected porosity of these NRFs is an important parameter that affects the mechanical properties [7]. This result can be explained by the cell density value. The cell density ($d_{cell}$) is calculated as follows [26]:

$$d_{cell} = \frac{3}{4\pi r^3}\left(1 - \frac{\rho}{\rho_s}\right) \qquad (10)$$

where $\rho$ is the foam density, $\rho_s$ is the solid phase density (NR 0.93 g/cm$^3$), and $r$ is the average cell radius.

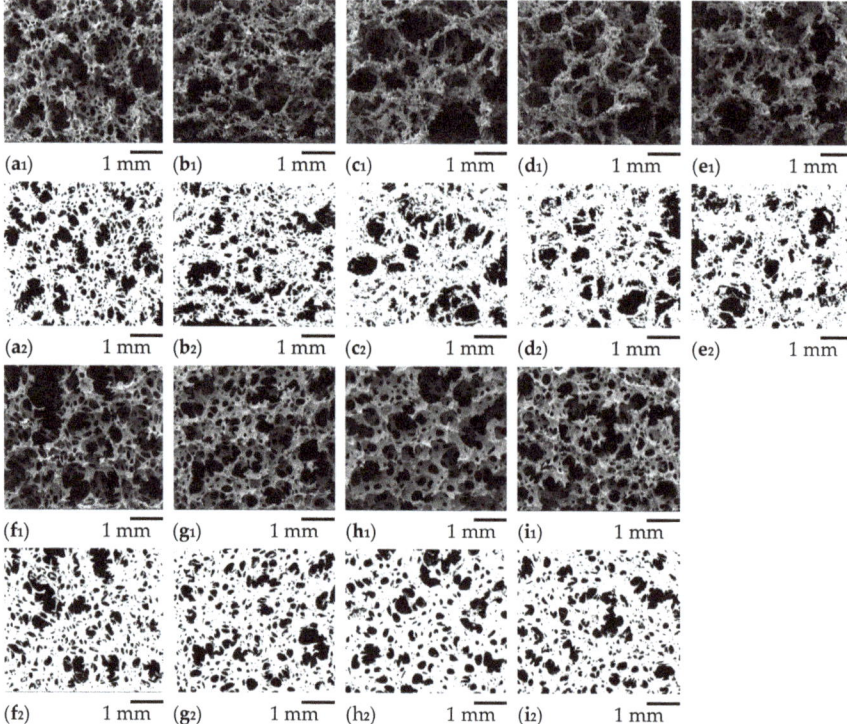

**Figure 6.** SEM images of: (**a1**) control NRF; (**b1**) NRF/2 Ch; (**c1**) NRF/4 Ch; (**d1**) NRF/6 Ch; (**e1**) NRF/8 Ch; (**f1**) NRF/2 Si; (**g1**) NRF/4 Si; (**h1**) NRF/6 Si; (**i1**) NRF/8 Si. The SEM images with ImageJ analysis of: (**a2**) control NRF; (**b2**) NRF/2 Ch; (**c2**) NRF/4 Ch; (**d2**) NRF/6 Ch; (**e2**) NRF/8 Ch; (**f2**) NRF/2 Si; (**g2**) NRF/4 Si; (**h2**) NRF/6 Si; (**i2**) NRF/8 Si.

As presented in Table 2, increasing charcoal loading increases the average NRF pore size, decreasing the porosity and cell density. On the other hand, increasing silica loading decreases the average pore size, increasing the porosity and cell density. Although the density of the NRF increases with increasing filler loading, the cell density is more complicated. For charcoal loading, the density slightly increases or remains almost constant with the addition of more than 4 phr of charcoal. The cell density of the NRF with charcoal loading decreases and becomes almost constant with the addition of more than 4 phr of charcoal. Since charcoal can act as the nucleating agent, which can promote foam growth, excess charcoal may stop acting as a filler and become a nucleating agent, resulting in an almost constant density with a larger average pore size and smaller porosity and cell density. For silica loading, the density increases as increasing silica loading, indicating the decreasing of the average pore size while the cell density increases.

**Table 2.** Average pore size, porosity, and cell density of the control NRF and NRF with filler loading.

| Sample Name | Average Pore Size (±0.300 mm) | Porosity (±1.00%) | Cell Density (±150 cm$^{-3}$) |
|---|---|---|---|
| Control NRF | 0.836 | 31.39 | 3041 |
| NRF/2 Ch | 0.860 | 31.69 | 2778 |
| NRF/4 Ch | 0.988 | 27.58 | 1807 |
| NRF/6 Ch | 1.081 | 25.46 | 1379 |
| NRF/8 Ch | 1.092 | 25.04 | 1333 |
| NRF/2 Si | 1.079 | 28.57 | 1412 |
| NRF/4 Si | 0.909 | 28.09 | 2316 |
| NRF/6 Si | 0.876 | 27.49 | 2546 |
| NRF/8 Si | 0.673 | 26.34 | 5606 |

### 3.3. Thermodynamic Aspects

Thermodynamic studies of the deformation process in uncrosslinked rubbers have already been discussed [27–29]. Most of the works showed the results of the temperature dependence of the stress in the extended state. In the present work, the mechanical compression properties of the crosslinked NRF samples are remarkable, especially for the %compression set and %recovery; thus, it is interesting to investigate the thermodynamic aspects. From the perspective of thermodynamics, the elasticity of rubber attributes to the changes in the conformations of rubber molecules from the unstrained molecules to the strained molecules. Such changes are related to the changes of internal energy and entropy associated with the deformation process as the following relationship [27,30,31]:

$$F = \left(\frac{\partial U}{\partial L}\right) - T\left(\frac{\partial S}{\partial L}\right) = F_u + F_s \tag{11}$$

$$F_u = \left(\frac{\partial U}{\partial L}\right) \tag{12}$$

$$F_s = -T\left(\frac{\partial S}{\partial L}\right) \tag{13}$$

where $F$ is the compression force causing changes in the length of NRF ($L$), $U$ is the internal energy of NRF, $T$ is the temperature used in the experiment, and $S$ is the entropy of NRF. When plotting the compression force graph as a function of the conducted temperature, the interception at 0 K is equal to $F_u$, and the slope multiplied by the temperature is equal to $F_s$.

To investigate the relationships between force (compression mode) and temperature, the NRF samples were compressed up to 20%strain, 30%strain, 40%strain, 50%strain, 60%strain, and 70%strain in the temperature controller chamber (at 298.15, 308.15, 318.15, 328.15, and 338.15 K). The relationships between compression force and temperature of the control NRF, NRF with 8 phr of charcoal, and NRF with 8 phr of silica are presented in Figures 7–9, respectively. The graphs of the other samples are presented in Figures S1–S6. The results reveal that the compression force to the sample increases with increasing %strain. At a given strain, the compression force decreases with increasing temperature. Moreover, the slope decreases at a higher strain due to a decrease of the rubber chains' degree of freedom in the NRF, which is unfavorable for entropy.

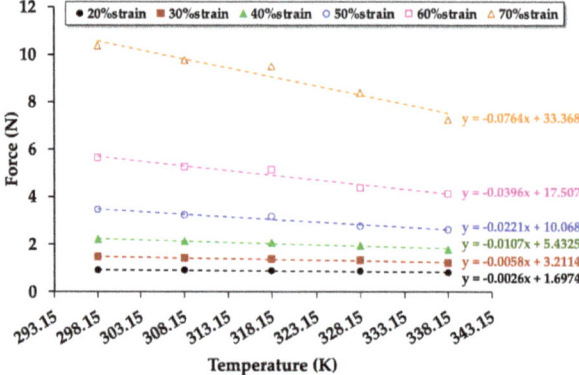

**Figure 7.** Force at a constant strain as a function of the temperature of the control NRF with a minimum of $R^2 = 0.9$ in each strain.

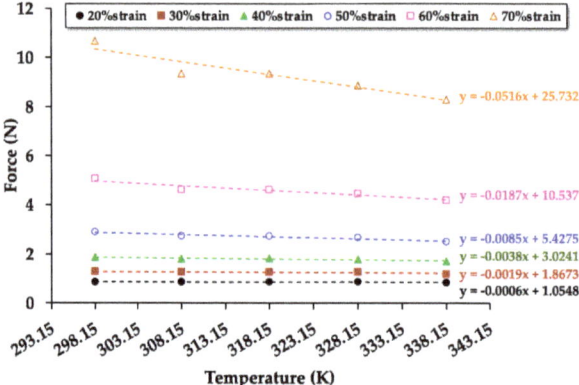

**Figure 8.** Force at a constant strain as a function of the temperature of the NRF with 8 phr of charcoal (NRF/8 Ch) with a minimum of $R^2 = 0.9$ in each strain.

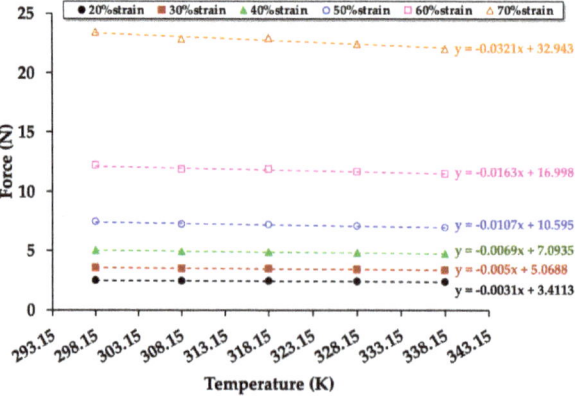

**Figure 9.** Force at a constant strain as a function of the temperature of the NRF with 8 phr of silica (NRF/8 Si) with a minimum of $R^2 = 0.9$ in each strain.

Figures 7–9 show the values of the compression force ($F$) and relative internal energy contributing to the compression force ($F_u/F$) at 298.15 K can be calculated as indicated in Table 3. The results of the other samples are presented in Table S1. The values of $F_u$ and $F$ increase with increasing compression strain, whereas the values of $F_u/F$ decrease. Since the internal energy ($U$) is varied by the compression force ($F$), the internal energy increases with increasing compression force. The entropy ($S$) can be varied by the length of the NRF, indicating the degree of freedom of rubber chains during the compression process. The compression causes a reduction in the length of the NRF, leading to a decrease in the rubber chains' degree of freedom. Thus, the entropy of compressed NRF is also reduced.

**Table 3.** Compression strain, compression limit, $F_u$, $F$, and $F_u/F$ values of the control NRF and NRF with various fillers at 298.15 K.

| Sample Name | Compression Strain (%) | Compression Limit ($\lambda$) | $F_u$ (N) | $F$ (N) | $F_u/F$ |
|---|---|---|---|---|---|
| Control NRF | 20 | 0.8 | 1.70 | 2.47 | 0.6865 |
|  | 30 | 0.7 | 3.21 | 4.94 | 0.6500 |
|  | 40 | 0.6 | 5.43 | 8.62 | 0.6300 |
|  | 50 | 0.5 | 10.07 | 16.66 | 0.6044 |
|  | 60 | 0.4 | 17.51 | 29.31 | 0.5972 |
|  | 70 | 0.3 | 33.37 | 56.15 | 0.5943 |
| NRF/8 Ch | 20 | 0.8 | 1.05 | 1.23 | 0.8550 |
|  | 30 | 0.7 | 1.87 | 2.43 | 0.7672 |
|  | 40 | 0.6 | 3.02 | 4.16 | 0.7275 |
|  | 50 | 0.5 | 5.43 | 7.96 | 0.6817 |
|  | 60 | 0.4 | 10.54 | 16.11 | 0.6540 |
|  | 70 | 0.3 | 25.73 | 41.12 | 0.6258 |
| NRF/8 Si | 20 | 0.8 | 3.41 | 4.34 | 0.7868 |
|  | 30 | 0.7 | 5.07 | 6.56 | 0.7727 |
|  | 40 | 0.6 | 7.09 | 9.15 | 0.7752 |
|  | 50 | 0.5 | 10.60 | 13.79 | 0.7686 |
|  | 60 | 0.4 | 17.00 | 21.86 | 0.7777 |
|  | 70 | 0.3 | 32.94 | 42.51 | 0.7749 |

The $F_u/F$ values of uncrosslinked rubber in the extension mode are typically in the range of 0.1–0.2 [27]. In this work, the $F_u/F$ values of the crosslinked NRF in the compression mode are in the range of 0.6–0.8, which are approximately three times higher than those of the literature review. The difference in $F_u/F$ values could come from the material structures (uncrosslinked rubber vs. crosslinked rubbers) and the test methods (extension mode vs. compression mode). The NRFs with fillers have higher $F_u/F$ values than the control NRF at a given strain level, possibly explained by the interaction of rubber and filler, which promotes changes of entropy in the deformation process. The slope direction of the $F_u/F$ values of the NRF with charcoal and NRF with silica are different (Figure 10). This is due to the control NRF and NRF with charcoal loading possess different degrees of freedom of rubber chains at different compression limits, and lower compression limit leads to lower $F_u/F$ values. However, the NRFs with silica loading possess a similar degree of freedom of rubber chains at different compression limits. This indicates that the stability of the degree of freedom of rubber chains at different compression strains or compression limits ($\lambda$) is related to the high mechanical property of the NRF with silica loading. Although the compression strength of the NRF with filler increased with the increment of filler, the ratio of $F_u/F$ indicates that the addition of filler affects the mechanical properties in the aspect of thermodynamics.

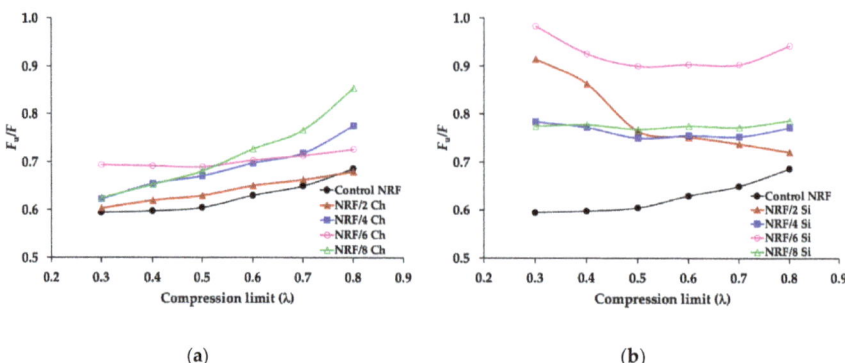

**Figure 10.** Relative internal energy contribution to the compression force ($F_u/F$): (**a**) NRF with charcoal loading compared to control NRF; (**b**) NRF with silica loading compared to control NRF.

We also investigated the change in Gibbs free energy ($\Delta G$) and entropy ($\Delta S$) in the NRF with fillers compared to the control NRF. These thermodynamic parameters can be calculated by the Flory–Huggins equation and statistical theory of rubber elasticity as follows [17,32]:

$$\Delta G = RT\left[\ln(1 - V_r) + V_r + V_r^2 \chi\right] \tag{14}$$

$$\Delta S = -\frac{\Delta G}{T} \tag{15}$$

where $R$ is the gas constant (8.3145 J/mol·K), and $T$ is the test temperature (300.15 K).

From the perspective of the crosslinking density, $\Delta G$ and $\Delta S$ are shown in Table 4. The volume fraction of rubber ($V_r$) with fillers is higher than the control NRF. The swelling behavior of the NRF with various filler loadings decreases with increasing filler loading, indicating that the filler enhances the rubber swelling resistance against the penetration of the solvent. Since the filler is the hard phase, which is impermeable to solvent molecules, there must be a higher interaction between phases and more rubber chains attached to the filler surface. Hence, the swelling ability of NR is reduced while increasing the volume fraction of rubber [17].

**Table 4.** Calculated parameters from crosslinking density results for the control NRF and NRF with various fillers.

| Sample Name | Swelling Ratio | Volume Fraction of Rubber ($V_r$) | $\Delta G$ (J/mol) | $\Delta S$ (J/mol·K) |
|---|---|---|---|---|
| Control NRF | 2.83 | 0.2377 | −29.15 | 0.0971 |
| NRF/2 Ch | 2.33 | 0.2755 | −42.85 | 0.1428 |
| NRF/4 Ch | 2.24 | 0.2823 | −45.74 | 0.1524 |
| NRF/6 Ch | 2.25 | 0.2790 | −44.33 | 0.1477 |
| NRF/8 Ch | 2.23 | 0.2843 | −46.61 | 0.1553 |
| NRF/2 Si | 2.33 | 0.2739 | −42.21 | 0.1406 |
| NRF/4 Si | 2.30 | 0.2732 | −41.91 | 0.1396 |
| NRF/6 Si | 2.24 | 0.2777 | −43.77 | 0.1458 |
| NRF/8 Si | 2.19 | 0.2839 | −46.41 | 0.1546 |

Table 4 shows that all samples present a negative $\Delta G$, which decreases with increasing filler and is a favorable spontaneous system. This is due to the restriction in the ability of the rubber chain motion in the presence of filler, resulting in a decrease in the Gibbs free energy, which can be attributed to good compatibility between polymer and filler [17,32]. ΔS increases with increasing filler loading, which is favorable in thermodynamics. This result is in good agreement with the result of the $F_u/F$ value.

Based on the dynamic mechanical properties of the sample, the storage modulus ($E'$) and tan δ results of the NRF with filler loading were determined by temperature sweep using dynamic mechanical analysis or DMA. The results are presented in Figure 11. The addition of filler, both charcoal and silica, affects the storage modulus and tan δ.

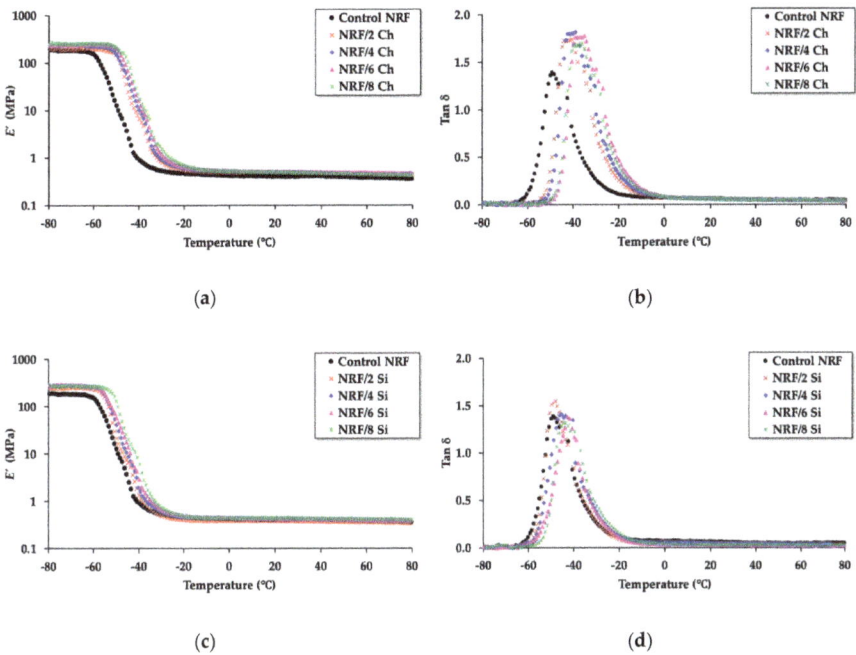

**Figure 11.** Storage modulus ($E'$) and tan δ as a function of the temperature of the NRF with various fillers: (**a**) storage modulus of the NRF with charcoal loading compared to the control NRF; (**b**) tan δ of the NRF with charcoal loading compared to the control NRF; (**c**) storage modulus of the NRF with silica loading compared to the control NRF; (**d**) tan δ of the NRF with silica loading compared to the control NRF.

The DMA results presented in Table 5 reveal that the storage modulus in both the glassy state and rubbery state of the NRF with filler loading is higher than the control NRF. The addition of filler decreases the free volume within the foam, which causes more rigidity, resulting in a higher storage modulus in the glassy state [33]. The storage modulus in the rubbery state of the control NRF is lower than the NRF with filler loading. This is due to the effect of the filler on the relaxation time of rubber chains. Increasing filler loading increases the volume fraction of filler ($\Delta V_f$), which causes a higher stress relaxation rate of rubber molecules where they require more time to unload the applied force [17,33]. This affects the degree of freedom of the rubber molecules to be more pronounced, i.e., when there is a greater number of interactions between the rubber chains and filler, the stress relaxation rate is increased, resulting in an increase in entropy [33,34]. Therefore, we can propose a model of the control NRF (Figure 12a) compared to the NRF with filler loading (Figure 12b). The thermodynamic meaning of this work can be explained as follows: the change in entropy (ΔS) of the NRF with filler loading is more pronounced compared to the control NRF due to the stress relaxation rate of rubber chains from the amplification of chain deformation between the rubber and filler interaction, as shown in Equation (16).

$$\Delta S = \frac{\Delta V_f \cdot \Delta St}{T} \quad (16)$$

where $\Delta S$ is the change of entropy, $\Delta V_f$ is the change of volume fraction of filler, $\Delta St$ is the change of stress relaxation rate of rubber molecules, and $T$ is the temperature.

**Table 5.** Obtained parameters by the dynamic mechanical analysis (DMA) test for the control NRF and NRF with various fillers.

| Sample Name | $E_g$ @ −70 °C (MPa) | $E_r$ @ 0 °C (MPa) | $T_g$ (°C) | tan δ max | $t_A$ | $(\Delta H_a)_{avg}$ (kJ·K/mol) |
|---|---|---|---|---|---|---|
| Control NRF | 184.05 | 0.42 | −49.08 | 1.39 | 31.08 | 128.33 |
| NRF/2 Ch | 193.35 | 0.49 | −40.75 | 1.75 | 39.04 | 107.81 |
| NRF/4 Ch | 248.15 | 0.51 | −37.17 | 1.82 | 40.87 | 109.92 |
| NRF/6 Ch | 240.76 | 0.54 | −35.17 | 1.78 | 40.24 | 112.16 |
| NRF/8 Ch | 251.01 | 0.52 | −36.17 | 1.69 | 38.13 | 118.76 |
| NRF/2 Si | 250.74 | 0.39 | −48.17 | 1.54 | 28.25 | 151.17 |
| NRF/4 Si | 283.10 | 0.44 | −45.50 | 1.41 | 30.51 | 143.53 |
| NRF/6 Si | 294.05 | 0.45 | −43.00 | 1.39 | 25.73 | 174.06 |
| NRF/8 Si | 280.63 | 0.44 | −44.67 | 1.32 | 27.60 | 159.44 |

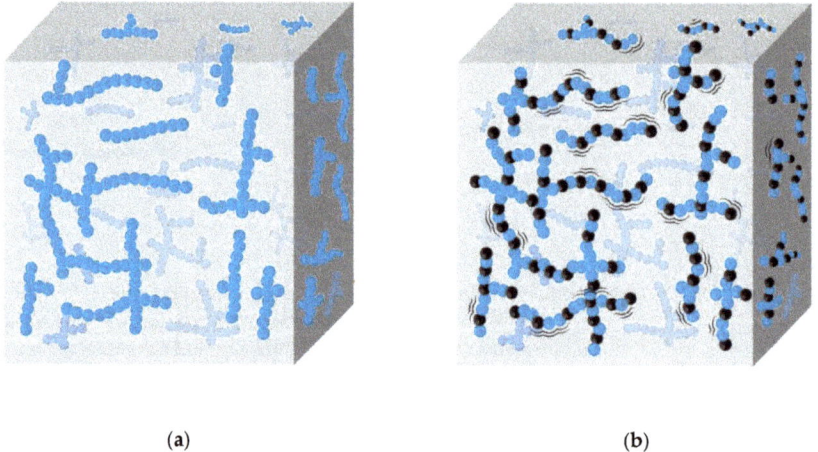

(a)          (b)

**Figure 12.** Model of the NRF with rubber chains (blue spot) and fillers (black spot): (a) control NRF; (b) NRF with filler, which increases the volume fraction of filler ($\Delta V_f$) induces more stress relaxation ($\Delta St$), resulting in a more pronounced entropy ($\Delta S$).

Table 5 also presents the $T_g$ value or peak of tan δ of the NRF with various fillers. The addition of filler causes this value to shift toward higher temperatures when compared to the control NRF. The shift of the $T_g$ value toward the higher temperatures indicates ionic and hydrogen bonding interactions between the rubber chains and filler [17]. However, the nonpolar charcoal might not disperse well in the concentrated natural latex or agglomerate and, instead, form filler–filler networks within the foam. This may cause a synergy effect where the filler–filler networks might defeat the movement of the free chains of rubber. Therefore, the addition of charcoal affects higher hysteresis with increasing tan δ max and $t_A$, resulting in lower activation enthalpy ($\Delta H_a$) than the control NRF.

At the same time, NRF is well-reinforced with silica. The rubber chains within the NRF with silica are hindered to freely move, and there are interactions between rubber–filler within the NRF.

Thus, it has a higher activation enthalpy than the control NRF. The tan δ max of the NRF with filler has a higher value than control NRF refers to more dissipation energy of the NRF in the existence of filler. The values of $\Delta H_a$ in this work are in the same order as in the works of Sadeghi Ghari and Jalali-Arani [17].

## 4. Conclusions

In this work, natural rubber foam (NRF) was prepared in two conditions: NRF with charcoal loading and NRF with silica loading. The results showed that increasing filler loading increases the density and mechanical properties of rubber foam. Since rubber chains may be occluded in the voids of filler causing the expansion of rubber chains, leading to increasing crosslinking density, somehow, at the same loading of vulcanizing chemicals, increasing filler loading is affected less in the crosslinking density.

Increasing filler loading increases the compression stress of NRF. The compression strength of the NRF with silica loading is higher than NFR with charcoal loading due to the better interaction within the foam structure caused by the smaller silica size, which presents a more specific surface area compared to charcoal filler. Since charcoal can act as a nucleating agent and promote foam growth, the excess of charcoal might change from being filler and become the nucleating agent, resulting in a larger average pore size and smaller porosity and cell density. Increased silica loading results in a decrease in the average pore size while the cell density increases. Thermodynamic aspects showed the relationships between the force and temperature of the NRF samples. The compression force ($F$) and internal energy force ($F_u$) values of the NRF samples increase with increasing compression strain. The NRF with various fillers has higher $F_u/F$ values than the control NRF. The slope direction of the $F_u/F$ value of the NRF with charcoal and NRF with silica are different, which comes from the different degrees of freedom of rubber chains in the NRF samples.

The swelling behavior of the NRF with filler loading decreases with increasing filler loading compared to the control NRF. The $\Delta G$ decreases while $\Delta S$ increases with increasing filler loading, which demonstrates a favorable thermodynamic system. The $\Delta H_a$ of the control NRF is lower than NRF with silica due to the movement limitation of rubber chains, whereas NRF filled with charcoal is more complicated. Increasing filler loading increases the volume fraction of filler, causing a higher stress relaxation rate due to the attempt to relax itself of molecules; therefore, the entropy is more pronounced. All the results from the theory to applications indicate that NRF, which normally behaves like a shape-memory material, can be developed into an anti-shape-memory material by the addition of silica loading, which is favorable in thermodynamics.

**Supplementary Materials:** The following are available online at http://www.mdpi.com/2073-4360/12/11/2745/s1, Figure S1. Force at a constant strain as a function of the temperature of the NRF with 2 phr of charcoal (NRF/2 Ch) with a minimum of $R^2$ = 0.9 in each strain. Figure S2. Force at a constant strain as a function of the temperature of the NRF with 4 phr of charcoal (NRF/4 Ch) with a minimum of $R^2$ = 0.9 in each strain. Figure S3. Force at a constant strain as a function of the temperature of the NRF with 6 phr of charcoal (NRF/6 Ch) with a minimum of $R^2$ = 0.9 in each strain. Figure S4. Force at a constant strain as a function of the temperature of the NRF with 2 phr of silica (NRF/2 Si) with a minimum of $R^2$ = 0.9 in each strain. Figure S5. Force at a constant strain as a function of the temperature of the NRF with 4 phr of silica (NRF/4 Si) with a minimum of $R^2$ = 0.9 in each strain. Figure S6. Force at a constant strain as a function of the temperature of the NRF with 6 hr of silica (NRF/6 Si) with a minimum of $R^2$ = 0.9 in each strain. Table S1. Compression strain, compression limit, $F_u$, $F$, and $F_u/F$ values of the control NRF and NRF with various fillers at 298.15 K.

**Author Contributions:** Conceptualization, W.S.; methodology, W.S.; software, T.P.; validation, W.S.; formal analysis, T.P.; investigation, W.S. and T.P.; resources, W.S.; data curation, W.S. and T.P.; writing—original draft preparation, T.P.; writing—review and editing, W.S.; visualization, W.S. and T.P.; supervision, W.S.; project administration, W.S.; funding acquisition, W.S. Both authors have read and agreed to the published version of the manuscript. All authors have read and agreed to the published version of the manuscript.

**Funding:** This research received no external funding.

**Acknowledgments:** This research was supported by a graduate study development scholarship from the National Research Council of Thailand as of the 2020 fiscal year. This research was also supported by the Specialized Center of Rubber and Polymer Materials in Agriculture and Industry (RPM), Faculty of Science, Kasetsart University, Bangkok, Thailand.

**Conflicts of Interest:** The authors declare no conflict of interest in this research. The funders had no role in the design of the study; in the collection, analyses, or interpretation of data; in the writing of the manuscript, or in the decision to publish the results.

## References

1. Chollakup, R.; Suwanruji, P.; Tantatherdtam, R.; Smitthipong, W. New approach on structure-property relationships of stabilized natural rubbers. *J. Polym. Res.* **2019**, *26*, 37. [CrossRef]
2. Oliveira-Salmazo, L.; Lopez-Gil, A.; Silva-Bellucci, F.; Job, A.E.; Rodriguez-Perez, M.A. Natural rubber foams with anisotropic cellular structures: Mechanical properties and modeling. *Ind. Crop. Prod.* **2016**, *80*, 26–35. [CrossRef]
3. Kim, D.Y.; Park, J.W.; Lee, D.Y.; Seo, K.H. Correlation between the Crosslink Characteristics and Mechanical Properties of Natural Rubber Compound via Accelerators and Reinforcement. *Polymers* **2020**, *12*, 2020. [CrossRef]
4. Nawamawat, K.; Sakdapipanich, J.T.; Ho, C.C.; Ma, Y.; Song, J.; Vancso, J.G. Surface nanostructure of Hevea brasiliensis natural rubber latex particles. *Colloids Surf. A Physicochem. Eng. Asp.* **2011**, *390*, 157–166. [CrossRef]
5. Suethao, S.; Shah, D.U.; Smitthipong, W. Recent Progress in Processing Functionally Graded Polymer Foams. *Materials* **2020**, *13*, 4060. [CrossRef]
6. Phomrak, S.; Nimpaiboon, A.; Newby, B.-m.Z.; Phisalaphong, M. Natural Rubber Latex Foam Reinforced with Micro- and Nanofibrillated Cellulose via Dunlop Method. *Polymers* **2020**, *12*, 1959. [CrossRef]
7. Suksup, R.; Sun, Y.; Sukatta, U.; Smitthipong, W. Foam rubber from centrifuged and creamed latex. *J. Polym. Eng.* **2019**, *39*. [CrossRef]
8. Blackley, D.C. *Polymer Latices: Science and Technology Volume 2: Types of Latices*, 2nd ed.; Springer: Dordrecht, The Netherlands, 1997.
9. Kamila, S. Introduction, Classification and Applications of Smart Materials: An Overview. *Am. J. Appl. Sci.* **2013**, *10*, 876–880. [CrossRef]
10. Cavicchi, K.A. Shape Memory Polymers from Blends of Elastomers and Small Molecule Additives. *Macromol. Symp.* **2015**, *358*, 194–201. [CrossRef]
11. Gunes, I.S.; Cao, F.; Jana, C.S. Evaluation of Nanoparticulate Fillers for Shape Memory Polyurethane Nanocomposites. *Polymer* **2008**, *49*, 2223–2234. [CrossRef]
12. Bashir, A.S.M.; Manusamy, Y.; Chew, T.L.; Ismail, H.; Ramasamy, S. Mechanical, thermal, and morphological properties of (eggshell powder)-filled natural rubber latex foam. *J. Vinyl Addit. Technol.* **2017**, *23*, 3–12. [CrossRef]
13. Ramasamy, S.; Ismail, H.; Munusamy, Y. Effect of Rice Husk Powder on Compression Behavior and Thermal Stability of Natural Rubber Latex Foam. *BioResources* **2013**, *8*, 4258–4269. [CrossRef]
14. Flory, P.J.; Rehner, J. Statistical Mechanics of Cross-Linked Polymer Networks II. Swelling. *J. Chem. Phys.* **1943**, *11*, 521–526. [CrossRef]
15. Smitthipong, W.; Nardin, M.; Schultz, J.; Suchiva, K. Adhesion and self-adhesion of rubbers, crosslinked by electron beam irradiation. *Int. J. Adhes. Adhes.* **2007**, *27*, 352–357. [CrossRef]
16. Nimpaiboon, A.; Sakdapipanich, J. Properties of peroxide cured highly purified natural rubber. *Kautsch. Gummi Kunstst.* **2012**, *65*, 55–59.
17. Sadeghi Ghari, H.; Jalali-Arani, A. Nanocomposites based on natural rubber, organoclay and nano-calcium carbonate: Study on the structure, cure behavior, static and dynamic-mechanical properties. *Appl. Clay Sci.* **2016**, *119*, 348–357. [CrossRef]
18. Kudori, S.N.I.; Ismail, H. The effects of filler contents and particle sizes on properties of green kenaf-filled natural rubber latex foam. *Cell. Polym.* **2019**, *39*, 57–68. [CrossRef]
19. Zhang, Q.; Lin, X.; Chen, W.; Zhang, H.; Han, D. Modification of Rigid Polyurethane Foams with the Addition of Nano-SiO$_2$ or Lignocellulosic Biomass. *Polymers* **2020**, *12*, 107. [CrossRef]
20. Rodrigue, D.; Souici, S.; Twite-Kabamba, E. Effect of wood powder on polymer foam nucleation. *J. Vinyl Addit. Technol.* **2006**, *12*, 19–24. [CrossRef]

21. Aussawasathien, D.; Jariyakun, K.; Pomrawan, T.; Hrimchum, K.; Yeetsorn, R.; Prissanaroon-Ouajai, W. Preparation and properties of low density polyethylene-activated carbon composite foams. *AIP Conf. Proc.* **2017**, *1914*, 060003. [CrossRef]
22. Thongsang, S.; Sombatsompop, N. Effect of filler surface treatment on properties of fly ash/NR blends. In *ANTEC 2005—Volume 8, Proceedings of the Annual Technical Conference—ANTEC, Boston, MA, USA, 1–5 May 2005*; Society of Plastics Engineers: Brookfield, WI, USA, 2005; pp. 3278–3282.
23. Promhuad, K.; Smitthipong, W. Effect of Stabilizer States (Solid vs. Liquid) on Properties of Stabilized Natural Rubbers. *Polymers* **2020**, *12*, 741. [CrossRef] [PubMed]
24. Li, S.; Wang, H.; Chen, C.; Li, X.; Deng, Q.; Gong, M.; Li, D. Size effect of charcoal particles on the properties of bamboo charcoal/ultra-high molecular weight polyethylene composites. *J. Appl. Polym. Sci.* **2017**, *134*, 45530. [CrossRef]
25. Ain, Z.N.; Azura, A.R. Effect of different types of filler and filler loadings on the properties of carboxylated acrylonitrile–butadiene rubber latex films. *J. Appl. Polym. Sci.* **2011**, *119*, 2815–2823. [CrossRef]
26. Forest, C.; Chaumont, P.; Cassagnau, P.; Swoboda, B.; Sonntag, P. Polymer nano-foams for insulating applications prepared from $CO_2$ foaming. *Prog. Polym. Sci.* **2015**, *41*, 122–145. [CrossRef]
27. Treloar, L.R.G. *The Physics of Rubber Elasticity*; Oxford University Press: New York, NY, USA, 1975.
28. Rubinstein, M.; Colby, R.H. *Polymer Physics*; Oxford University Press: New York, NY, USA, 2003.
29. Hiemenz, P.C.; Lodge, T.P. *Polymer Chemistry*; CRC Press: Boca Raton, FL, USA, 2007.
30. Roe, R.J.; Krigbaum, W.R. The contribution of internal energy to the elastic force of natural rubber. *J. Polym. Sci.* **1962**, *61*, 167–183. [CrossRef]
31. Pakornpadungsit, P.; Smitthipong, W.; Chworos, A. Self-assembly nucleic acid-based biopolymers: Learn from the nature. *J. Polym. Res.* **2018**, *25*, 45. [CrossRef]
32. Pojanavaraphan, T.; Magaraphan, R. Prevulcanized natural rubber latex/clay aerogel nanocomposites. *Eur. Polym. J.* **2008**, *44*, 1968–1977. [CrossRef]
33. Phuhiangpa, N.; Ponloa, W.; Phongphanphanee, S.; Smitthipong, W. Performance of Nano- and Microcalcium Carbonate in Uncrosslinked Natural Rubber Composites: New Results of Structure–Properties Relationship. *Polymers* **2020**, *12*, 2002. [CrossRef]
34. Maria, H.J.; Lyczko, N.; Nzihou, A.; Joseph, K.; Mathew, C.; Thomas, S. Stress relaxation behavior of organically modified montmorillonite filled natural rubber/nitrile rubber nanocomposites. *Appl. Clay Sci.* **2014**, *87*, 120–128. [CrossRef]

**Publisher's Note:** MDPI stays neutral with regard to jurisdictional claims in published maps and institutional affiliations.

© 2020 by the authors. Licensee MDPI, Basel, Switzerland. This article is an open access article distributed under the terms and conditions of the Creative Commons Attribution (CC BY) license (http://creativecommons.org/licenses/by/4.0/).

Article

# Adherence Kinetics of a PDMS Gripper with Inherent Surface Tackiness

Umut D. Çakmak [1,*], Michael Fischlschweiger [2], Ingrid Graz [3] and Zoltán Major [1]

[1] Institute of Polymer Product Engineering, Johannes Kepler University Linz, Altenbergerstrasse 69, 4040 Linz, Austria; zoltan.major@jku.at
[2] Chair of Technical Thermodynamics and Energy Efficient Material Treatment, Clausthal University of Technology, Agricolastrasse 4, 38678 Clausthal-Zellerfeld, Germany; michael.fischlschweiger@tu-clausthal.de
[3] School of Education, Johannes Kepler University Linz, Altenbergerstrasse 69, 4040 Linz, Austria; ingrid.graz@jku.at
* Correspondence: umut.cakmak@jku.at; Tel.: +43-732-2468-6596; Fax: +43-732-2468-6593

Received: 18 September 2020; Accepted: 19 October 2020; Published: 22 October 2020

**Abstract:** Damage and fiber misalignment of woven fabrics during discontinuous polymer processing remain challenging. To overcome these obstacles, a promising switchable elastomeric adherence gripper is introduced here. The inherent surface tackiness is utilized for picking and placing large sheets. Due to the elastomer's viscoelastic material behavior, the surface properties depend on loading speed and temperature. Different peeling speeds result in different adherence strength of an interface between the gripper and the substrate. This feature was studied in a carefully designed experimental test set-up including dynamic thermomechanical, as well as dynamic mechanical compression analyses, and adherence tests. Special emphases were given to the analyses of the applicability as well as the limitation of the viscoelastic gripper and the empirically modeling of the gripper's pulling speed-dependent adherence characteristic. Two formulations of poly(dimethylsiloxane) (PDMS) with different hardnesses were prepared and analyzed in terms of their applicability as gripper. The main insights of the analyses are that the frequency dependency of the loss factor tan$\delta$ is of particular importance for the application along with the inherent surface tackiness and the low sensitivity of the storage modulus to pulling speed variations. The PDMS-soft material formulation exhibits the ideal material behavior for an adhesive gripper. Its tan$\delta$ varies within the application relevant loading speeds between 0.1 and 0.55; while the PDMS-hard formulation reveals a narrower tan$\delta$ range between 0.09 and 0.19. Furthermore, an empirical model of the pulling speed-dependent strain energy release rate G(v) was derived based on the experimental data of the viscoelastic characterizations and the probe tack tests. The proposed model can be utilized to predict the maximum mass (weight-force) of an object that can be lifted by the gripper

**Keywords:** silicon rubber (PDMS); dynamic thermomechanical analyses; storage modulus; mechanical loss factor; viscoelastic gripper

## 1. Introduction

The picking and placing of limp solids (textile, woven fabrics, soft, and flexible electronics) during assembling in manufacturing processes remain challenging. The transfer to the desired position has to be achieved without damaging or misaligning the limp solid (i.e., fiber unravel, pull out, contamination by the gripper, etc.). A variety of technologies are available to overcome this challenge including vacuum suction, ingressive and adhesive grippers. The first two methods directly interact with the substrate's surface to be transferred and deform it to some extent: vacuum suction results in a local lift up of the limp substrate and ingressive picking in penetrating as well as interlocking with fibers. An adhesive gripper, on the other hand, adheres on the surface, and the substrate itself is neither deformed nor misaligned. The adhesive part is usually an elastomer (e.g., copolymers and plasticized elastomers) with permanently tacky surface [1,2]. A major drawback is that the adherence cannot be switched off and so the release of a substrate is critical for a textile or soft electronic part. The modification of surface properties and specifically the enhancement of the adsorption abilities (e.g., for catalysts) [3–7] and can help to achieve switchable grippers. Alternatively, transfer printing is utilized in the assembly of micro/nanofabrication, where a silicon elastomer stamp with a viscoelastic surface behavior is used to deliver a part from a donor to a receiver substrate (see, e.g., Cheng et al. (2012) [8]). At the interface of two elastic materials the adhesion force is rather a constant than tunable [9]. However, the adhesion can depend on the peeling speed (high → lifting and low → release), the mechanical loading protocol (directional shearing induced delamination), temperature (laser pulse) or could be controlled by the surface relief structure as it was pointed out by Cheng et al. (2012) [8], Li et al. (2012) [10], and Chen et al. (2013) [9]. Extensive research efforts were presented to exploit the switchable adhesion in kinetically controlled [9,11,12], shear-assisted [8,13], direction-controlled [14], laser-driven non-contact [10,15], and microstructure enabled transfer printing [16–19]. Furthermore, in robotics various designs for gripper are presented in order to pick up and handle arbitrary objects (see, e.g., Brown et al. (2010) [20]). This influence of external physical loadings on the elastomeric gripper is exploited in order to achieve a controlled picking and placing of a substrate. The pronounced viscoelastic behavior of the gripper with a high loading rate sensitivity of the mechanical properties is therefore important.

However, up to now no detailed study on the viscoelastic characteristics of an elastomeric adhesive gripper is presented. A throughout understanding of the interplay between loading rate sensitivity of the gripper's bulk property and the inherent surface tackiness is of particular necessity to optimize the performance of an adhesive gripper. This paper presents an approach to gain insights to the temperature and loading rate dependent storage and dissipative (loss) properties by dynamic thermomechanical analyses of adhesive grippers as well as probe tack tests. The grippers under investigation are made of two different formulations of poly(dimethylsiloxane). Two woven fabrics (limp substrate) and two woven reinforced thermoplastic matrix composites (stiff substrate) are used to examine the gripper capabilities by the probe tack test. In "Material and Specimens", the material selection is briefly discussed and the specimen geometries for the mechanical characterization ("Experimental") presented. The loading rate dependent bulk and interface properties are modeled by simple linear viscoelastic theory assumptions and linear elastic fracture mechanics approach ("Experimental"). Based on the comprehensive experimental characterization the main mechanical features required for a viscoelastic adhesive gripper is summarized and presented in "Results and Discussion".

## 2. Materials and Specimens

The silicone rubbers (poly(dimethylsiloxane); PDMS) under investigation were mixed and cured with two different base polymer:catalyst ratios (PDMS-soft → 20:1 and PDMS-hard → 10:1). The volume fraction of catalyst, curing temperature, and time to achieve a solid material are thereby critical, whose surface exhibits tackiness [21] high enough to pick up fabrics from a stack; the higher the tackiness the higher is the risk of contamination in a repetitive assembly process. Details about the fabrication of the material can be found in Cakmak et al. (2014a) [22].

Two categories of specimens were prepared. One category was dedicated for the general mechanical characterization of PDMS by dynamic thermomechanical analysis (DTMA); the other was for component tests under an application-related measurement protocol as well as dynamic mechanical analysis (DMA). DTMA was performed with the "barbell" specimen and the component tests were investigated with a cylindrical stamp. Figure 1 shows the specimen geometries and the respective dimensions.

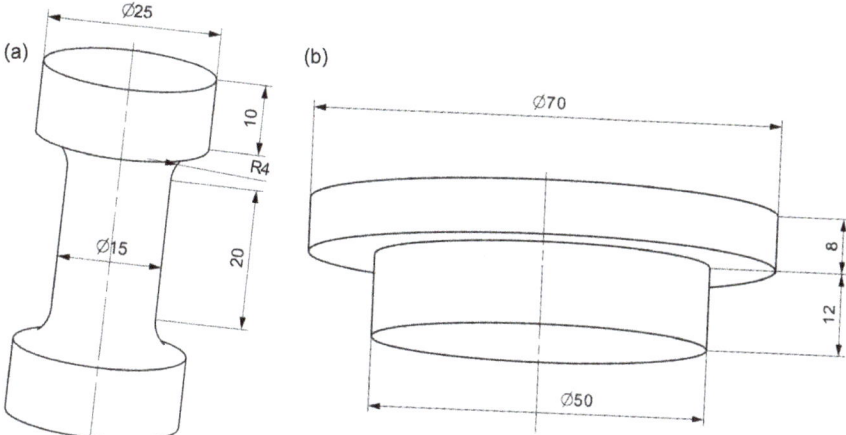

**Figure 1.** Specimen geometries: (**a**) barbell specimen and (**b**) gripper stamp specimen. Dimensions are in mm; R: radius.

In the following section, the experimental procedure ("Experimental") for the characterization of the employed PDMS is presented. Furthermore, the design and the application-related experiments of the gripper stamp are reported there.

## 3. Experimental

After the initiation of the idea to adopt the rate and delamination loading-dependent (switchable) adherence from transfer printing to a large-scale application of limp solids (fabrics), some promising preliminary experiments were performed. Motivated by these preliminary results, an experimental scheme limited to mechanical characterizations only was defined. All experiments were performed with an electro-mechanic/dynamic actuator (TestBench, Bose Corp., ElectroForce Systems Group, MN, USA).

### 3.1. Dynamic (Thermo-) Mechanical Analysis (D(T)MA)

In order to gain more information concerning the viscoelastic nature of the PDMS, D(T)MA experiments were performed with the barbell specimen as well as the gripper stamp. A classical D(T)MA test procedure was defined including frequency dependent experiments from 1 Hz to 16 Hz under isothermal condition at seven different temperatures (−30 to +30 °C). The loading mode was uniaxial tensile with a sinusoidal excitation of a mean strain level of 1% and a dynamic amplitude of 0.5%. A similar test procedure was performed with the gripper stamp in order to determine the viscoelastic behavior of the PDMS-soft material under component relevant condition. Component relevant condition means, thereby, a uniaxial compression dynamic mechanical analysis (DMA) just before delamination from the substrate. Compression DMA was examined at room temperature (22 °C) with a mean level of −0.5 mm and amplitude of 0.5 mm excitation from 0.1 Hz to 47 Hz. The data of the D(T)MA experiments were analyzed in WinTest software (Bose Corp., ElectroForce Systems Group, MN, USA) and the storage ($E'$) as well as the transient ($E''$ and $\tan\delta = E''/E'$) mechanical material properties were exported for further analyses. The complex modulus $E^*$ is given by Equation (1) and can be modeled by the well-known Prony series in the frequency ($\omega = 2\pi f$) domain [23]. In the series

$E_0$ refer to the instantaneous modulus, $\tau_i$ is the relaxation time and $g_i = (E_{i-1}-E_i)/E_0$ the Prony series coefficient corresponding to the relaxation time.

$$E^*(\omega) = E'(\omega) + j \cdot E''(\omega) = E_0 \cdot \left(1 - \sum_i \frac{g_i}{1+(\tau_i \cdot a_T \omega)^2} + j \cdot \sum_i \frac{g_i \tau_i a_T \omega}{1+(\tau_i \cdot a_T \omega)^2}\right) \quad (1)$$

If the material's thermorheological behavior is simple, then the time–temperature superposition can be applied and resulting in master curve at a reference temperature [22,24]. The temperature dependent shift factor $a_T$ can be given by the well-known WLF function [25] at a certain temperature $T$ where $T_g$ is the glass transition temperature:

$$|log_{10}(a_T)| = 17.44 \cdot \frac{T - T_g}{51.6 + T - T_g} \quad (2)$$

By inverse Laplace transformation of $E^*$ into the time-domain, the relaxation modulus in time is obtained:

$$E(t) = E_0 \cdot \left(1 - \sum_i g_i \cdot \left(1 - e^{-t/\tau_i}\right)\right) = E_\infty \cdot \frac{1 - \sum_i g_i \cdot \left(1 - e^{-t/\tau_i}\right)}{1 - \sum_i g_i} \quad (3)$$

where $E_\infty = E_0 \left(1 - \sum_i g_i\right)$ is the Young's modulus after infinite long time.

For further analyses of the rate-dependent adhesion, Equation (2) will be more convenient. From an experimental point of view, the DMA tests are preferable, as the storage and the transient viscoelastic behaviors are determined at once.

### 3.2. Probe Tack Test

The gripper stamp made of PDMS-soft was investigated in order to determine the rate dependent adherence properties for different substrate surfaces. Basically, the adherence test was a tensile delamination test (see Figure 2) at various pulling speeds (0.1 mm/s up to 100 mm/s) in accordance to the surface tack measurement set-up of Çakmak et al. (2011) [21]. The stamp was pressed onto the substrate, held for 10 s at −10 N, followed by the pulling and delamination from the substrate surface, while the force F was measured. In a real application, the stamp will lift up the substrate, but here the adherence was of particular interest and therefore the substrate was fixed. The substrates (100 × 100) mm² were attached to the support by means of double-side adhesive tape. To avoid inertia effects at high pulling speeds, the force was measured at the fixed side of the test setup. The maximum force corresponds to the fully delaminated stamp and is the limiting factor in-service for the considered pulling speed. If components with higher weight forces are applied, the stamp is not capable to pick them up. Application-relevant substrates were investigated and comprised dry carbon and glass fiber woven fabrics as well as poly(amide) matrix organo sheets reinforced with carbon and glass fiber weave.

The delamination process is assumed to be driven mainly by the crack propagation from the edge of the stamp when a tractive force $F$ was applied. For the sake of completeness, it should be mentioned that the other mechanisms would be the crack propagation starting from inside at some interfacial defects and decohesion at theoretical contact strength [26]. The real delamination process will be rather a combination of the aforementioned mechanisms. However, visual observations of the experiments revealed that the crack initiation and propagation at the edge was more dominant. From fracture mechanical point of view, the stress singularity at the crack tip can be sufficiently described by the stress intensity factor $K_I$. When the stamp is fully propagated along the interface, the crack length is equal to the radius $R$ of the stamp and $K_I$ is given by

$$K_I = \frac{F}{\pi \cdot R^2} \cdot \sqrt{\pi \cdot R} \cdot f \quad (4)$$

where $F$ is the tractive force ($F/\pi R^2$ is the tractive stress σ) and $f$ is the geometry coefficient and for a pillar shaped geometry $f = \frac{1}{2}$. The energy needed to create an area surface during a plain strain loading is connected to the stress intensity factor $K_I$ by

$$G = \frac{K_I^2 \cdot (1 - v^2)}{2 \cdot E} \quad (5)$$

where $v$ is the Poisson's ratio and $E$ is the Young's modulus. Since the material under investigation can be considered as incompressible ($v = 0.5$) and by combining Equation (4) with (5), the following relation can be found.

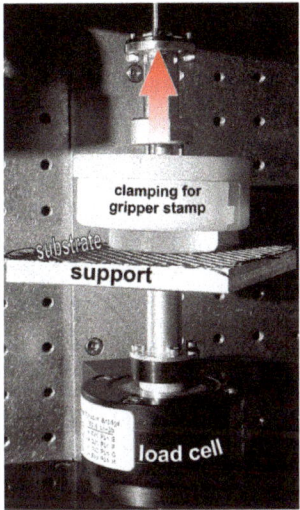

**Figure 2.** Component test set-up.

$$G_c(v, t_s(v)) = \frac{3 \cdot \sigma(v)^2 \cdot \pi \cdot R}{32 \cdot E(t_s)} \quad (6)$$

Equation (6) shows the critical energy release rate per area depending on the pulling speed v and the separation time $t_s$. The separation time is in turn a function of the pulling speed, meaning that the higher v, the lower $t_s$. This dependency is mainly driven by the viscoelastic nature of the stamp material and is also examined experimentally. To determine the actual value of the Young's Modulus $E(t_s)$, knowledge of the separation time is crucial. The pulling velocity-dependent tractive stress σ(v) has to be obtained by the component tests and is mainly determined by the surface energies of each solid in contact and the surface interaction energy of them in direct contact (cf. Duprè energy of adhesion). It is expected that the investigated substrates have a different trend in terms of measured tractive stresses. However, the velocity dependency of σ(v) is a respond given by the stamp's viscoelastic nature and can be described by the following square root function sufficiently enough.

$$\sigma(v) = \sigma_0 \cdot \sqrt{1 + \left(\frac{v}{v_0}\right)^n} \quad (7)$$

When the square root function is determined by simply fitting the measured data and combining Equations (3) and (7) with (6) lead to the pulling speed-dependent critical energy release rate:

$$G_c(v, t_s(v)) = G_0 \cdot \frac{1 + \left(\frac{v}{v_0}\right)^n \cdot \left(1 - \sum_i^3 g_i\right)}{1 - \sum_i^3 g_i \cdot (1 - e^{-t_s/\tau_r})} \tag{8}$$

where $G_0$ is the critical energy release rate near zero pulling speed v and infinite separation time $t_s$, $v_0$ is the reference pulling speed and n is the scaling exponent. Equation (8) is in accordance with the empirical form shown and utilized by Shull (2002) [27] and Feng et al. (2007) [11]. The modification is related to the time dependent Young's modulus and so the Prony series is incorporated. Parameter identification of the function in Equation (8) is carried out based on the experimental data of the component tests as well as DMA. The model will be used to predict the critical energy release rate for the desired in-service pulling speed.

## 4. Results and Discussion

As was mentioned in the "Materials and Specimens" section, two different formulations were considered for the application of the gripper stamp. These two materials will be herein after referred to as PDMS-soft and PDMS-hard. Basically, only the PDMS-soft material was suitable for the use as gripper; however, basic viscoelastic characterization (D(T)MA) was performed for both of the materials in order to show the range of possible mechanical behavior. Only PDMS-soft was investigated with the component testing procedure.

### 4.1. Dynamic (Thermo-) Mechanical Analysis (D(T)MA)

Figure 3 shows the DTMA results of the investigated materials. Each color corresponds to the results of isothermal testing at various frequencies, which are indicated as single points.

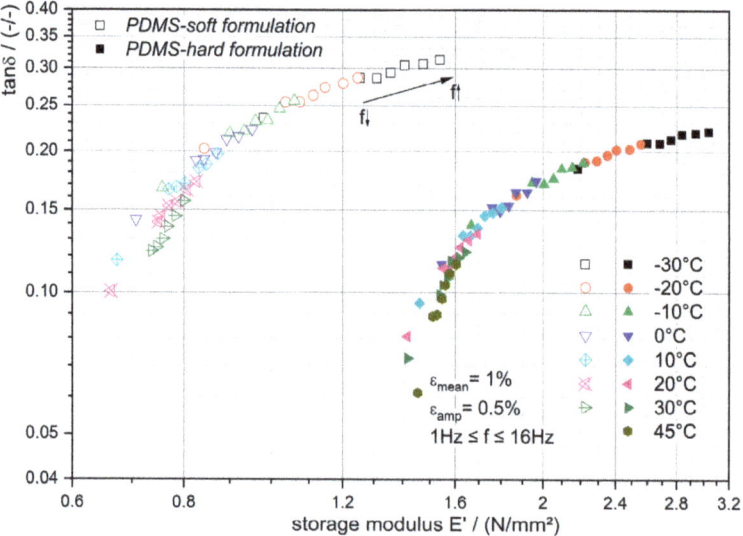

**Figure 3.** Loss factor tanδ over storage modulus $E'$ characteristic of the investigated materials at various temperatures.

This diagram is very convenient for analyzing the trend of the inherent respond time's temperature dependency as a result of the external mechanical loading. If all examined data reveal that tanδ is a unique function of $E'$, then the respond times (here the relaxation times) are equally temperature-dependent and time–temperature superposition (TTSP) is applicable (cf. Çakmak and Major (2014b) [28] and Tschoegl et al. (2002) [24]). Besides that, PDMS-soft formulation's characteristic is located within the diagram rather at the top right than the trend of PDMS-hard as expected. Important is that for both cases the TTSP can be applied, shift factors determined to construct the master curves at reference temperatures and these factors could be modeled with WLF function, among others. This is illustrated in Figure 4 for PDMS-soft (the candidate material for the application).

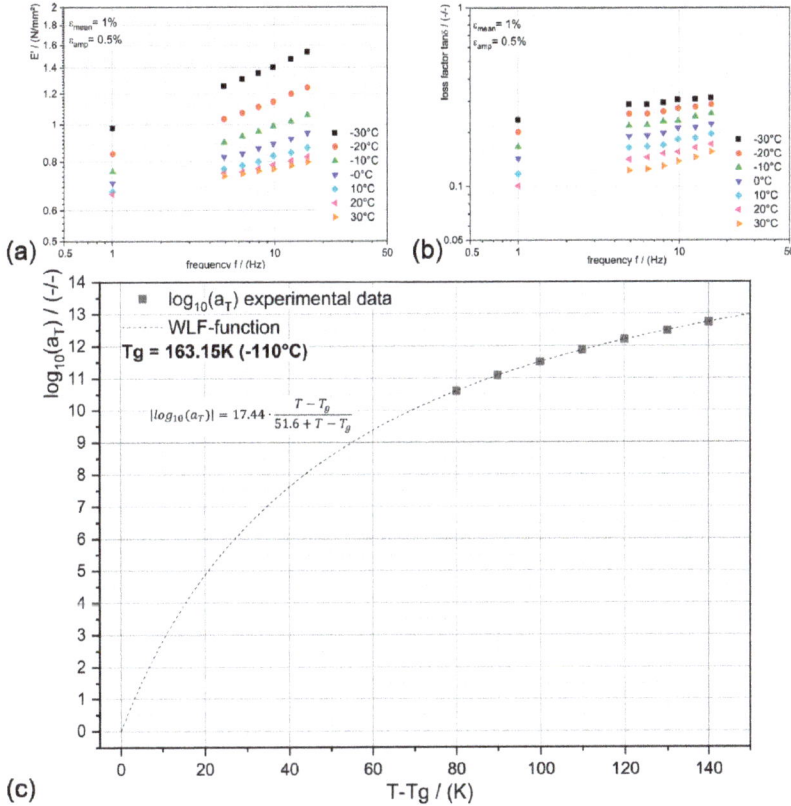

**Figure 4.** Panels (**a**,**b**) show the temperature and frequency dependent storage modulus and loss factor, respectively. In panel (**c**), the obtained time-temperature shift factors $a_T$ are presented as points and the WLF function as dashed line.

If the shift factor function (WLF) is applicable, then the temperature influence on the contact mechanical behavior can be easily predicted (e.g., Feng et al. (2007) [11]). This is of particular interest when the environmental conditions in-service of the gripper (i.e., during manufacturing of products) are fluctuating.

Figure 5 demonstrates the compression DMA results as before illustrated for the DTMA results. Now the gripper stamps made of the PDMS-formulations are investigated under contact and mechanically excited before delamination occurs. As mentioned earlier, the surface tackiness is an important requirement for the gripper stamp. However, the analyses of the viscoelastic behavior reveal that the loss factor characteristics are equally significant. It is believed that a high frequency (loading rate) dependency is essential to exploit the rate dependent adhesion. If the material has almost no rate-dependent tanδ, which is the case for PDMS-hard formulation, then the material will not be suitable as gripper with rate dependent adhesion. The storage modulus and its rate dependency are of minor importance. It should be low and remain low with increasing loading rate.

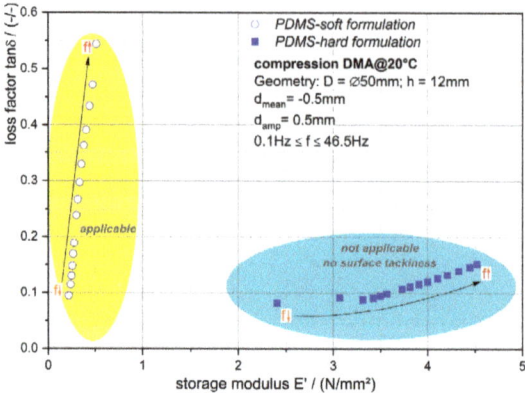

**Figure 5.** Loss factor tan$\delta$ over storage modulus $E'$ characteristic of the investigated formulation at 20 °C under compression DMA loading procedure.

The storage modulus of the PDMS-soft material is modeled with the 3rd order Prony series (cf. Equation (1)) and the respective Prony parameters are listed in Table 1. Figure 6 shows the experimental data of the frequency dependent storage modulus and the mentioned fit curve.

**Table 1.** Prony parameters of PDMS-soft formulation.

| Gripper Stamp | $E_0$ | $E_\infty$ | $g_i$ | $\tau_i$ |
|---|---|---|---|---|
| | /(MPa) | /(MPa) | - | /(s) |
| | | | $8.35 \times 10^{-2}$ | $2.50 \times 10^{-1}$ |
| PDMS-soft | 0.517 | 0.238 | $1.12 \times 10^{-1}$ | $5.23 \times 10^{-2}$ |
| | | | $3.44 \times 10^{-1}$ | $9.22 \times 10^{-3}$ |

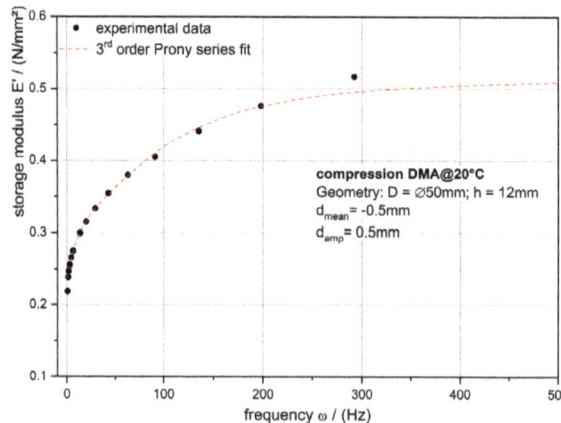

**Figure 6.** Linear plot of the frequency ($\omega = 2\pi$)-dependent storage modulus $E'$ of PDMS-soft formulation (black points) with the 3rd order Prony series fit of the data (cf. Table 1).

With the Prony series, the time dependent storage modulus needed for computing and modeling the critical energy release rate is obtained.

### 4.2. Probe Tack Test

From the tack tests the tractive stresses are evaluated, which are measured as the maximum stresses before delamination at various pulling speeds and are shown in Figure 7. The experimental data confirm the assumption of the square root relation between the stresses and the pulling speeds shown in Equation (7). This function's fit parameters for the investigated substrates are given in Table 2.

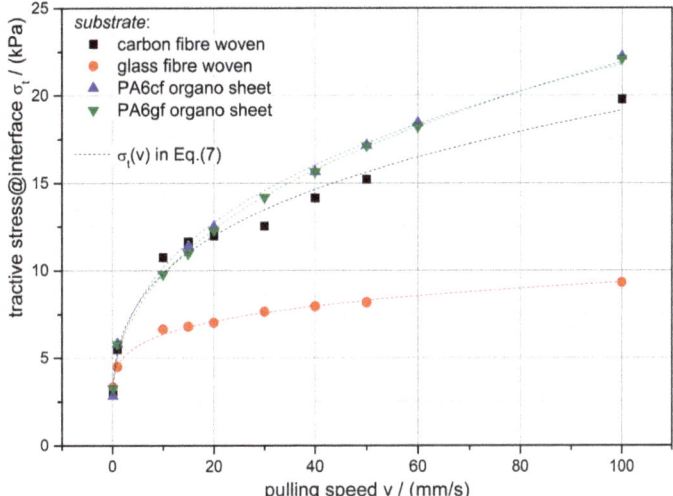

**Figure 7.** Measured tractive stresses $\sigma_t$ between the PDMS-soft gripper stamp and the investigated substrates at various pulling speeds (points) with the fit function (Equation (7)) as dashed lines.

**Table 2.** Fit parameter of the tractive stress and the corresponding calculated $G_0$.

| Substrate | $\sigma_0$ | n | $G_0$ |
|---|---|---|---|
| $v_0 = 0.1$ mm/s | /(kPa) | - | /(J/m$^2$) |
| carbon fiber woven | 2.4 | 0.60 | 0.36 |
| glass fiber woven | 2.5 | 0.38 | 0.38 |
| PA6 carbon fiber | 2.1 | 0.68 | 0.27 |
| PA6 glass fiber | 2.0 | 0.68 | 0.26 |

With $E_\infty$ and $\sigma_0$, the critical energy release rate $G_0$ near zero pulling speed and infinite separation time can be calculated. Table 2 shows the computed values and it can be observed, that the fabrics have a similar $G_0$ as well as the organo sheets. The surfaces of the fabrics are comprised by the fibers and the empty space in between and so the surface adhesion is determined by this meso-structure; on the other hand, the organo sheets' surfaces are dominated by the thermoplastic PA6.

Moreover, the time to separate the gripper stamp from the fixed substrates is evaluated (see Figure 8) in order to determine the actual storage modulus according to Equation (3). Figure 8 reveals that the separation time is with a good approximation independent of the substrate in contact with the gripper stamp. The red line shows the exponential decay trend over the pulling speed.

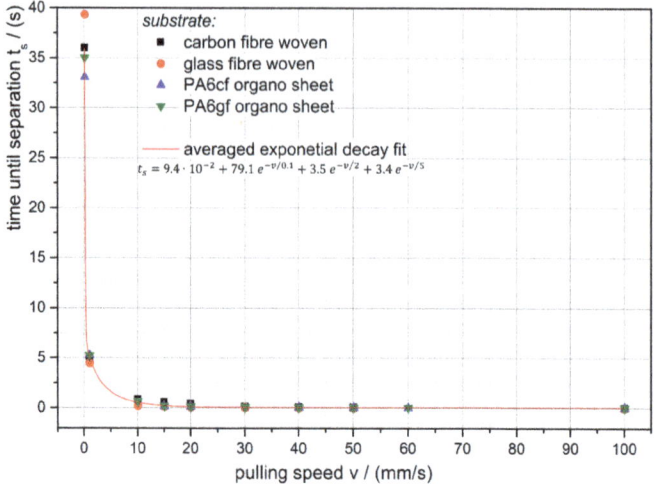

**Figure 8.** Pulling speed related time to separate the gripper stamp from the substrates; red line shows the trend of all measurement points.

From the data presented above, the critical energy release rates G are computed according to Equation (6) and are shown to be pulling speed-dependent in Figure 9. Also the modeled energy release rates are shown with good agreement with the experimental data. The model seems to be appropriate to model and predict the critical energy release rate for in-service use. From this, the pulling speed needed to lift up/release an object with a certain mass could be calculated as well.

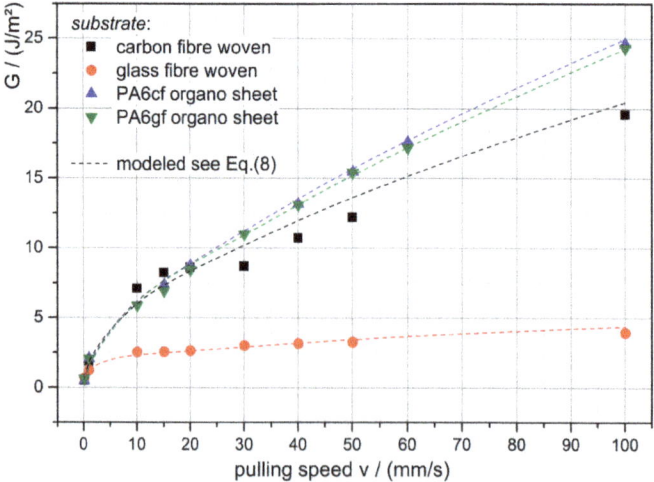

**Figure 9.** Critical energy release rate G versus the pulling speed v including the model of Equation (8) for each substrate.

## 5. Conclusions

The presented methodology to evaluate the viscoelastic as well as surface tack characteristics revealed that the gripper made of the soft PDMS formulation is capable for grasping, holding, transporting, and finally assembling (draping) of fabrics as well as semi-finished thermoplastic parts (organo sheets with PA6 matrix). Based on the observations and the experimental data, the essential viscoelastic as well as adherence kinetic behaviors are determined. A high loading speed dependency of the tanδ characteristic is needed (i.e., with higher loading speed, higher tanδ) to exploit the rate-dependent adherence. The tanδ of PDMS-soft varies between 0.1 (@0.1 Hz) and 0.55 (@46.5 Hz), while the PDMS-hard formulation reveals a narrower tanδ range between 0.09 and 0.19. If the material has almost no rate-dependent tanδ, which is the case for PDMS-hard formulation (see Figure 5), then the material is inapplicable as a gripper. The storage modulus and its loading speed dependency are of minor importance (PDMS-soft: 0.25 MPa (@0.1 Hz) to 0.5MPa (@46.5 Hz); PDMS-hard: 2.4 MPa (@0.1 Hz) to 4.5 MPa (@46.5 Hz)). It should be low and remain low with increasing loading speed.

Grasping and assembling are achieved by the utilization of the rate-dependent adhesion. For the application, the knowledge of the gripper's pulling speed dependent adherence characteristic is essential. Here, a model (see Equation (8)) is derived based on the experimental results and describes sufficiently the rate dependency of the gripper. It can be applied for the prediction of the critical energy release rate and, more importantly, the maximum mass (weight force) to lift and hold. We believe that the proposed model is of particular interest for the application and operation of a robot handling system with adhesive grippers. In the presented work, the aging, and therefore the altering of the gripper material's mechanical inherent viscoelastic properties, are omitted. For the prediction of the service (replacement) interval times, the aging behavior has to be known and has to be verified in future.

**Author Contributions:** Conceptualization, U.D.Ç.; Investigation, U.D.Ç. and M.F.; Methodology, U.D.Ç. and M.F.; Project administration, U.D.Ç.; Software, U.D.Ç.; Supervision, Z.M.; Validation, U.D.Ç. I.G. and Z.M.; Writing–original draft, U.D.Ç.; Writing–review & editing, U.D.Ç., M.F., I.G. and Z.M. All authors have read and agreed to the published version of the manuscript.

**Funding:** This research was funded by the Austrian Research Promotion Agency (FFG) in the framework of the Competence Headquarter Program operated by Engel Austria GmbH and by the LIT "ADAPT" grant number LIT2016-2-SEE-008.

**Acknowledgments:** Open Access Funding by the University of Linz.

**Conflicts of Interest:** The authors declare no conflict of interest.

## References

1. Monkman, G.J.; Shimmin, C. Permatak adhesives for robot grippers. *Assem. Autom.* **1991**, *11*, 17–19. [CrossRef]
2. Taylor, P.M. Presentation and gripping of flexible materials. *Assem. Autom.* **1995**, *15*, 33–35. [CrossRef]
3. Wang, M.; Xie, R.; Chen, Y.; Pu, X.; Jiang, W.; Yao, L. A novel mesoporous zeolite-activated carbon composite as an effective adsorbent for removal of ammonia-nitrogen and methylene blue from aqueous solution. *Bioresour. Technol.* **2018**, *268*, 726–732. [CrossRef] [PubMed]
4. Wang, F.; Xie, Z.; Liang, J.; Fang, B.; Piao, Y.A.; Hao, M.; Wang, Z. Tourmaline-modified FeMnTiO x catalysts for improved low-temperature NH3-SCR performance. *Environ. Sci. Technol.* **2019**, *53*, 6989–6996. [CrossRef] [PubMed]
5. Piao, Y.; Jiang, Q.; Li, H.; Matsumoto, H.; Liang, J.; Liu, W.; Pham-Huu, C.; Liu, Y.; Wang, F. Identify Zr promotion effects in atomic scale for co-based catalysts in Fischer–Tropsch synthesis. *ACS Catal.* **2020**, *10*, 7894–7906. [CrossRef]
6. Ouyang, J.; Zhao, Z.; Yang, H.; Zhang, Y.; Tang, A. Large-scale synthesis of sub-micro sized halloysite-composed CZA with enhanced catalysis performances. *Appl. Clay Sci.* **2018**, *152*, 221–229. [CrossRef]
7. Corrado, A.; Polini, W. Measurement of high flexibility components in composite material by touch probe and force sensing resistors. *J. Manuf. Process.* **2019**, *45*, 520–531. [CrossRef]
8. Cheng, H.; Li, M.; Wu, J.; Carlson, A.; Kim, S.; Huang, Y.; Kang, Z.; Hwang, K.-C.; Rogers, J.A. A viscoelastic model for the rate effect in transfer printing. *J. Appl. Mech.* **2013**, *80*, 041019. [CrossRef]
9. Chen, H.; Feng, X.; Huang, Y.; Huang, Y.; Rogers, J.A. Experiments and viscoelastic analysis of peel test with patterned strips for applications to transfer printing. *J. Mech. Phys. Solids* **2013**, *61*, 1737–1752. [CrossRef]

10. Li, R.; Li, Y.; Lü, C.; Song, J.; Saeidpouraza, R.; Fang, B.; Zhong, Y.; Ferreira, P.M.; Rogers, J.A.; Huang, Y. Thermo-mechanical modeling of laser-driven non-contact transfer printing: Two-dimensional analysis. *Soft Matter* **2012**, *8*, 7122–7127. [CrossRef]
11. Feng, X.; Meitl, M.A.; Bowen, A.M.; Huang, Y.; Nuzzo, A.R.G.; Rogers, J.A. Competing fracture in kinetically controlled transfer printing. *Langmuir* **2007**, *23*, 12555–12560. [CrossRef]
12. Cheng, H.; Wu, J.; Yu, Q.; Kim-Lee, H.-J.; Carlson, A.; Turner, K.T.; Hwang, K.-C.; Huang, Y.; Rogers, J.A. An analytical model for shear-enhanced adhesiveless transfer printing. *Mech. Res. Commun.* **2012**, *43*, 46–49. [CrossRef]
13. Carlson, A.; Kim-Lee, H.-J.; Wu, J.; Elvikis, P.; Cheng, H.; Kovalsky, A.; Elgan, S.; Yu, Q.; Ferreira, P.M.; Huang, Y.; et al. Shear-enhanced adhesiveless transfer printing for use in deterministic materials assembly. *Appl. Phys. Lett.* **2011**, *98*, 264104. [CrossRef]
14. Chen, H.; Feng, X.; Chen, Y. Directionally controlled transfer printing using micropatterned stamps. *Appl. Phys. Lett.* **2013**, *103*, 151607. [CrossRef]
15. Saeidpourazar, R.; Li, R.; Li, Y.; Sangid, M.D.; Lu, C.; Huang, Y.; Rogers, J.A.; Ferreira, P.M. Laser-driven micro transfer placement of prefabricated microstructures. *J. Microelectromechanical Syst.* **2012**, *21*, 1049–1058. [CrossRef]
16. Bogue, R. Smart materials: A review of capabilities and applications. *Assem. Autom.* **2014**, *34*, 16–22. [CrossRef]
17. Kim, S.; Wu, J.; Carlson, A.; Jin, S.H.; Kovalsky, A.; Glass, P.; Liu, Z.; Ahmed, N.; Elgan, S.L.; Chen, W.; et al. Microstructured elastomeric surfaces with reversible adhesion and examples of their use in deterministic assembly by transfer printing. *Proc. Natl. Acad. Sci. USA* **2010**, *107*, 17095–17100. [CrossRef]
18. Wu, J.; Kim, S.; Chen, W.; Carlson, A.; Hwang, K.-C.; Huang, Y.; Rogers, J.A. Mechanics of reversible adhesion. *Soft Matter* **2011**, *7*, 8657–8662. [CrossRef]
19. Kim, S.; Carlson, A.; Cheng, H.; Lee, S.; Park, J.-K.; Huang, Y.; Rogers, J.A. Enhanced adhesion with pedestal-shaped elastomeric stamps for transfer printing. *Appl. Phys. Lett.* **2012**, *100*, 171909. [CrossRef]
20. Brown, E.; Rodenberg, N.; Amend, J.R.; Mozeika, A.; Steltz, E.; Zakin, M.R.; Lipson, H.; Jaeger, H.M. Universal robotic gripper based on the jamming of granular material. *Proc. Natl. Acad. Sci. USA* **2010**, *107*, 18809–18814. [CrossRef]
21. Cakmak, U.D.; Grestenberger, G.; Major, Z. A novel test method for quantifying surface tack of polypropylene compound surfaces. *Express Polym. Lett.* **2011**, *5*, 1009–1016. [CrossRef]
22. Cakmak, U.D.; Hiptmair, F.; Major, Z. Applicability of elastomer time-dependent behavior in dynamic mechanical damping systems. *Mech. Time-Depend. Mater.* **2013**, *18*, 139–151. [CrossRef]
23. Emri, I.; Tschoegl, N. Determination of mechanical spectra from experimental responses. *Int. J. Solids Struct.* **1995**, *32*, 817–826. [CrossRef]
24. Tschoegl, N.; Knauss, W.G.; Emri, I. The effect of temperature and pressure on the mechanical properties of thermo-and/or piezorheologically simple polymeric materials in thermodynamic equilibrium-A critical review. *Mech. Time-Depend. Mater.* **2002**, *6*, 53–99. [CrossRef]
25. Williams, M.L.; Landel, R.F.; Ferry, J.D. The Temperature dependence of relaxation mechanisms in amorphous polymers and other glass-forming liquids. *J. Am. Chem. Soc.* **1955**, *77*, 3701–3707. [CrossRef]
26. Carbone, G.; Pierro, E.; Gorb, S.N. Origin of the superior adhesive performance of mushroom-shaped microstructured surfaces. *Soft Matter* **2011**, *7*, 5545–5552. [CrossRef]
27. Shull, K.R. Contact mechanics and the adhesion of soft solids. *Mater. Sci. Eng. R: Rep.* **2002**, *36*, 1–45. [CrossRef]
28. Cakmak, U.D.; Major, Z. Experimental thermomechanical analysis of elastomers under uni-and biaxial tensile stress state. *Exp. Mech.* **2013**, *54*, 653–663. [CrossRef]

**Publisher's Note:** MDPI stays neutral with regard to jurisdictional claims in published maps and institutional affiliations.

© 2020 by the authors. Licensee MDPI, Basel, Switzerland. This article is an open access article distributed under the terms and conditions of the Creative Commons Attribution (CC BY) license (http://creativecommons.org/licenses/by/4.0/).

*Article*

# Elastic Properties of Polychloroprene Rubbers in Tension and Compression during Ageing

Rami Bouaziz [1,2], Laurianne Truffault [2], Rouslan Borisov [3], Cristian Ovalle [1], Lucien Laiarinandrasana [1], Guillaume Miquelard-Garnier [2,*] and Bruno Fayolle [2,*]

1. Centre des Matériaux, Mines ParisTech, PSL University, CNRS UMR 7633 BP 87, F-91003 Evry, France; rami.bouaziz@mines-paristech.fr (R.B.); cristian.ovalle_rodas@mines-paristech.fr (C.O.); lucien.laiarinandrasana@mines-paristech.fr (L.L.)
2. PIMM, Arts et Metiers Institute of Technology, CNRS, Cnam, HESAM University, 151 boulevard de l'Hôpital, 75013 Paris, France; lauriane.truffault@ensam.eu
3. EDF, Nuclear New Build Engineering & Projects, Design and Technology Branch, La Grande Halle, 19 rue Pierre Bourdeix, 69007 Lyon, France; rouslan.borisov@edf.fr
* Correspondence: guillaume.miquelardgarnier@lecnam.net (G.M.-G.); bruno.fayolle@ensam.eu (B.F.)

Received: 18 September 2020; Accepted: 11 October 2020; Published: 14 October 2020

**Abstract:** Being able to predict the lifetime of elastomers is fundamental for many industrial applications. The evolution of both tensile and compression behavior of unfilled and filled neoprene rubbers was studied over time for different ageing conditions (70 °C, 80 °C and 90 °C). While Young's modulus increased with ageing, the bulk modulus remained almost constant, leading to a slight decrease in the Poisson's ratio with ageing, especially for the filled rubbers. This evolution of Poisson's ratio with ageing is often neglected in the literature where a constant value of 0.5 is almost always assumed. Moreover, the elongation at break decreased, all these phenomena having a similar activation energy (~80 kJ/mol) assuming an Arrhenius or pseudo-Arrhenius behavior. Using simple scaling arguments from rubber elasticity theory, it is possible to relate quantitatively Young's modulus and elongation at break for all ageing conditions, while an empirical relation can correlate Young's modulus and hardness shore A. This suggests the crosslink density evolution during ageing is the main factor that drives the mechanical properties. It is then possible to predict the lifetime of elastomers usually based on an elongation at break criterion with a simple hardness shore measurement.

**Keywords:** polychloroprene; ageing; mechanical properties; rubber elasticity theory; crosslinking

---

## 1. Introduction

Rubbers (also called elastomers) are loosely crosslinked networks of amorphous polymers having a very low glass transition temperature [1]. Due to their macromolecular structure, these materials are relatively soft (Young's modulus on the order of 1–10 MPa) but display a large deformability (up to about 10 times their original length), mostly reversible, due to an elasticity that is entropic in nature [2]. Amongst this class of materials, polydiene-based elastomers such as polyisoprene or polychloroprene (neoprene) have remarkable mechanical properties (especially a large elongation at break) along with a good stability towards chemicals [3]. However, their use is often limited by the fact that they harden and become brittle over the long term. These changes in mechanical properties are most often caused by an oxidation phenomenon [4]. Indeed, the oxidation process leads to the production of free alkyl/peroxyde radicals along the chain which can then react with the double bonds by addition [5]. These inter-macromolecular addition reactions then contribute to an increase in the crosslinking density [6].

The classical way to predict the lifetime of elastomers is to consider that the time to reach an end-life criterion in terms of elongation at break (such as 50% of initial elongation at break) follows a pseudo-Arrhenius behavior. Based on accelerated ageing time test, an extrapolation at room temperature can then be done to predict lifetime. However, this kind of extrapolation is highly questionable since there are many experimental results showing that end life does not always follow such an empirical law [7].

To avoid such extrapolation, a possible way is to develop an approach in two steps. The first step is to build a kinetic model taking into account all the chemical mechanisms involved during ageing: according to this model, it is then possible to simulate the chemical state and the chain scission/crosslinking events whatever the exposure conditions in terms of time or temperature of exposure (see for instance for the polychloroprene [8]). The second step is to relate the chemical state and the chain scission/crosslinking events to the relevant property such as elongation at break or hardness. This kind of structure–property relationship has to be valid whatever the accelerated ageing conditions to be used in real conditions.

The prediction of the lifetime of elastomer parts first requires reliable constitutive models to be established in 3D, i.e., taking the hydrostatic pressure into account. The basic Mooney–Rivlin deformation energy $W$ commonly used to assess the lifetime of engineering elastomeric structures is expressed as Equation (1) [9]:

$$W = C_{10}(I_1 - 3) + C_{01}(I_2 - 3) + 0.5K(J - 1)^2 \tag{1}$$

where $C_{10}$, $C_{01}$ and $K$ are the material parameters, $K$ being in particular the bulk modulus of the elastomer; $I_1$ and $I_2$ the first and second invariants of the deformation gradient tensor; $J$ the Jacobian which can be written as the volume ratio $V/V_0$, with $J = 1$ for perfectly incompressible materials.

Deriving the uniaxial Cauchy stress $\sigma$ from Equation (1) leads to:

$$\sigma = \frac{1}{J}\left(2C_{10} + \frac{2C_{01}}{\lambda}\right)\left(\lambda^2 - \frac{1}{\lambda}\right) + K(J-1) \tag{2}$$

with $\lambda$ the elongation, defined as $\lambda = l/l_0$ where $l_0$ is the initial sample length.

The relationship between hyperelastic (Mooney–Rivlin) material parameters ($C_{10}$, $C_{01}$), bulk modulus ($K$) and elastic properties at small strain (Young's modulus $E$ and Poisson's ratio $\nu$) can then be summarized as follows:

$$E = 3(2C_{10} + 2C_{01}) \tag{3}$$

$$K = \frac{E}{3(1 - 2\nu)} \tag{4}$$

In this context, relating the evolution of both hyperelastic and elastic material properties to the progress of the crosslinking process is one of the objectives of this study, since it is required for lifetime prediction of the elastomers.

In the case of unfilled elastomers, the value of the Poisson's ratio is often considered to be equal to 0.5, corresponding to an incompressible state of the material, i.e., a deformation that occurs at constant volume [10,11]. However, elastomers are actually "ever so slightly" compressible and it is noteworthy that measuring the Poisson's ratio is not straightforward. In the case of filled elastomers, the value of the Poisson's ratio has been typically measured between 0.46 and 0.49 [12–14]. By using Equation (4) and considering a typical elastomeric Young's modulus of 10 MPa, such variation of the Poisson's ratio leads to an increase of the bulk modulus $K$ from 42 MPa to 167 MPa. These values are in contradiction with the supposed incompressibility assumption letting the bulk modulus $K$ tends to infinity, and a small uncertainty in the Poisson's ratio measured value can easily lead to an uncertainty of as much as an order of magnitude for $K$.

To the best of our knowledge, there are no results available allowing us to characterize such a pair of hyperelastic and elastic parameters during an oxidative crosslinking, which is then required to perform finite elements simulations of industrial parts. In the literature, the changes in mechanical behavior during the crosslinking process induced by oxidation are characterized mainly by tensile tests to assess $E$ [15] or toughness in mode I [16]. However, although different loading conditions are investigated through these experiments, none of them gives access simultaneously to two parameters as described above. Therefore, a relevant mechanical characterization is required for a proper assessment of $E$ and $v$ (or $K$), which can monitor not only the changes in terms of Young's modulus during the crosslinking process due to ageing, but also the possible changes in terms of compressibility modulus or Poisson's ratio.

In this study, the objective is to characterize the mechanical behavior of filled and unfilled crosslinked polychloroprene through tensile and oedometric compression tests to assess both the change of $E$ and $v$ (or $K$) over time at different oxidative conditions. The choice of several oxidative conditions has been driven to check possible relationships between these values independently of oxidation rate, i.e., temperature of exposure. A specific attention is paid to promote spatial homogeneous oxidation for the samples considered here in order to assess only intrinsic material relationships.

Finally, the evolution of the elastic properties $E$ and $K$ during ageing is compared to elongation at break or hardness which are classically used to follow oxidative ageing of elastomers. The second objective of the study is then to propose correlations between all these mechanical properties, which shall be valid not only during the whole oxidation process but also for both unfilled and filled elastomers. As a result, a critical value for a specific property (easy to measure, such as hardness shore A) could be used as a proxy for critical values of properties that are much harder to assess (such as elongation at break), for lifetime prediction purposes.

## 2. Materials and Methods

Thin rectangular plates (thickness = 2 mm) of unfilled neoprene rubber and of neoprene filled with 40 phr (~30 wt %) of carbon black were provided by Nuvia (Villeurbanne, France). To study the impact of the thermal oxidative ageing on the mechanical properties, the plates were aged at three different temperatures (70 °C, 80 °C and 90 °C) during specific ageing times. It was verified with microindentation tests that ageing was homogeneously distributed throughout the samples (see Appendix A). All samples were thoroughly characterized after each ageing time. Specifically, differential scanning calorimetry (DSC) was used to follow the consumption of antioxidants/stabilizers under oxidative ageing. Mechanical tests, namely tensile, hardness, and oedometric tests, were developed to follow the evolution of the mechanical properties, namely Young's modulus, elongation at break and bulk modulus, with the ageing time. All mechanical tests were carried out at ambient temperature.

### 2.1. DSC/OIT

The antioxidants/stabilizers consumption was analyzed through the determination of the oxidation induction time (OIT) measured with a Q20 DSC from TA Instruments (New Castle, DE, USA). Polychloroprene samples with a mass ranging between 7.0 and 8.0 mg were placed in aluminum standard pans (TA Instruments) without a lid. Reference consisted of an empty aluminum pan, also without a lid. After equilibrating the temperature at 50 °C and an isothermal of five minutes, the samples were heated to 180 °C at 10 °C/min under nitrogen flow at 50 mL/min. Temperature was then maintained at 180 °C for five minutes before substituting nitrogen for oxygen for 250 min. OIT corresponds to the intersection of the tangent of the baseline at the time of atmosphere change with the tangent at the beginning of the oxidation peak.

### 2.2. Tensile Tests

The tensile tests were conducted on H3 dog-bone specimens (4 mm wide, 2 mm thick and 20 mm long obtained with a cutting-die) with a 5966 Instron tensile machine (Instron, Élancourt, France). Two samples were tested per ageing condition. Forces were monitored using a load cell of 10 kN. The specimens were stretched up to failure with a crosshead speed set to 20 mm/min. Due to the ability of the samples to undergo very large strains, pneumatic clamping jaws were used to avoid slippage.

Young's modulus was determined by fitting the stress–elongation curves in the λ = [1; 1.45] region with a Mooney–Rivlin incompressible model in uniaxial tension, i.e., following Equations (2) (adapted for engineering stress) and (3).

*2.3. Hardness Tests (Shore A)*

The hardness of the material was measured using a shore A durometer (model LX-A-Y from RS PRO, Corby, UK). The measured values indicate the resistance to indentation of the tested material on a scale between 0 (depth of indentation equal to the sample thickness) and 100 (no indentation).

*2.4. Oedometric Compression Test*

The oedometric compression tests were conducted on discs (diameter = 25 mm and thickness = 2 mm) obtained from the rubber plates with a cutting die. A stack of three discs was inserted in a lab-made apparatus consisting in a steel die with a compression applied by means of a rigid stamp. A small gap (less than 0.5 mm) between the samples and the steel die was initially set to allow the specimens to be positioned inside the die. The compression load was applied with a tensile/compression Instron machine (model 5982, Instron France) with a load cell of 100 kN. The specimens were compressed to about 95 kN with a displacement speed set to 1 mm/min.

## 3. Results

*3.1. Chemical Stability: OIT Changes*

Figure 1a reports the changes of OIT as a function of ageing time for filled and unfilled polymers at the three temperatures of exposure. The initial OIT gap between unfilled and filled samples can be attribute d to different processing conditions leading to a higher antioxidants loss in the case of the filled samples, such as a supplementary blending step at high temperature, using an extruder or an internal mixer.

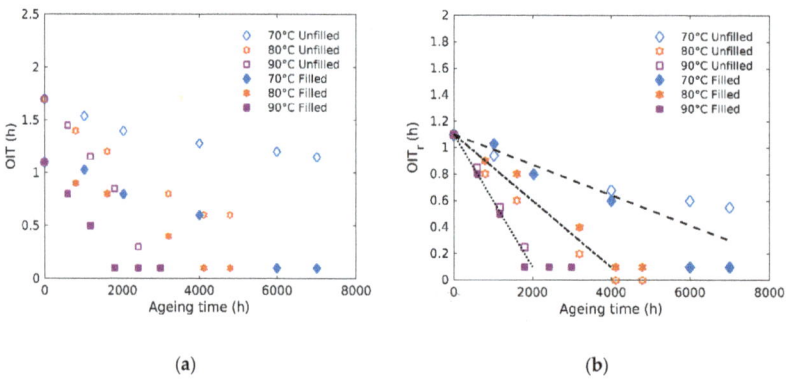

**Figure 1.** (a) Oxidation induction time (OIT) and (b) relative OIT ($OIT_r$) evolution as a function of ageing time.

As expected, OIT decreases with ageing time, indicating a consumption of the antioxidants/stabilizers during the samples exposure. OIT reaches a plateau value close to 0 for the filled rubbers indicating an almost total consumption of the antioxidants. This plateau is reached at 2000–3000 h for samples aged at 90 °C and after about 6000 h for the filled samples aged at 70 °C, while it was not obtained for the unfilled samples aged at the same temperature during the total time of the experiment (8000 h).

In order to take into account the fact that unfilled sample shows a higher initial OIT value, we plot here in Figure 1b "relative OIT" ($OIT_r$) which is defined equal to the actual OIT for the filled samples, and for the unfilled values as OIT − 0.6 (which is the OIT difference at t = 0 h between unfilled (1.7) and filled (1.1) samples). Thus, Figure 1b shows the relative OIT for the three temperatures of exposure and allows to compare both elastomers in terms of OIT decrease. It clearly appears that the relative OIT decrease rate (i.e., the slope of the "linear" region shown as dotted lines in Figure 1b) follows the same trend for both elastomers whatever the exposure temperature. As a result, filled and unfilled elastomers show the same ageing kinetic and differ only by their initial state.

A classical way to quantify the relative OIT change during ageing as a function of temperature is to consider the slope $x$ of the linear region follows a pseudo-Arrhenius behavior with temperature [7], i.e., $x = x_0 e^{\frac{E_a}{RT}}$ with $E_a$ the activation energy (J/mol), R the gas constant = 8.314 J/mol/K and $T$ the temperature (in kelvin). $E_a$ is then considered as a characteristic of the process and its activation by temperature. Here, the temperature of exposure activates the OIT drop rate with an activation energy close to 78 kJ/mol considering such behavior.

### 3.2. Tensile Tests

Figure 2 displays representative tensile curves for unfilled (a) and filled (b) samples aged at 90 °C and tested at different ageing times. The plots show the engineering stress (the applied load divided by the unstretched cross-sectional area) as a function of the elongation $\lambda$.

In Figure 2a, it can be observed that the unfilled rubber undergoes a very large strain, leading to a pronounced hardening before failure. For polychloroprene rubbers, it has been reported that this hardening is due to strain-induced crystallization (SIC) [17]. As proposed in [18], SIC-related stress hardening during tensile test appears amplified by the presence of fillers (see Figure 2b). This, along with the filler properties, leads to the samples initially stiffer but with a smaller elongation at break than the unfilled rubbers. Both SIC onset associated to the stress–elongation curve upturn and elongation at break decrease monotonously with ageing. Furthermore, the upturn phenomenon is only clearly visible for the nonaged sample and disappears for the aged samples. However, all tensile curves show a similar trend with ageing: stiffening and elongation at break decrease, both trends being attributed to extra crosslinking induced by ageing.

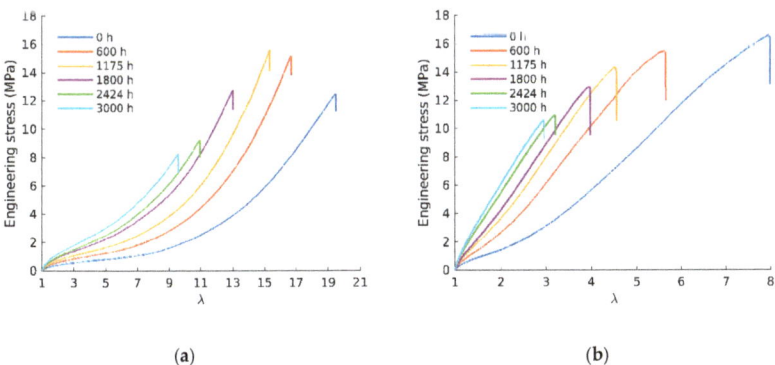

(a) (b)

**Figure 2.** (a) Tensile curves to failure of samples aged at 90 °C for unfilled and (b) filled rubbers.

Moduli changes can be extracted from the stress–elongation curves for both rubbers at all exposure conditions. Figure 3 shows the evolution of the Young's modulus ($E$) with ageing time for unfilled (a) and filled (b) rubbers. This confirms that $E$ continuously increases with ageing and that this increase is more pronounced at higher temperatures of exposure for a given ageing time. Similar trends have been obtained on a carbon black filled butadiene in [19]. Since modulus is driven by the crosslink density, it is thus possible to follow chemical modification through this quantity. Activation energy for the modulus (calculated using the same procedure as the one described for the OIT assuming here an Arrhenius behavior) increase rate is close to 90 kJ/mol for the unfilled and 84 kJ/mol for the filled rubber.

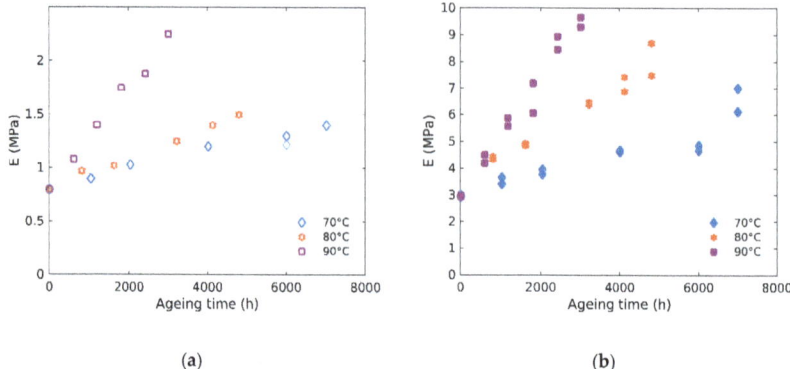

**Figure 3.** (a) Evolution of Young modulus ($E$) with ageing time for unfilled and (b) filled rubbers at 70 °C, 80 °C and 90 °C.

In order to study the fillers' effect on the modulus evolution, the ratio of the filled and unfilled moduli is plotted as a function of ageing time at the three temperatures in Figure 4. It appears this ratio is almost constant for all temperatures and ageing time, with a value of about 4.5. It is worth noting that this value is higher than the Guth and Gold equation [20,21] given by Equation (5):

$$\frac{E_{filled}}{E_{unfilled}} = 1 + 2.5\varphi + 14.1\varphi^2, \tag{5}$$

where $\varphi$ is the effective volume fraction of filler.

Assuming a density of 2 g/cm³ for carbon black and 1.2 g/cm³ for the elastomer [22], the volume fraction of fillers here is about 0.2, which yields a value of 2 for the modulus ratio using Equation (5).

A modified equation by Guth [20] takes into account the shape factor $f$ of the carbon black (see Equation (6)):

$$\frac{E_{filled}}{E_{unfilled}} = 1 + 0.67 f\varphi + 1.62 f^2 \varphi^2 \tag{6}$$

With a shape factor $f = 6.5$ as determined by Mullins and Tobin [23], Equation (6) gives a value of 4.6 for modulus ratio, consistent with the experimental data, as can be seen in Figure 4.

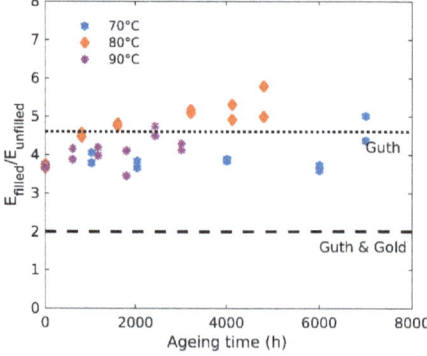

**Figure 4.** Ratio of the filled and unfilled moduli as a function of ageing time for three ageing temperatures.

Finally, if we consider modulus changes as a tracer of the chemical degradation, a correlation between chemical stability (i.e., OIT) and the evolution of normalized Young's modulus (defined as $E(t)/E(t = 0)$) shall be evidenced. Figure 5 presents normalized modulus as a function of normalized OIT ($OIT_n = OIT(t)/OIT(t = 0)$) for both filled and unfilled rubbers at the three ageing temperatures. All the data display a similar trend and fall within the same envelope (see dotted lines in Figure 5): starting from high normalized OIT values (unaged rubbers), normalized Young's modulus increases as normalized OIT decreases until a limit value of OIT which corresponds to a total consumption of stabilizers ($OIT_n$ reaching values close 0).

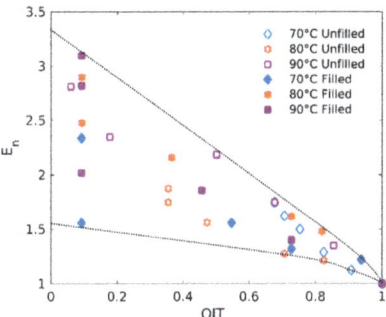

**Figure 5.** Normalized modulus $E_n$ as a function of normalized oxidation induction time $OIT_n$.

From Figure 2, it is also possible to extract the elongation at break evolution, which is a relevant parameter often used as a criterion to define end-life duration in industrial applications of such materials [24]. As expected, Figure 6 shows the elongation at break decreases with ageing, again in a much more pronounced way at higher temperatures.

This embrittlement process is necessarily associated with the crosslink density increase already witnessed through modulus changes. From a kinetic point of view, elongation at break decrease rate shows an activation energy close to 86 kJ/mol for the unfilled and 74 kJ/mol for the filled rubber, consistent with the values obtained for the moduli.

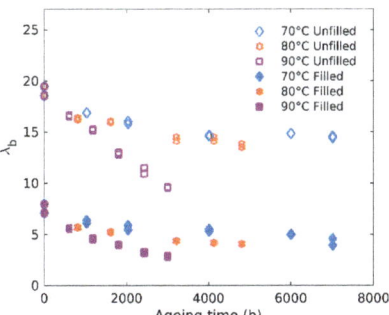

**Figure 6.** Evolution of the elongation at break with ageing time.

*3.3. Hardness Tests (Shore A)*

Although values obtained from hardness tests are a priori too qualitative to identify the elastic parameters, they are easy to obtain and as such often used in the industry. Figure 7 shows the evolution of the materials hardness with ageing time. It appears that the hardness of the rubbers increases with ageing similarly to the Young's modulus.

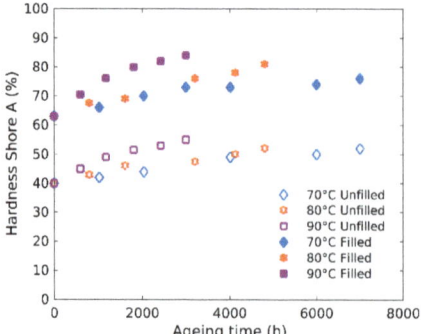

**Figure 7.** Hardness Shore A evolution with ageing time.

### 3.4. Bulk Modulus

The small-strain elastic mechanical behavior of isotropic materials is fully characterized by a couple of parameters, for example Young's modulus (*E*) and bulk modulus (*K*). If the evolution of the Young's modulus with ageing is often reported in the literature [15], no data concerning such evolution of the bulk modulus of elastomers are available according to our knowledge.

Figure 8 shows such evolution for unfilled (see Figure 8a) and filled (b) rubbers. The values obtained are in the GPa range, similar to previous reports for elastomers [25]. Contrary to what has been observed for Young's modulus (see Figure 3), the effect of ageing on the bulk modulus is small both in terms of time and temperature dependency (i.e., less than 5% variations). It can also be noticed that the bulk moduli of filled rubbers are slightly higher than the ones of unfilled rubbers (from about 1100 MPa to 1200 MPa).

Let us recall Young's modulus can be related to the crosslink density, whereas bulk modulus is essentially controlled by van der Waals forces [26,27]. Chemical degradation yielding a crosslinking process does not modify significantly the van der Waals' energy.

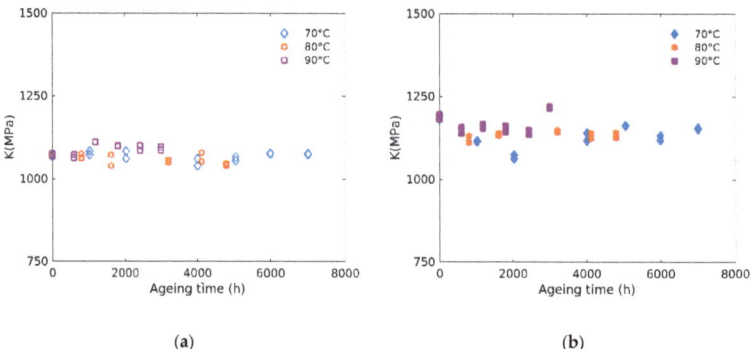

**Figure 8.** (a) Evolution of the bulk modulus (*K*) with ageing time for unfilled and (b) filled rubbers at 70 °C, 80 °C and 90 °C.

## 4. Discussion

### 4.1. Structure-Properties Relationships

The Young's modulus of rubbers can be related to the crosslink density following Equation (7):

$$E = 3N\rho RT. \tag{7}$$

with N the crosslink density in mol/kg and $\rho$ the elastomer density (kg/m$^3$) [28]. As a consequence, the crosslink density modifications which can be deduced from $E$ measurements may be linked to OIT, elongation at break and hardness changes. Although FTIR measurements only slightly evidence oxidation products in the timeframe of the ageing experiments (see infrared spectra in Appendix B), it is well known that oxidation leads to crosslinking in the case of polychloroprene rubbers during long-term exposure in air, due to addition reactions between alkyl/peroxyl radicals and double bonds [6,8]. This is consistent with the fact that OIT decreases when $E$ increases.

To confirm this, the activation energies estimated in the previous section for all these properties are listed and can be compared in Table 1: it appears that all activation energies are close to 80 kJ/mol, suggesting that all these properties changes have the same chemical origins and are driven by the same mechanisms. In [4], similar activation energy characterizing the degradation was found (~90 kJ/mol), the values for the unfilled neoprene also being slightly higher than those of the filled material. Finally, note that the hardness shore A activation energy value is a bit smaller for the unfilled samples but that is probably due to experimental uncertainty as the variations of hardness over ageing time are very small for these samples.

**Table 1.** Activation energy for all parameters understudied.

| $E_a$ (kJ/mol) | E | OIT | $\lambda_b$ | Hardness |
|---|---|---|---|---|
| Unfilled | 90 | 78 | 86 | 53 |
| Filled | 84 | 78 | 74 | 72 |

Let us also examine direct correlation between $E$ and hardness. Many relationships between $E$ and hardness shore A ($S_A$) are proposed in the literature [29,30]. In [30], a finite element simulation study leads to a linear relation between the logarithm of the elastic modulus and the hardness values (see Equation (8)):

$$log_{10}E = c_1 S_A - c_2 \qquad (8)$$

with $c_1$ and $c_2$ empirical constants determined from the best fit as $c_1 = 0.0235$ and $c_2 = 0.6403$ for $20 < S_A < 80$ [30].

In Figure 9, $E$ is plotted as a function of $S_A$. According to the previous approach, our experimental results lead to $c_1 = 0.0245$ and $c_2 = 1.075$ for both filled and unfilled samples, independently of ageing, in the $40 < S_A < 85$ range, which is in good agreement with the values reported previously [30].

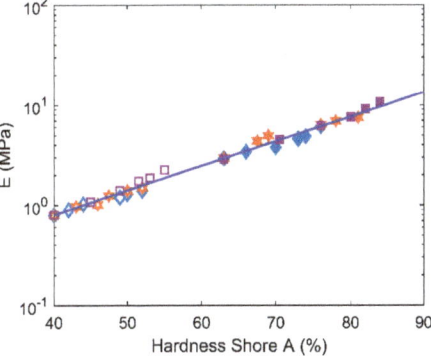

**Figure 9.** Young's modulus (E) versus hardness shore A for all samples (filled, unfilled) and temperatures.

Finally, we would like now to study possible correlations between $\lambda_b$ and $E$. As a first approximation, the elongation at break can be related to the chain dimensions [1]: assuming an initial Gaussian conformation $R_0 = \sqrt{n}l$ with $R_0$ the chain end-to-end distance, $n$ the average number of repeating units in the chain and $l$ the monomer typical size, and fracture occurring when the chain extension reaches a maximal value $R_{max} = nl$, we can write Equation (9):

$$\lambda_b \propto \frac{R_{max}}{R_0} = \sqrt{n} \qquad (9)$$

Similarly, the Young's modulus of rubbers can be related to the molar mass between crosslinks, according to Equation (7) rewritten under the form $E = 3\frac{\rho RT}{M_c}$ with $M_c$ the molar mass between crosslinks (in kg/mol), which is proportional to $n$. Hence, $\lambda_b$ should scale as $1/\sqrt{E}$ which is indeed observed in Figure 10. Interestingly, this simple scaling remains valid whatever the ageing conditions and for both filled and unfilled materials.

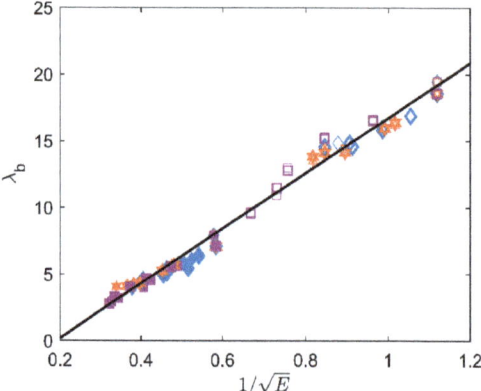

**Figure 10.** $\lambda_b$ versus $1/\sqrt{E}$ for all samples (filled, unfilled) and temperatures.

The linear fit gives the following value with E in MPa: $\lambda_b = 20.7/\sqrt{E} - 4$.

Assuming again a density of 1.2 g/cm³, and with a molar mass $M_0$ of 88 g/mol for the polychloroprene monomer, the calculated prefactor is $\sqrt{\frac{3\rho RT}{M_0}} \approx 10$ MPa$^{-1/2}$. This is in very good agreement with the value obtained from the fit. Note the calculated value is obtained by assuming the typical molar mass between crosslinks is the molar mass of the whole chain, which explains the slightly underestimated value obtained.

To conclude this part, correlations between $E$, $\lambda_b$ and hardness shore A show a direct relation according to simple physical relations from rubber elasticity theory. Interestingly, these correlations are valid for the filled and unfilled rubbers during ageing. From a practical purpose, the choice of a critical value for one of these three properties can be simply translated in terms of crosslink density changes during ageing.

*4.2. Poisson's Ratio*

All mechanical properties determined here have to allow the description of the behavior law which can be, for example, later used in a finite element code: $E$ and $v$ or the shear modulus $G$ and $K$ for instance. The final objective would be to perform simulations to predict strain and stress in an elastomer part under loading and after ageing.

Figure 11 shows the evolution of the Poisson's ratio ($v$) with ageing time of both unfilled (a) and filled (b) rubbers aged at three different temperatures. Values of $v$ were obtained using the following relationship that links $E$ and $K$ (see Equation (4), rewritten as follows): $v = 0.5\,(1 - E/3K)$.

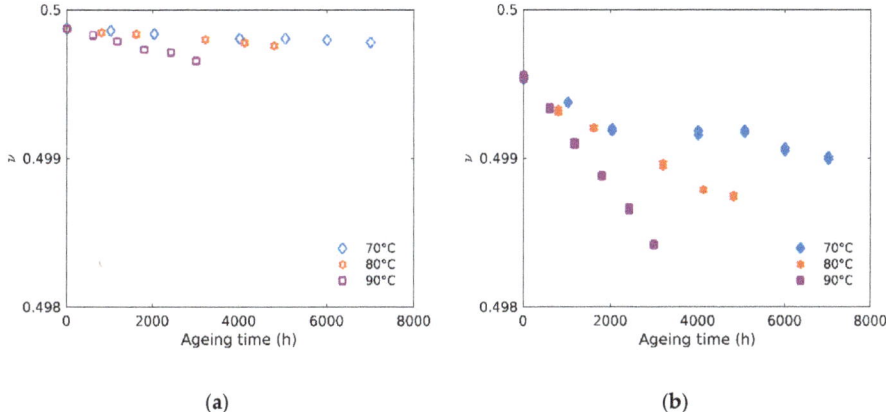

**Figure 11.** (**a**) Poisson's ratio versus ageing time for unfilled and (**b**) filled rubbers at 70 °C, 80 °C and 90 °C.

It can be noted again that the often used approximation for rubbers $v = 0.5$ is an upper limit (see Figure 11) obtained when $K$ tends to infinity. For this reason, this limit value cannot be used in finite elements simulations.

We can observe a small linear decrease of $v$ from 0.4996 to 0.4986 at 90 °C and 0.4988 at 80 °C in the case of filled rubbers. However, in the case of unfilled rubbers (and of filled rubbers at 70 °C), the decrease of $v$ may be considered as negligible, with a value of $0.4995 \pm 0.0004$. To the best of our knowledge, this is one of the first measurements of the evolution of the Poisson's ratio with ageing for elastomers.

## 5. Conclusions

In this article, we described the evolution of Young's modulus, elongation at break and hardness shore A as a function of ageing conditions. In this study, we also developed an experimental protocol allowing to estimate the Poisson's ratio evolution for the same sample. If the Poisson's ratio is only marginally affected by ageing for the unfilled rubbers, a slight decrease is witnessed in the case of the filled rubbers. This point, previously unseen in the literature, has to be taken into account for 3D element finite simulations of aged rubber industrial parts.

It is shown that Young's modulus and elongation at break can be quantitatively linked to each other using simple scaling arguments coming from the rubber elasticity theory, whatever the samples (filled or unfilled) and ageing time and temperature. This strongly suggests that the mechanical properties evolution is mainly driven by the extra crosslinking of the neoprene matrix, induced by ageing.

Since Young's modulus and hardness are also related quantitatively to each other, it can then be possible to predict the lifetime of elastomers based on an elongation at break criterion with a simple hardness shore measurement. This result may have numerous applications for industries where predicting such lifetime is of fundamental importance. This precise measurement of the Poisson's ratio could also open the way to more quantitative mechanical models.

**Author Contributions:** Conceptualization, L.L., G.M.-G. and B.F.; methodology, R.B. (Rami Bouaziz), L.T., G.M.-G.; C.O., R.B. (Rouslan Borisov), L.L., B.F. validation, R.B. (Rami Bouaziz), L.T., G.M.-G., L.L. and B.F.; formal analysis, R.B. (Rami Bouaziz), L.T., G.M.-G. and B.F.; investigation, R.B. (Rami Bouaziz), L.T. and C.O.; resources, L.L. and B.F.; data curation, R.B. (Rami Bouaziz), L.T. and R.B. (Rouslan Borisov); writing—original draft preparation, R.B. (Rami Bouaziz), L.T., G.M.-G. and B.F.; writing—review and editing, R.B. (Rami Bouaziz), C.O., L.L., R.B. (Rouslan Borisov) G.M.-G. and B.F.; visualization, R.B. (Rami Bouaziz) and L.T.; supervision, L.L., C.O., G.M.-G. and B.F.; project administration, R.B. (Rouslan Borisov); funding acquisition, L.L., G.M.-G., B.F. All authors have read and agreed to the published version of the manuscript.

**Funding:** This research was funded by EDF and Nuvia provided the rubber samples used in the study.

**Acknowledgments:** The authors would like to thank M. Kuntz (EDF), B. Basile (Freyssinet) and C. Guyonnet (Nuvia) for fruitful discussions. L. Truffault and R. Bouaziz also thank S. Baiz and C. Gorny from the PIMM lab for their help with the samples preparation and microindentation measurements.

**Conflicts of Interest:** The authors declare no conflict of interest. EDF participated in the design of the study and in the interpretation of the obtained data. EDF agreed to the publication of the results presented in the study.

## Appendix A

Small pieces of neoprenes (about 10 mm long, 5 mm large and 2 mm thick) were cut and embedded in epoxy resin. The epoxy embedded neoprenes were then polished in order to obtain flat areas. Local moduli were measured each 200 µm along the neoprenes thickness with a microhardness tester from CSM + Instruments equipped with a conospherical indenter. Hertz's model was used to determinate the modulus from the loading region of the strength–indentation depth curves.

The modulus profiles along the thickness measured by microindentation for three filled neoprenes aged at 90 °C during different times are presented in Figure A1. For the most aged sample (3000 h), the highest modulus increase between the center of the sample and its edge was around 17%. This value was slightly higher (33%) for the unaged sample. Our results demonstrate the absence of oxidation profile within the samples thickness. Indeed, these increases are both very small compared to the increases reported by Wise et al. [31] for neoprene aged at 100 °C (>100% increase) and by Le Gac et al. who reported a modulus approaching 1 GPa in the 100 µm superficial zone for neoprene aged at 120 °C [32].

Furthermore, it is interesting to note that the ratio between the average moduli measured by microindentation and the average moduli measured by traction is similar for the different exposition conditions (~1.5–2).

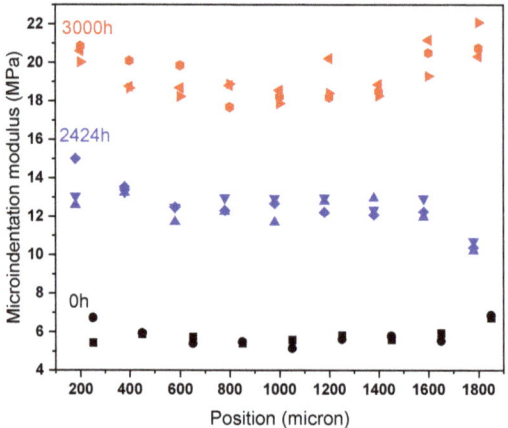

**Figure A1.** Evolution/profile of modulus through the thickness measured by microindentation of a 2 mm-thick sample aged at 90 °C for different durations.

## Appendix B

Unfilled neoprene plates were analyzed by Fourier transform infrared spectroscopy (FTIR) in attenuated total reflectance (ATR) mode using a Frontier FTIR spectrometer from Perkin Elmer. Spectra were obtained in the 4000–650 cm$^{-1}$ region with a resolution of 4 cm$^{-1}$ and 16 scans.

The spectra of neoprene aged at 90 °C for different times are presented in Figure A2. The spectral range between 2000 and 1000 cm$^{-1}$ is of particular interest when studying polychloroprene degradation due to the presence of bands attributed to oxidation products [4,5,32]. It should be noted that unfilled neoprene has been aged up to 6000 h for this analysis, while in the rest of the study the longest ageing time was 3000 h. The intensity of the band centered at 1590 cm$^{-1}$ clearly increases with ageing, but the change becomes only significant after 4900 h. This band can be attributed to carboxylate species. Their presence results from the conversion of acid chlorides and carboxylic acids during ageing in the presence of a small amount of metal oxides such as ZnO, classically used as an activator for the sulfur vulcanization of rubbers [33].

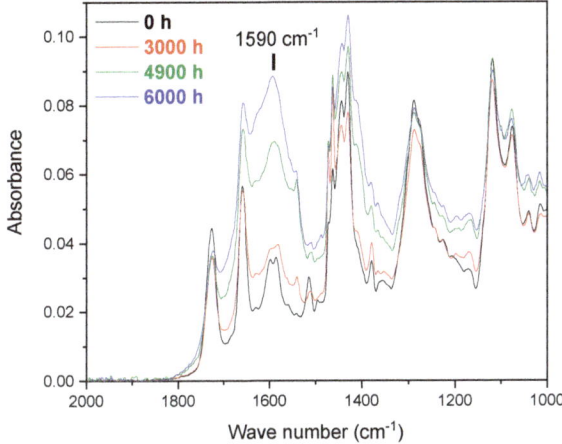

**Figure A2.** FTIR spectra of neoprene aged at 90 °C for different durations.

## References

1. Rubinstein, M.; Colby, R. *Polymer Physics*; Oxford University Press: Oxford, UK, 2003.
2. Mark, J. Rubber Elasticity. *J. Chem. Educ.* **1981**, *58*, 98. [CrossRef]
3. Nagdi, K. *Rubber as an Engineering Material: Guideline for User*; Hanser: Munich, Germany, 1993.
4. Celina, M.C.; Wise, J.; Ottesen, D.; Gillen, K.; Clough, R. Correlation of chemical and mechanical property changes during oxidative degradation of neoprene. *Polym. Degrad. Stabil.* **2000**, *68*, 171–184. [CrossRef]
5. Shelton, R. Aging and oxidation of elastomers. *Rubber Chem. Technol.* **1957**, *30*, 1251–1290. [CrossRef]
6. Delor, F.; Lacoste, J.; Lemaire, J.; Barrois-Oudin, N.; Cardinet, C. Photo-and thermal ageing of polychloroprene: Effect of carbon black and crosslinking. *Polym. Degrad. Stabil.* **1996**, *53*, 361–369. [CrossRef]
7. Celina, M.C. Review of polymer oxidation and its relationship with materials performance and lifetime prediction. *Polym. Degrad. Stabil.* **2013**, *98*, 2419–2429. [CrossRef]
8. Le Gac, P.-Y.; Roux, G.; Verdu, J.; Davies, P.; Fayolle, B. Oxidation of unvulcanized, unstabilized polychloroprene: A kinetic study. *Polym. Degrad. Stabil.* **2014**, *109*, 175–183. [CrossRef]
9. Bergström, J. *Mechanics of Solid Polymers*; Elsevier: Amsterdam, The Netherlands, 2015.
10. Peng, S.; Landel, R. Stored energy function and compressibility of compressible rubberlike materials under large strain. *J. Appl. Phys.* **1975**, *46*, 2599. [CrossRef]
11. Laufer, Z.; Diamant, Y.; Gill, M.; Fortuna, G. A Simple Dilatometric Method for Determining Poisson's Ratio of Nearly Incompressible Elastomers. *Int. J. Polym. Mater.* **1978**, *6*, 159–174. [CrossRef]
12. Kugler, H.; Stacer, R.; Steimle, C. Direct Measurement of Poisson's Ratio in Elastomers. *Rubber Chem. Technol.* **1990**, *63*, 473–487. [CrossRef]
13. Elektrova, L.; Melent'ev, P.; Zelenev, Y. Influence of fillers on the poisson ratios of rubber-like polymers. *Polym. Mech.* **1973**, *8*, 308–309.
14. Robertson, C.; Bogoslovov, R.; Roland, C. Effect of structural arrest on Poisson's ratio in nanoreinforced elastomers. *Phys. Rev. E* **2007**, *75*, 051403. [CrossRef] [PubMed]
15. Planes, E.; Chazeau, L.; Vigier, G.; Fournier, J. Evolution of EPDM networks aged by gamma-irradiation —Consequences on the mechanical properties. *Polymer* **2009**, *50*, 4028–4038. [CrossRef]

16. Le Gac, P.-Y.; Broudin, M.; Roux, G.; Verdu, J.; Davies, P.; Fayolle, B. Role of strain induced crystallization and oxidative crosslinking in fracture properties of rubbers. *Polymer* **2014**, *55*, 2535–2542. [CrossRef]
17. Le Gac, P.-Y.; Albouy, P.-A.; Petermann, D. Strain-induced crystallization in an unfilled polychloroprene rubber: Kinetics and mechanical cycling. *Polymer* **2018**, *142*, 209–217. [CrossRef]
18. Le Gac, P.-Y.; Albouy, P.-A.; Sotta, P. Strain-induced crystallization in a carbon-black filled polychloroprene rubber: Kinetics and mechanical cycling. *Polymer* **2019**, *173*, 158–165. [CrossRef]
19. Bouaziz, R.; Ahose, K.; Lejeunes, S.; Eyheramendy, D.; Sosson, F. Characterization and modeling of filled rubber submitted to thermal aging. *Int. J. Solids Struct.* **2019**, *169*, 122–140. [CrossRef]
20. Guth, E. Theory of Filler Reinforcement. *J. Appl. Phys.* **1945**, *16*, 596–604. [CrossRef]
21. Fukahori, Y.; Hon, A.A.; Jha, V.; Busfield, J.J.C. Modified GUTH–GOLD Equation for Carbon Black-Filled Rubbers. *Rubber Chem. Technol.* **2013**, *86*, 218–232. [CrossRef]
22. Mark, J. *Polymer Handbook*; Oxford University Press: Oxford, UK, 1999.
23. Mullins, L.; Tobin, N. Stress softening in rubber vulcanizates. Part I. Use of a strain amplification factor to describe the elastic behavior of filler-reinforced vulcanized rubber. *J. Appl. Polym. Sci.* **1965**, *9*, 2993–3009. [CrossRef]
24. Anandakumaran, K.; Seidl, W.; Castaldo, P. Condition Assesment of Cable Insulation Systems in Operating Nuclear Power Plants. *IEEE T. Dielect. El. In.* **1999**, *6*, 376–384. [CrossRef]
25. Burns, J.; Dubbelday, P.; Ting, R. Dynamic Bulk Modulus of Various Elastomers. *J. Polym. Sci. B Polym. Phys.* **1990**, *28*, 1187–1205. [CrossRef]
26. Diani, J.; Fayolle, B.; Gilormini, G. Study on the temperature dependence of the bulk modulus of polyisoprene by molecular dynamics simulations. *Mol. Simul.* **2008**, *34*, 1143–1148. [CrossRef]
27. Seitz, J. The estimation of mechanical properties of polymers from molecular structure. *J. Appl. Phys.* **1993**, *49*, 1331–1351. [CrossRef]
28. Flory, P. *Principles of Polymer Chemistry*; Cornell University Press: Ithaca, NY, USA; London, UK, 1953.
29. Gent, A.J. On the relation between indentation hardness and Young's modulus. *Rubber Chem. Technol.* **1958**, *31*, 896–906. [CrossRef]
30. Qi, H.J.; Joyce, K.; Boyce, M.C. Durometer hardness and the stress-strain behavior of elastomeric materials. *Rubber Chem. Technol.* **2003**, *76*, 419–435. [CrossRef]
31. Wise, J.; Gillen, K.; Clough, R. Quantitative model for the time development of diffusion-limited oxidation profiles. *Polymer* **1997**, *38*, 1229–1244. [CrossRef]
32. Le Gac, P.-Y.; Celina, M.; Roux, G.; Verdu, J.; Davies, P.; Fayolle, B. Predictive ageing of elastomers: Oxidation driven modulus changes for polychloroprene. *Polym. Degrad. Stabil.* **2016**, *130*, 348–355. [CrossRef]
33. Heideman, G.; Datta, R.; Noordermeer, J.; van Baarle, B. Influence of zinc oxide during different stages of sulfur vulcanization. Elucidated by model compound studies. *J. Appl. Polym. Sci.* **2005**, *96*, 1388–1404. [CrossRef]

**Publisher's Note:** MDPI stays neutral with regard to jurisdictional claims in published maps and institutional affiliations.

© 2020 by the authors. Licensee MDPI, Basel, Switzerland. This article is an open access article distributed under the terms and conditions of the Creative Commons Attribution (CC BY) license (http://creativecommons.org/licenses/by/4.0/).

Article

# Electrorheological Properties of Polydimethylsiloxane/TiO₂-Based Composite Elastomers

Alexander V. Agafonov [1,*], Anton S. Kraev [1], Alexander E. Baranchikov [2] and Vladimir K. Ivanov [2,3]

[1] Krestov Institute of Solution Chemistry, Russian Academy of Sciences, 153045 Ivanovo, Russia; a.s.kraev@mail.ru
[2] Kurnakov Institute of General and Inorganic Chemistry, Russian Academy of Sciences, 119991 Moscow, Russia; a.baranchikov@yandex.ru (A.E.B.); van@igic.ras.ru (V.K.I.)
[3] Higher School of Economics, National Research University, 101000 Moscow, Russia
* Correspondence: ava@isc-ras.ru

Received: 19 August 2020; Accepted: 16 September 2020; Published: 18 September 2020

**Abstract:** Electrorheological elastomers based on polydimethylsiloxane filled with hydrated titanium dioxide with a particle size of 100–200 nm were obtained by polymerization of the elastomeric matrix, either in the presence, or in the absence, of an external electric field. The viscoelastic and dielectric properties of the obtained elastomers were compared. Analysis of the storage modulus and loss modulus of the filled elastomers made it possible to reveal the influence of the electric field on the Payne effect in electrorheological elastomers. The elastomer vulcanized in the electric field showed high values of electrorheological sensitivity, 250% for storage modulus and 1100% for loss modulus. It was shown, for the first time, that vulcanization of filled elastomers in the electric field leads to a significant decrease in the degree of crosslinking in the elastomer. This effect should be taken into account in the design of electroactive elastomeric materials.

**Keywords:** crosslinking; TiO₂; nanomaterials; stimuli-responsive materials; smart materials

---

## 1. Introduction

The electrorheological effect is a reversible and rapid change in the physicomechanical properties of a composite dielectric material when an external electric field is applied [1,2]. Materials exhibiting the electrorheological effect, and their analogues–magnetorheological materials, are typical representatives of smart materials that are extremely promising for application in various industrial devices, including vibration absorbers, dampers, etc. [3,4].

Depending on the type of dielectric dispersion medium, electrorheological fluids (liquid dispersion medium) and electrorheological gels and elastomers (solid dispersion medium) are distinguished [5,6]. In electrorheological fluids, the electrorheological effect is caused by the movement of particles of the polarizable disperse phase in an electric field, with the formation of ordered structures and the transition of the dispersed system from a viscous flowing state to a viscoplastic state [7,8]. The main disadvantage of electrorheological fluids is the lack of long-term stability and gradual sedimentation of particles of the disperse phase. Despite the fact that a strong electrorheological effect was observed for suspensions containing particles of a disperse phase with a high density (up to 7 g/cm³) [9–12], providing sedimentation stability of electrorheological fluids usually requires the use of additional stabilizers or constant mixing of the suspension.

In electrorheological elastomers, the mobility of the particles of the disperse phase (filler particles) is hindered by the polymer matrix, so there is no problem of sedimentation stability in such systems. Unfortunately, almost complete immobility of the filler particles does not allow for restructuring of the electrorheological elastomer when an electric field is applied, and so in these materials the electrorheological effect arises due to dipole-dipole interactions between isolated particles of the disperse phase [13].

A number of studies have been devoted to the creation of electrorheological elastomers, covering the use of various fillers and polymer compositions, aimed at the development of electrorheological elastomers with improved dynamic characteristics and mechanical properties, and the development of model representations to describe the electrorheological and viscoelastic mechanical behavior of elastomers, as well as the search for possible areas of practical application of electrorheological elastomers [14–20]. Despite some advances in the field of electrorheological elastomers production, the design of electrorheological elastomers with large-scale changes in viscoelastic properties in electric fields is currently difficult, due to the lack of a generally accepted behavior theory for such disperse systems. As a first approximation, the polarization theory is applied, which is generally used for the interpretation of the electrorheological effect in electrorheological fluids. Based on this theory, the key parameters determining the electrorheological effect are the electric field strength, the distance between the filler particles, the dielectric relaxation time as well as the dielectric constant, the dielectric loss tangent and the electrical conductivity of the materials of the filler particles, and the dispersion medium [3,21–27]. The surface chemical composition of particles in the disperse phase [28–30], the presence of the regions with a reduced viscosity in the dispersion medium [31], as well as the nature of the distribution of filler particles in the matrix, can also make a significant contribution to the formation of the electrorheological effect.

The most well-known and simple method to organize the structure of electrorheological elastomers is the polymerization of dielectric suspensions in a constant electric field [13]. During the polymerization, ordered structures are preserved, resulting from the arrangement of polarized particles of the disperse phase in a liquid. As a rule, the electrorheological effect for structured elastomers is greater than for materials with a stochastic distribution of filler particles, which is associated with the smaller distance between the particles of the disperse phase. Anisotropically-filled elastomers are also characterized by a higher dielectric constant and Maxwell–Wagner dispersion [32–34].

At the same time, the factors determining the magnitude of the electrorheological response of filled elastomers remain largely unexplored. In our opinion, the strength of contact between the particles of the disperse phase and the dispersion medium is of great importance, as well as the presence of low molecular weight compounds on the surface of the particles, which act as activators of the electrorheological effect. In particular, during the curing of the elastomer in the electric field, regions with different degrees of polymerization of the elastomer may appear, which could unpredictably affect the physicomechanical properties of the material.

The aim of this work is to analyze the dielectric and electrorheological characteristics of polydimethylsiloxane elastomers containing amorphous hydrated titanium dioxide vulcanized in, and in the absence of, an electric field.

## 2. Experimental Part

Titanium dioxide was synthesized by the hydrolysis of titanium tetraisopropoxide (#87560, Sigma-Aldrich, Darmstadt, Germany) in non-absolute ethanol. A 250-mL flask containing 50 mL of 95.6% ethanol was placed in a dry box and, with vigorous stirring, 12.7 g of Ti (O$i$Pr)$_4$ was added dropwise to ethanol. The resulting white precipitate was kept in the mother liquor, under continuous stirring at room temperature, for 5 h, after which it was separated by centrifugation and dried to a constant mass in an oven at 60 °C.

To obtain a composite material containing titanium dioxide dispersed in a polydimethylsiloxane matrix, we used liquid siloxane rubber (polymerization degree, $n$ = 100–5000) with terminal silanol groups (Vixint PK-68A compound, NPP Khimprom LLC, Yekaterinburg, Russia). Titanium dioxide powder was mixed with a siloxane liquid to obtain a composite containing 30 wt.% (~17 vol.%) TiO$_2$, and the resulting suspension was mixed for 30 min, followed by the addition of 3 wt.% of the polymerization catalyst (a solution of aminopropyltriethoxysilane in ethyl silicate with a mass ratio of

1:4) [35] and vigorous stirring of the resulting mixture at 500 rpm for 5 min, after which the mixture was evacuated at $10^{-2}$ atm to remove air bubbles. The obtained suspension was placed in a cylindrical polymethylmethacrylate container with a diameter of 20 mm and depth of 2.5 mm, and was left to cure at room temperature. A mold with flat electrodes installed in its base was used for curing the suspension in the electric field. Within an hour of filling the mold, an alternating (10 Hz) voltage of 5 kV was applied to the electrodes. Complete curing of the elastomer was achieved within 5 h. Upon the vulcanization, the elastomers adhered firmly to the electrodes, eliminating possible electrode slip relative to the elastomer sample surface. Hereafter, a sample vulcanized in the absence of an electric field is designated ERE-0, and a sample vulcanized in an electric field at a voltage of 5 kV is designated ERE-5.

X-ray diffraction analysis (XRD) was carried out using a Bruker D8 Advance diffractometer (Karlsruhe, Germany) (CuK$\alpha$ radiation) in the angle range 5–80° 2$\theta$, with a step of 0.02° 2$\theta$ and an accumulation duration of 0.2 s/step.

Scanning electron microscopy (SEM) was performed using a Tescan Vega 3 SBH microscope (Kohoutovice, Czech Republic) at an accelerating voltage of 5 kV.

IR spectra were recorded using a Bruker VERTEX 80v IR Fourier spectrometer (Karlsruhe, Germany). The FTIR reflection spectra were recorded in the region of 400 to 4000 cm$^{-1}$, with a resolution of 2 cm$^{-1}$, at room temperature.

Low-temperature nitrogen adsorption analysis was performed using a Quantachrome NOVAtouch NT LX-specific surface and porosity analyzer (Boynton Beach, FL, USA).

The dependences of the dielectric constant and dielectric loss tangent on the frequency of the electric field of composite elastomers were measured in a capacitor-type cell with spring-loaded disk plane-parallel electrodes made of polished stainless steel, using a Solartron SI 1260 Impedance/Gain-Phase analyzer (Farnborough, United Kingdom) in the frequency range $25-10^6$ Hz at 1 V voltage. The effect of the electric field strength on the dielectric characteristics of filled elastomers was analyzed in the same cell, using an Electronpribor MEP-4CA Schering Bridge (Moscow, Russia) with a reference capacitor at a frequency of 50 Hz in the voltage range 0.2–1.0 kV. All measurements were performed at room temperature.

To conduct electrorheological measurements, a rheometer operating in the controlled shear deformation mode was used with a stepper motor with a controlled rotation speed and a torque measuring system. A cell with two parallel plates of polished brass with a diameter of 20 mm and an adjustable gap was used for all measurements. A voltage up to 5.0 kV was generated between the upper movable plate and the lower plate connected with the strain gauge; after 60 s, the rheometer drive was turned on and the measured torque values were recorded automatically every 0.1 s at a shear rate of 0.1 rad/s until shearing angle reached 0.192 rad. After the measurement, the rheometer plate was returned to a starting position. The absence of the electrode slip relative to the sample surface was checked by the coincidence of marks applied on the side faces of the electrodes and the elastomer sample.

## 3. Results and Discussion

Figure 1 shows the results of the analysis (XRD, SEM, low temperature nitrogen adsorption) of titanium dioxide powder obtained by hydrolysis of titanium isopropoxide in ethanol. The titanium dioxide powder was almost completely X-ray amorphous; the diffraction pattern (Figure 1a) contained weak reflections (in particular, ~25.7° 2$\theta$), which can be attributed to anatase [36]. The amorphous state is typical of titanium dioxide obtained by the sol-gel method without any further hydrothermal or thermal treatment [37,38]. At the same time, as previously noted in a number of studies, amorphous titanium dioxide can exhibit short-range order typical of crystalline polymorphic modifications of $TiO_2$ (in particular, anatase), which determines the mechanism of its crystallization in aqueous media at elevated temperatures [39].

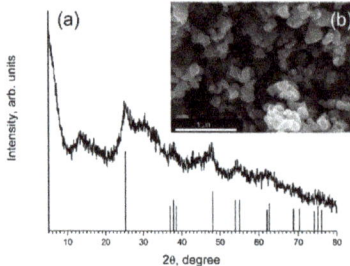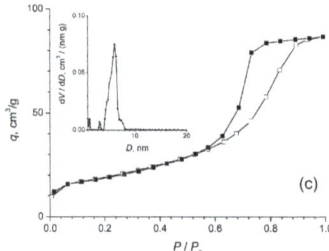

**Figure 1.** The analysis results of titanium dioxide powder obtained by hydrolysis of titanium isopropoxide in ethanol, using (**a**) XRD (Bragg positions correspond to anatase); (**b**) SEM; (**c**) low-temperature nitrogen adsorption. The inset shows the pore size distribution.

SEM data (Figure 1b) indicated a rather low degree of agglomeration of the obtained $TiO_2$ powder, which consisted of relatively large (100–200 nm) particles of an almost isotropic shape. The specific surface area of the powder (65 m$^2$/g) indicated that these particles were sufficiently dense, which was further confirmed by the fact that only small mesopores (4–8 nm) were present in the powder, while the larger mesopores, as well as micropores, were nearly absent. Capillary condensation hysteresis (Figure 1c), which belongs to type H2, according to the IUPAC classification (type E according to the de Boer classification) [40], indicated the presence of narrow, cylindrical pores.

The shear modulus [41] was used to describe the viscoelastic properties of elastomers, which is considered as the complex number $G^* = G' + iG''$, where $G'$ (storage modulus) is connected with the energy consumption for the deformation of the elastic structural elements in the elastomer, and $G''$ (loss modulus) characterises the energy dissipation into viscous losses.

Figure 2 shows the storage and loss moduli of the composite elastomer obtained by vulcanization in the absence or in the presence of an electric field, as functions of the relative deformation in external electric fields of various strengths. For comparison purposes, the storage and loss moduli of bare elastomer samples obtained by vulcanization in the absence or in the presence of an electric field, as functions of the relative deformation are presented in Figure S2. Bare elastomer samples vulcanized in the absence or in the presence of an electric field (see Figure S2), show virtually no dependence on the storage and loss moduli on the relative deformation. The $G'$ and $G''$ values for unfilled elastomers are lower than that of the filled ones, confirming higher elasticity and lower viscosity of the former under rotational strain.

The data presented in Figure 2 indicate that, in the absence of an electric field, the ERE-0 sample behaved as a highly elastic material; the storage modulus at a shear rate of 0.1 rad/s at low strain values was about 1.8 MPa. In turn, the storage modulus of the ERE-5 sample under the same conditions was about 0.5 MPa. Until a 0.5% degree of deformation was reached, the values of the storage modulus for both samples remained constant and decreased at large strains. This effect is similar to the well-known Payne effect (otherwise called the Fletcher-Gent effect) observed for filled elastomers [42,43] and which arises from the bonds breaking between filler particles [44]. The presence of the Payne effect indicates a high degree of interaction between the filler particles in the composite elastomer [45].

Previously, the Payne effect was detected in magnetorheological elastomers, and the magnitude of the effect increased when an external magnetic field was applied [46]. Our data confirm the results of a recent study [45] mentioning the Payne effect in polysiloxane-based electrorheological elastomers.

It is noteworthy that, for ERE-0 and ERE-5 samples, the values of the storage and loss moduli remained constant at low strain values in the entire range of external electric field strengths (Figure 2). To determine the effect of the electric field strength on the value of the Payne effect in electrorheological elastomers based on polydimethylsiloxane, the values of the storage and loss moduli were estimated at zero and infinitely large deformations [42] and the corresponding differences were calculated, $G'_0 - G'_\infty$ and $G''_0 - G''_\infty$ (Figure 3).

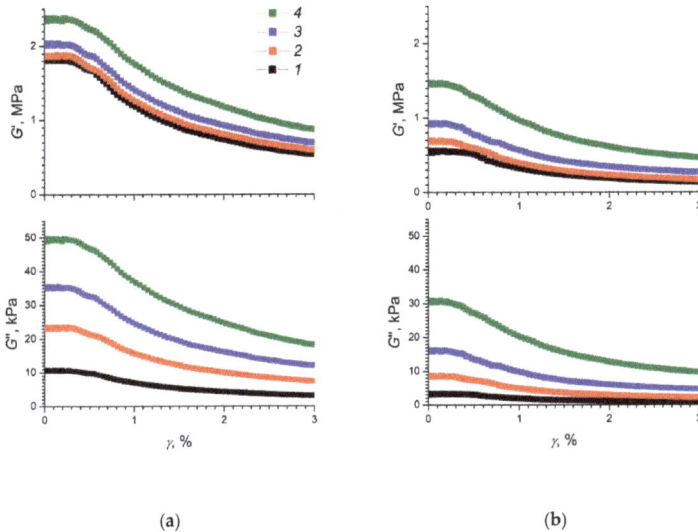

**Figure 2.** Storage modulus ($G'$) and loss modulus ($G''$) as functions of relative deformation in electric fields of various strengths (1–0 kV/mm; 2–0.4 kV/mm; 3–1.2 kV/mm; 4–2 kV/mm) for elastomer samples modified with titanium dioxide and vulcanized (**a**) in the absence of an electric field and (**b**) in an electric field. The shear rate was 0.1 rad/s.

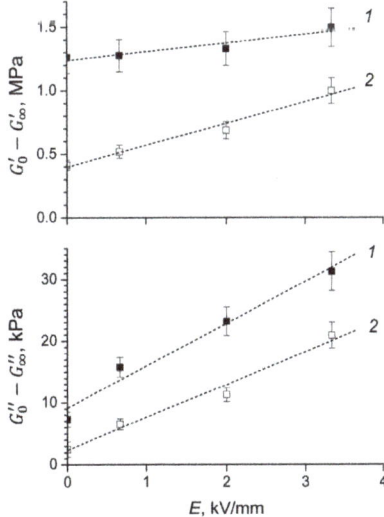

**Figure 3.** The effect of the external electric field strength on the value of the Payne effect for the storage modulus ($G'_0 - G'_\infty$) and loss modulus ($G''_0 - G''_\infty$) for the ERE-0 and ERE-5 samples.

One can note that the values of the Payne effect for the storage and loss moduli increased in an external electric field. This increase was apparently due to the fact that polarized filler particles, as a result of small spatial displacements (mainly rotation), contributed to an increase in the conformational mobility of macromolecules.

The electrorheological sensitivity of the elastomers (the efficiency of converting the energy of an electric field into mechanical energy, $\eta_E$) can be estimated using the relative increments of the storage and loss moduli and in electric fields of different strengths (Figure 4) [25]:

$$\eta'_E = \frac{G'_E - G'_0}{G'_0} \text{ and } \eta''_E = \frac{G''_E - G''_0}{G''_0} \qquad (1)$$

where $G_E$ is a storage modulus or loss modulus of the elastomer in an electric field; $G_0$ is a storage or loss modulus of the elastomer in the absence of an electric field.

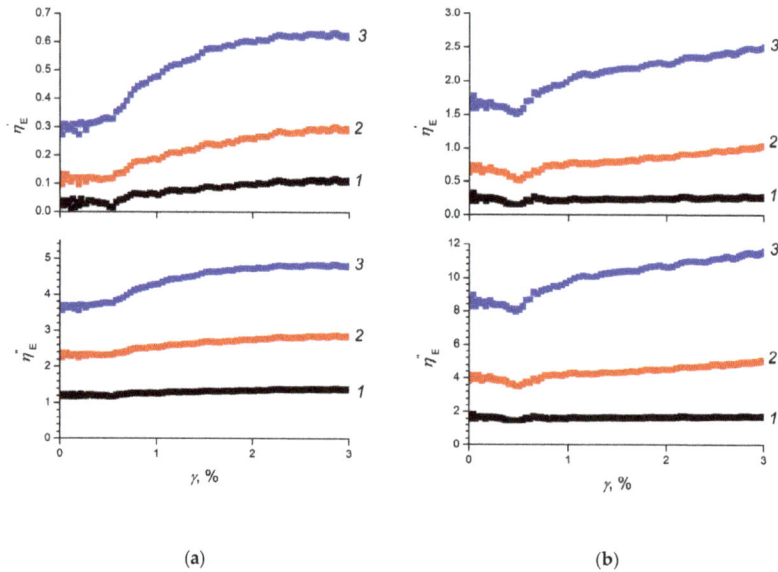

(a)         (b)

**Figure 4.** The electrorheological sensitivity of composite elastomers (**a**) ERE-0 and (**b**) ERE-5 with respect to storage ($\eta'_E$) and loss ($\eta''_E$) moduli as function of relative deformation at various strengths of an external electric field (1–0.4 kV/mm, 2–1.2 kV/mm, 3–2.0 kV/mm).

As follows from Figure 4a, the electrorheological sensitivity for the ERE-0 sample by storage modulus reached 63% in the 2.0 kV/mm electric field, which corresponds to the characteristics obtained for electrorheological composite elastomers filled with titanium dioxide [25,26]. Electrorheological sensitivity for the sample ERE-0 by the loss modulus reached 450% in the 3 kV/mm field.

The electrorheological sensitivity for the ERE-5 sample was significantly higher by both storage modulus and loss modulus (Figure 4b). In a 2.0 kV/mm electric field, the $\eta'_E$ value reached 250%, and the $\eta''_E$ value exceeded 1100%. High values of electrorheological sensitivity for the ERE-5 sample indicate the high mobility of macromolecules in its structure and a significant range of changes in its elasto-plastic properties when an electric field was applied, comparable with the control range of magnetorheological elastomers [47]. It should be noted that the high electrorheological sensitivity of the ERE-5 sample, in comparison with the ERE-0 sample, is consistent with significantly higher values of the dielectric characteristics of the ERE-5 sample (see below)–the permittivity difference $\varepsilon_0-\varepsilon_\infty$, higher tg$\delta$ values at the point of relaxation maximum and slower relaxation processes during polarization, which, according to general theories about the electrorheological sensitivity of electrorheological fluids, should contribute to a pronounced electrorheological effect.

A specific feature of the electrorheological elastomers ERE-0 and ERE-5 is their almost constant electrorheological sensitivity in the region of small deformations (up to $\gamma \approx 0.5\%$) at a fixed value of the external electric field. In the region of relative deformations $\gamma > 0.5\%$, an increase in deformation led to an increase in electrorheological sensitivity by 1.5–2 times. These observations indicate that, with a relative deformation $\gamma \approx 0.5\%$, the polymer matrix was being rearranged, forming a relatively mobile viscoplastic structure. This effect allows for expanding the range of regulation of the physicomechanical properties of the composite elastomers using external electric fields.

The results obtained indicate that, in order to achieve high electrorheological sensitivity of composite elastomeric systems, an elastomer material must be preliminarily turned to a stressed state by a slight mechanical load. This stress can be caused by either external or internal, factors. In an electric field, filler particles immobilized in a polymer matrix are affected by polarizing forces that contribute to the rotation and displacement of particles relative to electric field lines. The movement of particles leads to the deformation of polymer molecules, primarily due to their conformational mobility, which in turn leads to an increase in the forced elasticity of the elastomer. The summation of these phenomena determines the features of the electrorheological effect in filled elastomers.

At the same time, these features cannot clarify the differences in the electrorheological behavior observed for the ERE-0 sample (vulcanized in the absence of an electric field and presumably characterized by a stochastic distribution of particles of the disperse phase) and ERE-5 sample (vulcanized in an electric field and presumably characterized by an anisotropic structure). In particular, the significantly higher electrorheological sensitivity with respect to loss modulus observed for the ERE-5 elastomer compared to the ERE-0 elastomer may have been due to the presence of a viscous flowing component in the structure of the former.

To identify the nature of the differences in the electrorheological properties of the obtained composite elastomers, an analysis of their dielectric characteristics was performed (Figures 5 and 6).

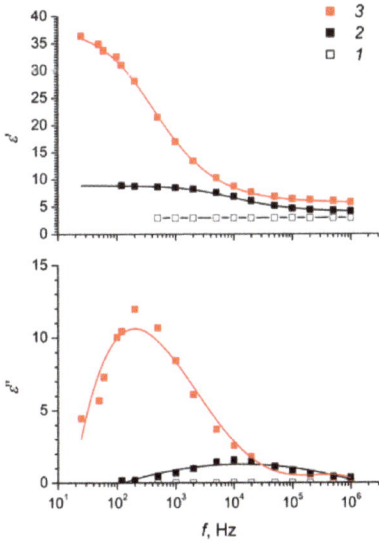

**Figure 5.** The dielectric constants ($\varepsilon'$, $\varepsilon''$) as functions of frequency for (1) unmodified polydimethylsiloxane and TiO$_2$-modified polydimethylsiloxane samples: (2) ERE-0; (3) ERE-5.

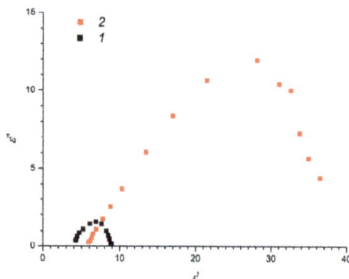

**Figure 6.** Cole–Cole diagrams for TiO$_2$-modified elastomers: (1) ERE-0; (2) ERE-5.

The dielectric characteristics of a filler-free elastomer (Figure 5) were frequency independent, while filled elastomers exhibited pronounced dependences of the dielectric constant and dielectric loss tangent on the frequency of the electric field. Comparison of the dielectric characteristics of filled elastomers (Figure 5) reveals that, for a sample vulcanized in the absence of an electric field, the nature of the relaxation processes was close to Debye and typical for systems with an isotropic structure. Conversely, for the asymmetry of the Cole–Cole diagrams (Figure 6), a larger range of changes in the dielectric constant and large values of the dielectric loss tangent for a composite elastomer vulcanized in an electric field indicates the presence of polarizable elements with different characteristic relaxation times.

Figure 5 and Table 1 show the results of fitting the dielectric spectra of the elastomer samples to the Havriliak–Negami Equation [48]:

$$\varepsilon'_\omega = \varepsilon_\infty + \frac{\varepsilon_S - \varepsilon_\infty}{\left(1 + (i\,\omega\tau)^\alpha\right)^\beta} \qquad (2)$$

where $\varepsilon'_\omega$ is a dielectric permittivity at a circular frequency $\omega$; $\varepsilon_S$–dielectric permittivity at zero frequency; $\varepsilon_\infty$–dielectric permittivity at infinite frequency; $\tau$–dielectric relaxation time; $\alpha$, $\beta$–parameters related to the distribution of relaxation times.

**Table 1.** The results of fitting the dielectric spectra of composite elastomer samples to Havriliak–Negami equation.

| Sample Name | $\varepsilon_S$ | $\varepsilon_\infty$ | $\varepsilon_\infty - \varepsilon_S$ | $\tau$ | $\alpha$ | $\beta$ |
|---|---|---|---|---|---|---|
| ERE-0 | 8.9 | 4.0 | 4.9 | $2.4\cdot10^{-5}$ | 0.85 | 0.80 |
| ERE-5 | 37.7 | 5.7 | 32.0 | $5.6\cdot10^{-4}$ | 0.85 | 0.80 |

Table 1 shows that both ($\varepsilon_\infty - \varepsilon_S$) and $\tau$ values for the ERE-5 sample are notably higher than those for the ERE-0 sample, which is probably due to the different contribution of interphase polarization. Figure S3 shows the frequency dependences of the composite elastomer conductivity, which are typical to hopping charge transport mechanism in titanium dioxide prepared by sol-gel method [49]. ERE-5 sample demonstrates higher conductivity at frequencies below 55 kHz, while at higher frequencies, the sample has lower conductivity. The different efficiency of charge transfer process in these composites may be due to the different environment of titanium dioxide particles in the elastomer matrix.

The ordered arrangement of the particles of the disperse phase in the elastomer is not the only factor determining the dielectric characteristics of polymer composites, as the crosslinking density of macromolecules is also an extremely important parameter. For temperatures above the glass transition temperature of the polymer, the dielectric constant and dielectric loss tangent increase with decreasing crosslink density of the macromolecules. A decrease in the crosslinking density of polydimethylsiloxane macromolecules (for example, due to the introduction of plasticizers [50]) can be used for the advanced design of composites with improved electrorheological properties.

From the analysis of the electric field strength effect on the dielectric characteristics of composite elastomers (Figure 7), it follows that in 0.1–0.4 kV/mm electric fields, the dielectric constant and the dielectric loss tangent were directly proportional to the field strength. With an increase in the field strength from 0.1 to 0.4 kV/mm, the change in $\varepsilon$ value for the ERE-0 and ERE-5 samples was almost the same, amounting to 0.5–0.9, while the change in the dielectric loss tangent for the ERE-5 sample (~0.14) significantly exceeded the similar characteristic for the ERE-0 sample (~0.05). Since the conductivity can be derived from the dielectric characteristics of the material [51]:

$$\sigma = \varepsilon_0 \cdot \varepsilon \cdot \omega \cdot \mathrm{tg}\delta \qquad (3)$$

where $\omega$ is a circular frequency, such a difference leads to a significantly larger increase in conductivity in the electric field of an ERE-5 sample, compared with an ERE-0 sample.

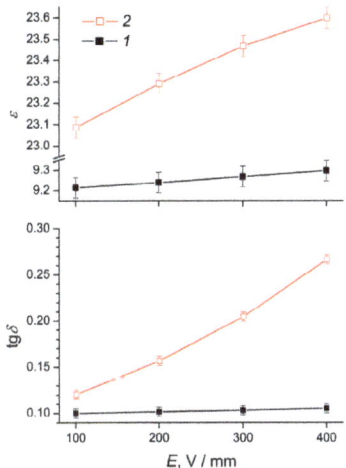

**Figure 7.** The dielectric constant ($\varepsilon$) and dielectric loss tangent (tg$\delta$) of composite elastomers (1) ERE-0 and (2) ERE-5 as functions of the electric field strength.

The higher conductivity of the composite elastomer ERE-5 can be caused by various factors. First, as a result of vulcanization of the composite elastomer in an external electric field, aggregates of disperse phase particles oriented in the direction of the electric field form in the volume of the polymer, along which charge transfer can occur [52]. Obviously, with an equal content of the disperse phase, the percolation threshold for composites containing oriented particle aggregates will be lower than for a composite with a stochastic distribution of filler particles. At the same time, high dielectric characteristics and, consequently, relatively high conductivity of elastomeric materials can be caused by the presence of mobile structural elements, for example, low molecular weight compounds or polymer molecules, with a lower average molecular weight compared to the main polymer component.

In order to identify the possible contribution of the latter factor to the electrorheological and dielectric properties of the obtained elastomers, we assessed a degree of polymer crosslinking in composites vulcanized in an electric field or without it. The crosslinking degree was evaluated gravimetrically by determining the degree of swelling [53,54].

For this, samples of elastomers filled with titanium dioxide were immersed in toluene and left to swell for one day, with stirring. After this, toluene was replaced by a different volume of toluene and the procedure was repeated for an additional day. After swelling, the samples were thoroughly washed with ethanol and dried under reduced pressure ($10^{-2}$ atm) at 50 °C.

The following equations were used for calculations:

$$\beta = \frac{m_H - m_t}{m_t} \cdot 100\% \qquad (4)$$

$$\omega_c = \frac{m_0 - m_t}{m_0} \cdot 100\% \qquad (5)$$

Here, $\beta$ is the swelling degree; $\omega_c$ is the mass fraction of components not polymerized into a common network; $m_0$ is the initial mass of the composite elastomer; $m_H$ is the mass of swelled composite elastomer; $m_t$ is the mass of elastomer after washing with ethanol and drying. The weight of titanium dioxide contained in the composite sample was subtracted from the values $m_0$, $m_H$, $m_t$ before calculations. The Flory–Rehner equation was used to determine the degree of crosslinking [55]:

$$\nu = -\frac{\ln(1 - \vartheta_{2m}) + \vartheta_{2m} + \chi_{12}\vartheta_{2m}^2}{V_1\left(\vartheta_{2m}^{\frac{1}{3}} - \frac{\vartheta_{2m}}{2}\right)}, \qquad (6)$$

where $\nu$ is a degree of crosslinking of the elastomer; $V_1$ is the molar volume of the solvent (106.5 cm$^3$/mol for toluene); $\vartheta_{2m}$ is the molar fraction of the polymer for the equilibrium swelling; $\chi_{12}$ is the Flory–Higgins coefficient, which characterizes the polymer-solvent interaction and is equal to 0.393 for the analyzed system [56,57]. The calculation results are presented in Table 2.

**Table 2.** The results of swelling and crosslinking degree calculations for the samples of composite elastomers based on polydimethylsiloxane and titanium dioxide, cured in, and in the absence of, an electric field.

| Sample Name | $\beta$, % | $\omega_c$, % | $\nu$, mol/g |
|---|---|---|---|
| ERE-0 | 170 | 3.5 | 0.0056 |
| ERE-5 | 296 | 10.3 | 0.0013 |

From the data obtained, it follows that a sample of composite elastomer cured in the absence of an external electric field has a lower content of unbound components, a lower degree of swelling, and a higher degree of crosslinking of polymer molecules. In turn, vulcanization of the composite polymer in an electric field results in a material with a relatively low degree of crosslinking and contains organic components that are not connected to a common polymer network. These features are consistent with data characterizing the electrorheological behavior of composite elastomers and their dielectric properties.

The analysis of ERE-0 and ERE-5 samples after determining the degree of swelling (washed with toluene and ethanol), by scanning electron microscopy (Figure 8), showed that the microstructure of the former sample was characterized by the presence of TiO$_2$ particles immobilized in a continuous polymer matrix. Conversely, the surface of the ERE-5 sample after washing acquired a loose structure, indicating a partial destruction of the polymer matrix.

Based on the combined results obtained, one can conclude that the ERE-0 sample was an elastomeric composite with filler particles uniformly distributed in a polymer matrix, characterized by a high degree of crosslinking. In addition to the crosslinked polymer, the polymer matrix of the ERE-5 composite included about 10% of the organic component that was not bound with the polymer matrix.

Apparently, during the curing of the elastomer in an external electric field, a partial segregation of the hardener, which has a higher dielectric constant compared to oligomers of polydimethylsiloxane, occurred. This resulted in its localization near the polarized particles of titanium dioxide, so the curing of the elastomer occurred non-uniformly and a fraction of the organic molecules was not bound into a polymer matrix. This effect can be used to obtain elastomeric composites with a controlled elasticity and creates the background for the design of highly effective electrorheological elastomers.

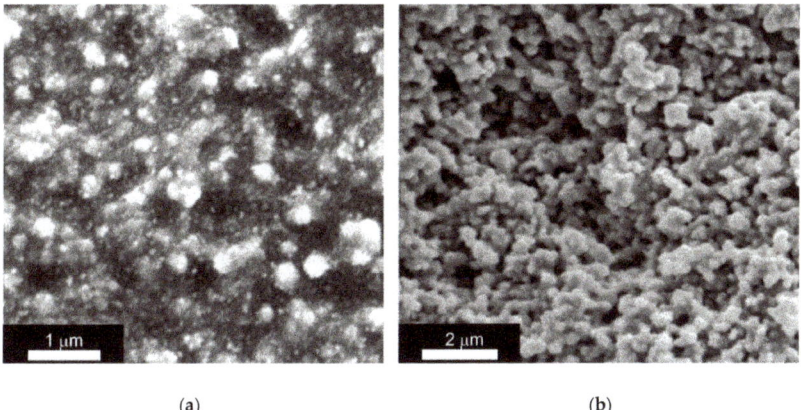

(a)  (b)

**Figure 8.** Scanning electron microscopy data for composite elastomers (**a**) ERE-0 and (**b**) ERE-5 after analysis of the degree of crosslinking of the polymer matrix (after washing in toluene and ethanol).

## 4. Conclusions

The viscoelastic and dielectric properties of silicone elastomers filled with isotropic particles of hydrated titanium dioxide with 100–200 nm size and cured in, and in the absence of, an electric field were compared. An analysis of the dependences of the storage modulus and the loss modulus of filled elastomers made it possible to reveal the effect of the electric field on the Payne effect in electrorheological elastomers. An elastomer vulcanized in an electric field showed high values of electrorheological sensitivity, which reached 250% for the storage modulus and 1100% for the loss modulus.

It was found that it is necessary to bring the elastomer to a pre-stressed state by applying an external tangential load in order to achieve high electrorheological sensitivity of the composite elastomeric system.

It has been shown, for the first time, that the vulcanization of filled elastomers in an electric field can lead to a significant decrease in the degree of crosslinking in the elastomer. This effect should be taken into account in the design of electroactive elastomeric materials.

**Supplementary Materials:** The following are available online at http://www.mdpi.com/2073-4360/12/9/2137/s1.

**Author Contributions:** Conceptualization, A.V.A. and V.K.I.; data curation, A.S.K.; investigation, A.V.A. and A.S.K.; methodology, A.V.A.; project administration, A.E.B.; resources, A.V.A.; supervision, V.K.I.; writing, original draft, A.V.A.; writing, review, and editing, A.E.B. and V.K.I. All authors have read and agreed to the published version of the manuscript.

**Funding:** This work was supported by the Russian Science Foundation (project 16-13-10399).

**Conflicts of Interest:** The authors declare no conflict of interest.

## References

1. Stangroom, J.E. Electrorheological fluids. *Phys. Technol.* **1983**, *14*, 290–296. [CrossRef]
2. Sheng, P.; Wen, W. Electrorheological Fluids: Mechanisms, Dynamics, and Microfluidics Applications. *Annu. Rev. Fluid Mech.* **2011**, *44*, 143–174. [CrossRef]
3. Hao, T. Electrorheological suspensions. *Adv. Colloid Interface Sci.* **2002**, *97*, 1–35. [CrossRef]
4. Moučka, R.; Sedlačík, M.; Kutálková, E. Magnetorheological elastomers: Electric properties versus microstructure. *AIP Conf. Proc.* **2018**, *2022*. [CrossRef]

5. Dong, Y.Z.; Seo, Y.; Choi, H.J. Recent development of electro-responsive smart electrorheological fluids. *Soft Matter* **2019**, *15*, 3473–3486. [CrossRef]
6. Shiga, T.; Ohta, T.; Hirose, Y.; Okada, A.; Kurauchi, T. Electroviscoelastic effect of polymeric composites consisting of polyelectrolyte particles and polymer gel. *J. Mater. Sci.* **1993**, *28*, 1293–1299. [CrossRef]
7. Hutter, K.; Ursescu, A.; van de Ven, A.A.F. Electrorheological fluids. *Lect. Notes Phys.* **2006**, *710*, 279–366. [CrossRef]
8. Parthasarathy, M.; Klingenberg, D.J. Electrorheology: Mechanisms and models. *Mater. Sci. Eng. R Rep.* **1996**, *17*, 57–103. [CrossRef]
9. Egorysheva, A.V.; Kraev, A.S.; Gajtko, O.M.; Kusova, T.V.; Baranchikov, A.E.; Agafonov, A.V.; Ivanov, V.K. High electrorheological effect in Bi1.8Fe1.2SbO7 suspensions. *Powder Technol.* **2020**, *360*, 96–103. [CrossRef]
10. Agafonov, A.V.; Kraev, A.S.; Kusova, T.V.; Evdokimova, O.L.; Ivanova, O.S.; Baranchikov, A.E.; Shekunova, T.O.; Kozyukhin, S.A. Surfactant-Switched Positive/Negative Electrorheological Effect in Tungsten Oxide Suspensions. *Molecules* **2019**, *24*, 3348. [CrossRef] [PubMed]
11. Agafonov, A.V.; Kraev, A.S.; Ivanova, O.S.; Evdokimova, O.L.; Gerasimova, T.V.; Baranchikov, A.E.; Kozik, V.V.; Ivanov, V.K. Comparative study of the electrorheological effect in suspensions of needle-like and isotropic cerium dioxide nanoparticles. *Rheol. Acta* **2018**, *57*, 307–315. [CrossRef]
12. Agafonov, A.V.; Krayev, A.S.; Davydova, O.I.; Ivanov, K.V.; Shekunova, T.O.; Baranchikov, A.E.; Ivanova, O.S.; Borilo, L.P.; Garshev, A.V.; Kozik, V.V.; et al. Nanocrystalline ceria: A novel material for electrorheological fluids. *RSC Adv.* **2016**, *6*, 88851–88858. [CrossRef]
13. Dong, X.; Niu, C.; Qi, M. Electrorheological Elastomers. In *Elastomers*; IntechOpen: London, UK, 2017. [CrossRef]
14. Shiga, T. Deformation and Viscoelastic Behavior of Polymer Gels in Electric Fields. In *Neutron Spin Echo Spectrosc. Viscoelasticity Rheol*; Springer: Berlin, Germany, 1997; pp. 131–163. [CrossRef]
15. Shiga, T.; Okada, A.; Kurauchi, T. Electroviscoelastic effect of polymer blends consisting of silicone elastomer and semiconducting polymer particles. *Macromolecules* **1993**, *26*, 6958–6963. [CrossRef]
16. Wei, K.; Bai, Q.; Meng, G.; Ye, L. Vibration characteristics of electrorheological elastomer sandwich beams. *Smart Mater. Struct.* **2011**, *20*, 055012. [CrossRef]
17. Niu, C.; Dong, X.; Qi, M. Damping mechanism and theoretical model of electrorheological elastomers. *Soft Matter* **2017**, *13*, 5409–5420. [CrossRef] [PubMed]
18. Ma, N.; Zhang, Z.; Dong, X.; Wang, Q.; Niu, C.; Han, B. Dynamic viscoelasticity and phenomenological model of electrorheological elastomers. *J. Appl. Polym. Sci.* **2017**, *134*, 45407. [CrossRef]
19. Ma, N.; Yao, Y.; Wang, Q.; Niu, C.; Dong, X. Properties and mechanical model of a stiffness tunable viscoelastic damper based on electrorheological elastomers. *Smart Mater. Struct.* **2020**, *29*, 045041. [CrossRef]
20. Choi, S.-B.; Wereley, N.M.; Li, W. *Controllable Electrorheological and Magnetorheological Materials*; Frontiers Media SA: Laussane, Switzerland, 2019. [CrossRef]
21. Klingenberg, D.J.; Zukoski, C.F. Studies on the steady-shear behavior of electrorheological suspensions. *Langmuir* **1990**, *6*, 15–24. [CrossRef]
22. Kim, Y.D.; Klingenberg, D.J. An interfacial polarization model for activated electrorheological suspensions. *Korean J. Chem. Eng.* **1997**, *14*, 30–36. [CrossRef]
23. Kunanuruksapong, R.; Sirivat, A. Electrical properties and electromechanical responses of acrylic elastomers and styrene copolymers: Effect of temperature. *Appl. Phys. A* **2008**, *92*, 313–320. [CrossRef]
24. Ludeelerd, P.; Niamlang, S.; Kunaruksapong, R.; Sirivat, A. Effect of elastomer matrix type on electromechanical response of conductive polypyrrole/elastomer blends. *J. Phys. Chem. Solids* **2010**, *71*, 1243–1250. [CrossRef]
25. Kossi, A.; Bossis, G.; Persello, J. Electro-active elastomer composites based on doped titanium dioxide. *J. Mater. Chem. C* **2015**, *3*, 1546–1556. [CrossRef]
26. Liu, B.; Shaw, M.T. Electrorheology of filled silicone elastomers. *J. Rheol.* **2001**, *45*, 641–657. [CrossRef]
27. Niu, C.; Dong, X.; Qi, M. Enhanced Electrorheological Properties of Elastomers Containing $TiO_2$/Urea Core–Shell Particles. *ACS Appl. Mater. Interfaces* **2015**, *7*, 24855–24863. [CrossRef]
28. Shen, R.; Wang, X.; Lu, Y.; Wang, D.; Sun, G.; Cao, Z.; Lu, K. Polar-Molecule-Dominated Electrorheological Fluids Featuring High Yield Stresses. *Adv. Mater.* **2009**, *21*, 4631–4635. [CrossRef]
29. Cheng, Y.; Guo, J.; Xu, G.; Cui, P.; Liu, X.; Liu, F.; Wu, J. Electrorheological property and microstructure of acetamide-modified $TiO_2$ nanoparticles. *Colloid Polym. Sci.* **2008**, *286*, 1493–1497. [CrossRef]

30. Cao, J.G.; Shen, M.; Zhou, L.W. Preparation and electrorheological properties of triethanolamine-modified $TiO_2$. *J. Solid State Chem.* **2006**, *179*, 1565–1568. [CrossRef]
31. Trusova, T.A.; Redozubov, A.A.; Agafonov, A.V.; Kraev, A.S.; Davydova, O.I. Effect of polydimethylsiloxane viscosity on the electrorheological activity of dispersions based on it. *Russ. J. Phys. Chem. A* **2016**, *90*, 1269–1273. [CrossRef]
32. Tsai, P.J.; Nayak, S.; Ghosh, S.; Puri, I.K. Influence of particle arrangement on the permittivity of an elastomeric composite. *AIP Adv.* **2017**, *7*, 015003. [CrossRef]
33. Kostrov, S.A.; Shamonin, M.; Stepanov, G.V.; Kramarenko, E.Y. Magnetodielectric Response of Soft Magnetoactive Elastomers: Effects of Filler Concentration and Measurement Frequency. *Int. J. Mol. Sci.* **2019**, *20*, 2230. [CrossRef]
34. Høyer, H.; Knaapila, M.; Kjelstrup-Hansen, J.; Helgesen, G. Microelectromechanical strain and pressure sensors based on electric field aligned carbon cone and carbon black particles in a silicone elastomer matrix. *J. Appl. Phys.* **2012**, *112*, 094324. [CrossRef]
35. Zhukov, A.V.; Mushenko, V.D.; Baratova, T.N. Catalytic Mixture for Hardening Siloxane. Rubber. Patent No. RU2424610, 20 July 2011.
36. Ramanavicius, S.; Tereshchenko, A.; Karpicz, R.; Ratautaite, V.; Bubniene, U.; Maneikis, A.; Jagminas, A.; Ramanavicius, A. $TiO_2$-x/$TiO_2$-Structure Based 'Self-Heated' Sensor for the Determination of Some Reducing Gases. *Sensors* **2019**, *20*, 74. [CrossRef] [PubMed]
37. Wang, C.-C.; Ying, J.Y. Sol−Gel Synthesis and Hydrothermal Processing of Anatase and Rutile Titania Nanocrystals. *Chem. Mater.* **1999**, *11*, 3113–3120. [CrossRef]
38. Li, Y.; White, T.; Lim, S. Low-temperature synthesis and microstructural control of titania nano-particles. *J. Solid State Chem.* **2004**, *177*, 1372–1381. [CrossRef]
39. Yanagisawa, K.; Yamamoto, Y.; Feng, Q.; Yamasaki, N. Formation mechanism of fine anatase crystals from amorphous titania under hydrothermal conditions. *J. Mater. Res.* **1998**, *13*, 825–829. [CrossRef]
40. Rojas, F.; Kornhauser, I.; Felipe, C.; Esparza, J.M.; Cordero, S.; Domínguez, A.; Riccardo, J.L. Capillary condensation in heterogeneous mesoporous networks consisting of variable connectivity and pore-size correlation. *Phys. Chem. Chem. Phys.* **2002**, *4*, 2346–2355. [CrossRef]
41. Mezger, T.G. *The Rheology Handbook: For Users of Rotational and Oscillatory Rheometers*, 4th ed.; Vincentz Network: Hanover, Germany, 2014.
42. Payne, A.R. The dynamic properties of carbon black-loaded natural rubber vulcanizates Part I. *J. Appl. Polym. Sci.* **1962**, *6*, 57–63. [CrossRef]
43. Payne, A.R. The dynamic properties of carbon black loaded natural rubber vulcanizates. Part II. *J. Appl. Polym. Sci.* **1962**, *6*, 368–372. [CrossRef]
44. Hentschke, R. The Payne effect revisited. *Express Polym. Lett.* **2017**, *11*, 278–292. [CrossRef]
45. Dong, X.; Niu, C.; Qi, M. Enhancement of electrorheological performance of electrorheological elastomers by improving $TiO_2$ particles/silicon rubber interface. *J. Mater. Chem. C* **2016**, *4*, 6806–6815. [CrossRef]
46. An, H.; Picken, S.J.; Mendes, E. Nonlinear rheological study of magneto responsive soft gels. *Polymer* **2012**, *53*, 4164–4170. [CrossRef]
47. Kwon, S.; Lee, J.; Choi, H. Magnetic Particle Filled Elastomeric Hybrid Composites and Their Magnetorheological Response. *Materials* **2018**, *11*, 1040. [CrossRef] [PubMed]
48. Kutalkova, E.; Plachy, T.; Sedlacik, M. On the enhanced sedimentation stability and electrorheological performance of intelligent fluids based on sepiolite particles. *J. Mol. Liq.* **2020**, *309*, 113120. [CrossRef]
49. Capan, R.; Ray, A.K. Dielectric Measurements on Sol–Gel Derived Titania Films. *J. Electron. Mater.* **2017**, *46*, 6646–6652. [CrossRef]
50. Yang, D.; Zhang, L.; Liu, H.; Dong, Y.; Yu, Y.; Tian, M. Lead magnesium niobate-filled silicone dielectric elastomer with large actuated strain. *J. Appl. Polym. Sci.* **2012**, *125*, 2196–2201. [CrossRef]
51. Geyer, R.D. *Dielectric Characterization and Reference Materials*; NIST Technical Note 1338; U.S. Government Printing Office: Washington, DC, USA, 1990. Available online: https://www.govinfo.gov/content/pkg/GOVPUB-C13-319af9ee141c44f1f5c7072208f56b6b/pdf/GOVPUB-C13-319af9ee141c44f1f5c7072208f56b6b.pdf (accessed on 17 September 2020).
52. Moucka, R.; Sedlacik, M.; Cvek, M. Dielectric properties of magnetorheological elastomers with different microstructure. *Appl. Phys. Lett.* **2018**, *112*, 122901. [CrossRef]

53. Manaila, E.; Stelescu, M.D.; Craciun, G.; Surdu, L. Effects of benzoyl peroxide on some properties of composites based on hemp and natural rubber. *Polym. Bull.* **2014**, *71*, 2001–2022. [CrossRef]
54. Zhang, H.; Cai, C.; Liu, W.; Li, D.; Zhang, J.; Zhao, N.; Xu, J. Recyclable Polydimethylsiloxane Network Crosslinked by Dynamic Transesterification Reaction. *Sci. Rep.* **2017**, *7*, 11833. [CrossRef]
55. Flory, P.J.; Rehner, J. Statistical Mechanics of Cross-Linked Polymer Networks II. Swelling. *J. Chem. Phys.* **1943**, *11*, 521–526. [CrossRef]
56. Kim, J.K.; Lee, S.H. New technology of crumb rubber compounding for recycling of waste tires. *J. Appl. Polym. Sci.* **2000**, *78*, 1573–1577. [CrossRef]
57. Valentín, J.L.; Carretero-González, J.; Mora-Barrantes, I.; Chasésé, W.; Saalwächter, K. Uncertainties in the Determination of Cross-Link Density by Equilibrium Swelling Experiments in Natural Rubber. *Macromolecules* **2008**, *41*, 4717–4729. [CrossRef]

© 2020 by the authors. Licensee MDPI, Basel, Switzerland. This article is an open access article distributed under the terms and conditions of the Creative Commons Attribution (CC BY) license (http://creativecommons.org/licenses/by/4.0/).

*Article*

# Micromechanics of Stress-Softening and Hysteresis of Filler Reinforced Elastomers with Applications to Thermo-Oxidative Aging

Jan Plagge and Manfred Klüppel *

Deutsches Institut für Kautschuktechnologie e. V., Eupener Str. 33, D-30519 Hannover, Germany; jan.plagge@dikautschuk.de
* Correspondence: manfred.klueppel@dikautschuk.de

Received: 25 May 2020; Accepted: 8 June 2020; Published: 15 June 2020

**Abstract:** A micromechanical concept of filler-induced stress-softening and hysteresis is established that describes the complex quasi-static deformation behavior of filler reinforced rubbers upon repeated stretching with increasing amplitude. It is based on a non-affine tube model of rubber elasticity and a distinct deformation and fracture mechanics of filler clusters in the stress field of the rubber matrix. For the description of the clusters we refer to a three-dimensional generalization of the Kantor–Webman model of flexible chain aggregates with distinct bending–twisting and tension deformation of bonds. The bending–twisting deformation dominates the elasticity of filler clusters in elastomers while the tension deformation is assumed to be mainly responsible for fracture. The cluster mechanics is described in detail in the theoretical section, whereby two different fracture criteria of filler–filler bonds are considered, denoted "monodisperse" and "hierarchical" bond fracture mechanism. Both concepts are compared in the experimental section, where stress–strain cycles of a series of ethylene–propylene–diene rubber (EPDM) composites with various thermo-oxidative aging histories are evaluated. It is found that the "hierarchical" bond fracture mechanism delivers better fits and more stable fitting parameters, though the evolution of fitting parameters with aging time is similar for both models. From the adaptations it is concluded that the crosslinking density remains almost constant, indicating that the sulfur bridges in EPDM networks are mono-sulfidic, and hence, quite stable—even at 130 °C aging temperature. The hardening of the composites with increasing aging time is mainly attributed to the relaxation of filler–filler bonds, which results in an increased stiffness and strength of the bonds. Finally, a frame-independent simplified version of the stress-softening model is proposed that allows for an easy implementation into numerical codes for fast FEM simulations

**Keywords:** filled elastomers; stress softening; filler-induced hysteresis; cluster mechanics; FEM simulation

## 1. Introduction

Nanoscopic fillers like carbon black or silica play an important role in the mechanical reinforcement of elastomers [1,2]. They make the elastomer stiffer and tougher, leading to a pronounced reduction of crack propagation rates and wear, which is accompanied by an increased life time of rubber goods [3]. But the incorporation of fillers results in a nonlinear dynamic-mechanical response, which is reflected e.g., by the amplitude-dependence of the dynamic moduli. This so called Payne effect was investigated by several authors like Payne [4] and Medalia [5]. A related phenomenon is the stress softening effect under quasi-static cyclic deformation, which is also called Mullins effect due to the intensive studies by Mullins [6]. Accordingly, a drop in stress appears if the loading goes beyond the previous maximum. Most of the stress drop at a certain strain occurs in the first cycle, and in the following cycles the specimen approaches a steady state stress–strain curve. A second characteristic effect caused by fillers is the pronounced hysteresis which is related to the dissipation of mechanical energy during every cycle.

Based on the investigations of filler network morphology by different experimental techniques, a model of rubber reinforcement by flexible filler clusters has been proposed that allows for a quantitative understanding of the complex mechanical response under quasi-static and dynamic deformations [7–10]. This model of rubber reinforcement refers to the kinetics of cluster–cluster aggregation (CCA) of filler networking in elastomers, which represents a reasonable theoretical basis for analyzing the linear viscoelastic properties of reinforced rubbers. According to this approach, filler networks consist of a space-filling configuration of fractal CCA-clusters with characteristic mass fractal dimension $df \approx 1.8$. The mechanical response of such filler networks at small strain depends purely on the fractal connectivity of the CCA-clusters. It can be evaluated by referring to the Kantor–Webman model [11] of flexible chains of filler particles that allows for a micromechanical description of the elastic properties of tender CCA-clusters in elastomers. The main contribution of the elastically stored energy in the strained filler clusters results from the bending–twisting deformation of filler–filler bonds, considered by an elastic constant $\overline{G}$. Based on this approach a power law behavior of the small strain modulus vs. filler concentration can be derived. The predicted exponent 3.5 is in good agreement with experimental data of Payne [4] for carbon black filled butyl rubber. This power law behavior has been confirmed by further investigations of carbon black and silica filled rubbers as well as composites with polymeric model fillers (microgels) [7–10].

Based on this approach, a micromechanical material model of filler reinforced elastomers has been put forward, denoted dynamic flocculation model (DFM) [7,12–16]. Similar models have been proposed by other authors [17–19]. The DFM describes the complex quasi-static stress–strain response of filler reinforced elastomers up to large strains in fair agreement with experimental data. It is based on a non-affine tube model of rubber elasticity and considers hydrodynamic amplification of the rubber matrix by a fraction of hard rigid filler clusters with filler–filler bonds in the unbroken, virgin state. The filler-induced hysteresis is described by the cyclic breakdown and re-aggregation of the residual fraction of more soft filler clusters with already broken, but reformed filler–filler bonds. The difference between hard and soft filler–filler bonds arises due to the softening of glassy-like polymer bridges between adjacent filler particles [7,20–22]. If they break under stress the recovery to a virgin bond will take time and/or high temperatures, as applied during vulcanization [23]. A finite element implementation of the DFM was established [24] by referring to the concept of representative directions introduced by Ihlemann [25]. Thereby also temperature and time dependent effects have been considered [26].

It is the aim of this study to consider the underlying fracture mechanics of filler clusters entering the DFM more closely, whereby we will investigate two different fracture criteria for filler–filler bonds. The differences between both approaches will be evaluated by comparing fitting results obtained for stress–strain cycles of ethylene–propylene–diene monomer rubber (EPDM) composites with various thermo-oxidative aging histories, which will be analyzed by the variation of fitting parameters. This is in close relation to recent investigations of the structure–property relationships of silica/silane formulations in different rubber composites [27] delivering microscopic information about the material properties of the systems. Finally, we will introduce a frame-invariant version of the stress-softening part of the DFM that allows for an easy implementation into a finite element code for fast finite element (FEM) applications of the isotropic discontinuous damage effects in engineering rubber science.

## 2. Modeling Approach

### 2.1. Basic Assumptions of the Dynamic Flocculation Model

The (microscopic) free energy density of the dynamic flocculation model (DFM) consists of two contributions, which are weighted by the effective filler volume fraction $\Phi_{\text{eff}}$:

$$W(\varepsilon_\mu) = (1 - \Phi_{\text{eff}})W_R(\varepsilon_\mu) + \Phi_{\text{eff}}\, W_A(\varepsilon_\mu) \tag{1}$$

The first addend is the equilibrium energy density stored in the extensively strained rubber matrix, which includes hydrodynamic amplification by a fraction of rigid filler clusters. The second addend considers the energy stored in the strained soft filler clusters and is responsible for the filler-induced hysteresis. The symbol $\varepsilon_\mu$ is defined in this work as the macroscopic strain in direction $\mu$. The rubber elastic part is modeled by the free energy density of the extended non-affine tube model [12,28]:

$$W_R(\varepsilon_\mu) = \frac{G_c}{2}\left\{\frac{\left(\sum_{\mu=1}^{3}\lambda_\mu^2 - 3\right)\left(1 - \frac{T_e}{n_e}\right)}{1 - \frac{T_e}{n_e}\left(\sum_{\mu=1}^{3}\lambda_\mu^2 - 3\right)} + \ln\left[1 - \frac{T_e}{n_e}\left(\sum_{\mu=1}^{3}\lambda_\mu^2 - 3\right)\right]\right\} + 2G_e\left(\sum_{\mu=1}^{3}\lambda_\mu^{-1} - 3\right) \tag{2}$$

with $n_e$ being the number of statistical chain segments between neighboring entanglements and $T_e$ is the trapping factor ($0 < T_e < 1$) characterizing the fraction of elastically active entanglements. We define $\lambda_\mu$ to be the microscopic strain ratio (or stretch) on the nanoscale in direction $\mu$. Thus, for unfilled rubbers the usual relation $\lambda_\mu = 1 + \varepsilon_\mu$ holds. The first addend in Equation (2) considers the constraints due to interchain junctions, with an elastic modulus $G_c$ proportional to the density of network junctions. The second addend considers topological constraints in densely packed polymer networks, whereby $G_e$ is proportional to the entanglement density of the rubber. The parenthetical expression in the first addend considers the finite chain extensibility of polymer networks by referring to an approach of Edwards and Vilgis [29]. For the limiting case $n_e/T_e = \sum \lambda_\mu^2 - 3$ a singularity is obtained for the free energy density $W_R$, indicating the maximum extensibility of the network. This is reached when the chains between successive trapped entanglements are fully stretched out. In the limit $n_e \to \infty$ the original Gaussian formulation of the non-affine tube model, derived by Heinrich et al. [30] for infinite long chains, is recovered.

The presence of tightly bonded (virgin bonds) rigid filler clusters gives rise to hydrodynamic reinforcement of the rubber matrix. This is specified by the strain amplification factor $X$ as proposed by Mullins and Tobin [6], which relates the external, macroscopic strain $\varepsilon_\mu$ of the sample to the internal, microscopic strain ratio $\lambda_\mu$ of the rubber matrix, $\lambda_\mu = 1 + X(\varepsilon_{\mu,\text{max}})\varepsilon_\mu$. For strain amplified rubbers this strain has to be used in the free energy density Equation (2). The microscopic stress of the rubber matrix is then obtained by differentiation with respect to the internal strain $\lambda_\mu$:

$$\sigma_{R,\nu} \equiv \frac{\partial W_R}{\partial \lambda_\nu} = G_c \sum_{\mu=1}^{3}\frac{\partial \lambda_\mu}{\partial \lambda_\nu}\lambda_\mu\left\{\frac{\left(1 - \frac{T_e}{n_e}\right)}{\left(1 - \frac{T_e}{n_e}\left(\sum_{\mu=1}^{3}\lambda_\mu^2 - 3\right)\right)^2} - \frac{\frac{T_e}{n_e}}{1 - \frac{T_e}{n_e}\left(\sum_{\mu=1}^{3}\lambda_\mu^2 - 3\right)}\right\} - 2G_e\sum_{\mu=1}^{3}\frac{\partial \lambda_\mu}{\partial \lambda_\nu}\lambda_\mu^{-2} \tag{3}$$

This is the microscopic stress between the filler clusters that can be identified with the macroscopically measured engineering stress (1. PK stress) in equilibrium. For uniaxial deformations ($\lambda_2 = \lambda_3 = \lambda_1^{-1/2}$ and $\partial\lambda_2 = \partial\lambda_3 = -1/2\,\lambda_1^{-3/2}\partial\lambda_1$) we obtain for the engineering stress in stretching direction:

$$\sigma_{R,1} = G_c(\lambda_1 - \lambda_1^{-2})\left\{\frac{\left(1 - \frac{T_e}{n_e}\right)}{\left(1 - \frac{T_e}{n_e}(\lambda_1^2 + 2/\lambda_1 - 3)\right)^2} - \frac{\frac{T_e}{n_e}}{1 - \frac{T_e}{n_e}(\lambda_1^2 + 2/\lambda_1 - 3)}\right\} + 2G_e(\lambda_1^{-\frac{1}{2}} - \lambda_1^{-2}) \tag{4}$$

In the case of a preconditioned sample and for strains smaller than the previous straining ($\varepsilon_\mu < \varepsilon_{\mu,\max}$), the materials microscopic structure is already adjusted to the maximum load and the strain amplification factor $X$ is independent of strain. In that case it is determined by $\varepsilon_{\mu,\max}$ ($X = X(\varepsilon_{\mu,\max})$). We relate this to the irreversible fracture of filler clusters (see below). A relation for the strain amplification factor of overlapping fractal clusters of size $\xi$ was derived by Huber and Vilgis [31]. By using path integral methods they found $X = 1 + c\Phi^{2/(3-d_f)}\xi^{d_w - d_f}$ where $\Phi$ is the filler volume fraction and $c$ is a constant of order one. With this, $X(\varepsilon_{\mu,\max})$ can be evaluated by averaging over the size distribution of hard clusters in all space directions. In the case of preconditioned samples this yields:

$$X(\varepsilon_{\mu,\max}) = 1 + c\Phi_{\text{eff}}^{\frac{2}{3-d_f}} \sum_{\mu=1}^{3} \frac{1}{d}\left[\int_0^{\xi_{\mu,\min}} \left(\frac{\xi'_\mu}{d}\right)^{d_w - d_f} \varphi(\xi'_\mu) d\xi'_\mu + \int_{\xi_{\mu,\min}}^{\infty} \varphi(\xi'_\mu) d\xi'_\mu\right] \quad (5)$$

Here, $d$ is the particle size, $\xi_\mu$ is the cluster size in spatial direction $\mu$ and $\xi_{\mu,\min}$ is the minimum cluster size which will be calculated later on. The fractal exponents are determined as $d_f \approx 1.8$ for the mass fractal dimension and $d_w = 3.1$ for the anomalous diffusion exponent of CCA-clusters [2]. Note that the effective filler volume fraction $\Phi_{\text{eff}} > \Phi$ is used in Equations (1) and (5), which considers the effective volume of the rigid phase of structured filler particles, e.g., carbon black or silica, according to the "occluded rubber concept" of Medalia [32]. Occluded rubber is defined as the rubber part of the rubber matrix that penetrates into the voids of the particles, which partially shields it from deformation. The second addend of Equation (5) takes into account that also fully broken clusters contribute to the strain amplification factor by the remaining particles. $\varphi(\xi_\mu)$ is the normalized cluster size distribution:

$$\varphi(\xi_\mu) = 4d\frac{\xi_\mu}{\langle\xi_\mu\rangle^2}\exp\left(-2\frac{\xi_\mu}{\langle\xi_\mu\rangle}\right) \quad (6)$$

This is a peaked cluster size distribution with $\langle\xi_\mu\rangle$ being the ensemble average in spatial direction $\mu$. It is motivated by analytical results referring to Smoluchowski's equation for the kinetics of cluster–cluster aggregation of colloids [33–35] (comp. also [7]). In the undeformed state it is assumed to be isotropic, i.e., $\varphi(\xi_1) = \varphi(\xi_2) = \varphi(\xi_3) \equiv \varphi(\xi)$.

The model of stress softening and hysteresis assumes that the breakdown of filler clusters during the first deformation of the virgin samples is reversible, though the initial virgin state of filler–filler bonds is not recovered. This implies that, on the one side, the fraction of hard (virgin) filler clusters decreases with increasing pre-strain, leading to pronounced stress softening after the first deformation cycle. On the other side, the fraction of soft (reaggregated) filler clusters increases with rising pre-strain, which affects the filler-induced hysteresis. A schematic view of the decomposition of filler clusters in hard and soft units for preconditioned samples is shown in Figure 1.

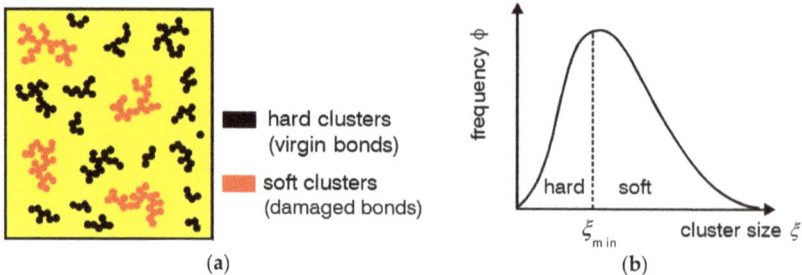

**Figure 1.** (a) Schematic view of the decomposition of filler clusters in hard and soft units for preconditioned samples and (b) cluster size distribution with the pre-strain dependent boundary size $x_{\min}$ between hard and soft clusters.

The second addend of Equation (1) describes the filler-induced hysteresis. It considers the energy stored in the substantially strained soft filler clusters, which break under stress and reaggregate on retraction

$$W_A(\varepsilon_\mu) = \sum_\mu^{\partial \varepsilon/\partial t > 0} \frac{1}{2d} \int_{\xi_{\mu,\min}}^{\xi_\mu(\varepsilon_\mu)} G_A(\xi'_\mu) \varepsilon^2_{A,\mu}(\xi'_\mu, \varepsilon_\mu) \varphi(\xi'_\mu) \, d\xi'_\mu \tag{7}$$

$G_A$ is the elastic modulus and $\varepsilon_{A,\mu}$ is the strain of the fragile filler clusters in spatial direction $\mu$. These quantities and their dependence on cluster size $x_\mu$ and external strain $\varepsilon_\mu$ will be specified in the next sections. In addition. The integral boundaries of Equations (5) and (7) have to be described more closely, which requires the consideration of elasticity and fracture of filler clusters in stretched elastomers.

### 2.2. Elasticity and Fracture of Filler Clusters in Stretched Elastomers

For consideration of filler network breakdown in stretched rubbers, the elasticity and failure properties of tender filler clusters have to be evaluated in dependence of cluster size. This will be obtained by referring to the two-dimensional Kantor–Webman model of flexible chains of arbitrary connected filler particles [11] as represented in Figure 2a. We apply here a simplified generalization of this model to three dimensions, where on-plane bending, and off-plain twisting deformations of bonds are considered by a single bending–twisting term [20]. By identifying the three-dimensional flexible chain with the backbone of a CCA-cluster, the model can be applied for modeling the small-strain modulus of fractal filler networks, consisting of a space-filling configuration of CCA-clusters [7–10]. Here, we use it for a micromechanical description of CCA-clusters that are deformed in the stress field of a strained rubber matrix. Note that this is possible because the CCA-cluster backbone is not branched on large length scales, which is a typical result of cluster–cluster aggregation.

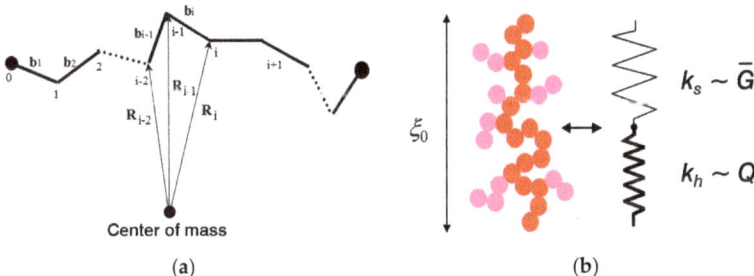

**Figure 2.** Schematic view of the Kantor–Webman model of flexible chains of arbitrary connected filler particles (**a**) and mechanical equivalence between a filler cluster and a series of soft and stiff molecular springs (**b**). The two springs with force constants $k_s$ and $k_h$ are related to bending–twisting- and tension deformations of filler–filler bonds referring to elastic constants $\bar{G}$ and $Q$, respectively.

In our model two kinds of deformations of filler–filler bonds are considered, bending–twisting- and tension deformations. This corresponds to a mechanical equivalence between a filler cluster and a series of two molecular springs depicted schematically in Figure 2b. We will see that the bending–twisting deformation governs the elasticity while the tension deformation is sensitive for fracture. The total force constant of a cluster of size $\xi_0$ with $N_B$ particles in the backbone reads:

$$k_\xi = \left( \frac{1}{k_h} + \frac{1}{k_s} \right)^{-1} \tag{8}$$

with the tension part given by:

$$k_h = \frac{Q}{\sum_{i=1}^{N_B}\left(\vec{\frac{F}{F}}\cdot\vec{b}_i\right)^2} = \frac{Q}{N_B d^2 \oint\left(\vec{\frac{F}{F}}\cdot\vec{\frac{b}{d}}\right)^2 dS} = \frac{Q}{d^2 g}\left(\frac{d}{\xi_0}\right)^{d_{f,B}} \tag{9}$$

The bending–twisting part reads:

$$k_s = \frac{\overline{G}}{\sum_{i=1}^{N_B}\left[\left(\vec{\frac{F}{F}}\times\vec{z}\right)\left(\vec{R}_{i-1}-\vec{R}_{N_B}\right)\right]^2} = \frac{\overline{G}}{N_B S_\perp^2} = \frac{\overline{G}}{d^2 g'}\left(\frac{d}{\xi_0}\right)^{2+d_{f,B}} \tag{10}$$

Here, $d$ is the bond length (particle size), $\vec{F}$ is the force and $g \equiv \oint\left(\vec{\frac{F}{F}}\cdot\vec{\frac{b}{d}}\right)^2 dS$ is the average projection of bond vectors $\vec{b}_i$ on the direction of the force ($0 < g < 1$). $S_\perp^2 \equiv \frac{1}{N_B}\sum_{i=1}^{N_B}\left[\left(\vec{\frac{F}{F}}\times\vec{z}\right)\left(\vec{R}_{i-1}-\vec{R}_{N_B}\right)\right]^2$ is the average squared radius of gyration in direction perpendicular to the force and includes a unit vector $\vec{z}$ pointing perpendicular to the connecting vector $\left(\vec{R}_{i-1}-\vec{R}_{N_B}\right)$. It scales with the squared cluster size, $S_\perp^2 = g'\xi_0^2$, with a scaling factor $0 < g' < 1$. $\overline{G}$ and $Q$ are elastic constants due to bending–twisting—and tension deformations, respectively. For the particle number the scaling relation $N_B = (\xi_0/d)^{d_{f,B}}$ was used with $d_{f,B} = 1.3$ being the backbone fractal dimension of CCA-clusters.

By comparing the exponents of Equations (9) and (10) one finds that the force constant $k_s$ decreases much more rapidly with cluster size $\xi_0$ than the force constants $k_h$. Accordingly, Equation (8) implies $k_\xi \approx k_s$ for sufficient large clusters, i.e., the stiffness of the cluster is determined by the bending–twisting deformations of bonds. This determines the following scaling law for the elastic modulus entering Equation (7):

$$G_A(\xi_0) \approx \xi_0^{-1} k_s = \frac{\overline{G}}{d^3 g'}\left(\frac{d}{\xi_0}\right)^{3+d_{f,B}} \simeq \frac{\overline{G}}{d^3}\left(\frac{d}{\xi_0}\right)^{3+d_{f,B}} \tag{11}$$

This approximation without the tension term can be applied for sufficient large clusters with $(\xi_0/d)^{2+d_{f,B}} \gg (\xi_0/d)^{d_{f,B}}$. For the evaluation of the scaling factor $g'$ in Equation (11) we have to consider the ensemble average of clusters. However, this will not be considered here, because we are mainly interested in the scaling exponents.

The stretching of the clusters can be evaluated in the same approximation [7]:

$$\Delta l_\xi = \Delta l_s + \Delta l_h = \Delta l_b\left(\frac{k_b}{k_s(\xi_0)} + gN_B\right) \approx \Delta l_b \frac{g'Q}{\overline{G}}\left(\frac{\xi_0}{d}\right)^{2+d_{f,B}} \tag{12}$$

Here, we have introduced the stretching $\Delta l_b$ and force constant $k_b \equiv Q/d^2$ related to stretching of the single bonds. In addition. we used the equilibrium conditions for the force $\Delta l_s/\Delta l_b = k_b/k_s(\xi_0)$ and $\Delta l_h/\Delta l_b = k_b/k_h(\xi_0) = gN_B$. In the next section we will use Equation (12) for describing the fracture of filler clusters by relating it to the fracture of bonds under tension.

Examples:

The cluster mechanics described by Equations (8)–(10) shall be illustrated by two simple examples which are depicted in Figure 3. In Figure 3a a linear chain is considered with the force $\vec{F}$ pointing perpendicular or alternatively in direction of the chain. In the first case we have $g = 0$ and $\vec{F}\times\vec{z}$ points into the direction of the chain. This implies for the force constant and the deformation:

$$k_\xi = k_s = 12\frac{\overline{G}}{N_B(N_B d)^2} \rightarrow \Delta\vec{l}_s = \frac{\vec{F}}{k_s} = \frac{\vec{F}}{12d\overline{G}}(N_B d)^3 \tag{13}$$

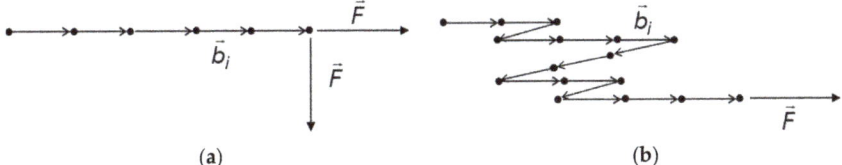

**Figure 3.** Illustration of the cluster mechanics by simple examples: (**a**) Linear chain and (**b**) one dimensional random walk.

With $g' = 1/12$ being the ratio between the squared radius of gyration and the squared length $L^2 = (N_B d)^2$ of a linear chain. Accordingly, the force constants $k_s$ drops with the 3rd power of the length.

In the second case, where $\vec{F}$ points parallel to the chain, we have $g = 1$ and $\vec{F} \times \vec{z}$ points perpendicular to the chain. This implies $\bar{S}_\perp = 0$ yielding:

$$k_\xi = k_h = \frac{Q}{N_B d^2} \rightarrow \Delta \vec{l}_h = \frac{\vec{F}}{k_h} = N_B \frac{d^2 \vec{F}}{Q} \tag{14}$$

As expected, the deformation increases linear with the number of bonds.

In Figure 3b we consider the case where a random walk structure of the chain is realized, corresponding to three 1-dimensional random walks with $N_B/3$ particles. The average projection $g$ is then given by $g = \oint \left( \frac{\vec{F}}{F} \cdot \frac{\vec{b}}{d} \right)^2 dS = 1/3$. The ratio between the ensemble average of the squared radius of gyration $\langle R_g^2 \rangle$ and the squared end-to-end distance $R^2 = N_B d^2$ is evaluated as $g' = 1/6$ (see e.g., Chapter 2 in [36]). This implies for the force constant:

$$k_\xi = \left( \frac{N_B^2 d^2}{6G} + \frac{N_B d^2}{3Q} \right)^{-1} \tag{15}$$

For the deformation under tension of the 1-dimensional random walk shown in Figure 3b one obtains:

$$\Delta \vec{l}_h = \frac{\vec{F}}{k_h} = \frac{N_B}{3} \frac{d^2 \vec{F}}{Q} \tag{16}$$

For the more general case that the force points in arbitrary direction also the bending–twisting deformations of bonds must be taken into account by referring to the full force constant $k_\xi$ of Equation (15).

### 2.3. Evaluation of Boundary Cluster Size and Cluster Stress

In view of introducing a fracture criterion for strained clusters, we assume that the tension of bonds is a much more critical deformation compared to bending and twisting, since it separates the filler particles from each other. Equation (12) relates the total stretching of a cluster to the stretching of the bonds and can therefore be used for evaluation of the failure strain of the cluster by defining a fracture criterion for the bonds. We will introduce here two different fracture criteria, which will be denoted "monodisperse" and "hierarchical".

In the first approach all bonds are considered to be equal (monodisperse) having the same strength. Then, the failure strain $\varepsilon_{f,b}$ of the bonds is given by the critical stretch of the bonds $\Delta l_{f,b}$ in relation to the bond length: $\varepsilon_{f,b}^m \equiv \Delta l_{f,b}/d$. This implies for the failure strain of the cluster:

$$\varepsilon_{f,\xi} \equiv \frac{\Delta l_{f,\xi}}{\xi_0} \approx \varepsilon_{f,b}^m \frac{g'Q}{G} \left( \frac{\xi_0}{d} \right)^{1+d_{f,B}} \tag{17a}$$

In contrast, the hierarchical model takes into account that a hierarchy of bond strengths develops during cluster–cluster aggregation, because the mobility of the clusters decreases with cluster size. Accordingly, the first bond formed between two particles is the strongest while successive bonds formed between the growing sub-clusters become weaker and weaker. The last bond formed in the cluster is the weakest and will break first under tension. This effect is taken into account by the hierarchical fracture criterion, where the failure strain $\varepsilon_{f,b}$ of the bonds is defined in relation to the cluster size, which is the only relevant length scale in our model: $\varepsilon_{f,b}^h \equiv \Delta l_{f,b}/\xi_0$. This implies that the failure strain of the cluster increases more rapidly with cluster size compared to the monodisperse case:

$$\varepsilon_{f,\xi} \equiv \frac{\Delta l_{f,\xi}}{\xi_0} \approx \varepsilon_{f,b}^h \frac{g'Q}{G}\left(\frac{\xi_0}{d}\right)^{2+d_{f,B}} \tag{17b}$$

For the evaluation of the boundary cluster size between broken and unbroken clusters in stretched rubbers, we assume that a stress equilibrium is realized between the strain amplified rubber matrix and the clusters $\sigma_{R,\mu}(\varepsilon_\mu) = G_A \varepsilon_{A,\mu}(\varepsilon_\mu)$. With the scaling relation Equation (11) for the elastic modulus of the clusters this delivers for the cluster strain:

$$\varepsilon_{A,\mu}(\varepsilon_\mu) = G_A^{-1}\hat{\sigma}_{R,\mu}(\varepsilon_\mu) \approx \frac{g'd^3}{G}\left(\frac{\xi_0}{d}\right)^{3+d_{f,B}} \hat{\sigma}_{R,\mu}(\varepsilon_\mu) \tag{18}$$

Here we have replaced the rubber stress by a relative stress with respect to the minimum strain:

$$\hat{\sigma}_{R,\mu}(\varepsilon_\mu) \equiv \left(\sigma_{R,\mu}(\varepsilon_\mu) - \sigma_{R,\mu}(\varepsilon_{\mu,\min})\right) \tag{19}$$

This ensures that the stretching of clusters in spatial direction $\mu$ starts at the minimum strain $\varepsilon_{\mu,\min}$ for each cycle. Here, we assume that clusters reaggregate into a stress-free state at minimum strain.

A comparison of the exponents in Equations (18) and (17) makes clear that the strain of the clusters under external strain increases faster with cluster size than the failure strain, in both cases. This implies that large clusters break first followed by smaller ones, i.e., the boundary cluster size between broken and unbroken clusters $\xi_\mu(\varepsilon_\mu)$ moves from larger to smaller values with increasing strain. It is obtained by equating the cluster strain to the failure strain. This yields for the two fracture criteria:

$$\left(\frac{\xi_\mu(\varepsilon_\mu)}{d}\right)^2 = \frac{Q\varepsilon_{f,b}^m}{d^3\hat{\sigma}_{R,\mu}(\varepsilon_\mu)} \equiv \frac{s_d}{\hat{\sigma}_{R,\mu}(\varepsilon_\mu)} \tag{20a}$$

And

$$\frac{\xi_\mu(\varepsilon_\mu)}{d} = \frac{Q\varepsilon_{f,b}^h}{d^3\hat{\sigma}_{R,\mu}(\varepsilon_\mu)} \equiv \frac{s_d}{\hat{\sigma}_{R,\mu}(\varepsilon_\mu)} \tag{20b}$$

Here, $s_d$ is defines as the fracture stress under tension of bonds, i.e., the tensile strength of damaged filler–filler bonds. The boundary cluster size $\xi_\mu(\varepsilon_\mu)$ applies for the integral boundaries of Equation (7), describing the filler-induced hysteresis due to the successive breakdown of soft filler clusters with damaged filler–filler bonds.

Similar expressions are found for the upper boundary of Equation (5), but now the tensile strength $s_v$ of virgin filler–filler bonds is entering:

$$\left(\frac{\xi_{\mu,\min}}{d}\right)^2 = \frac{\tilde{Q}\varepsilon_{f,b}^m}{d^3\hat{\sigma}_{R,\mu}(\varepsilon_{\mu,\max})} \equiv \frac{s_v}{\hat{\sigma}_{R,\mu}(\varepsilon_{\mu,\max})} \tag{21a}$$

And

$$\frac{\xi_{\mu,\min}}{d} = \frac{\tilde{Q}\varepsilon_{f,b}^h}{d^3\hat{\sigma}_{R,\mu}(\varepsilon_{\mu,\max})} \equiv \frac{s_v}{\hat{\sigma}_{R,\mu}(\varepsilon_{\mu,\max})} \tag{21b}$$

The elastic constant and failure strains are denoted by $\tilde{Q}$ and $\tilde{\varepsilon}_{f,b}$, respectively. Note that the tensile strength of virgin filler–filler bonds must be larger than the tensile strength of damaged bonds, i.e., $s_v > s_d$. Equation (21a) or (21b) together with Equation (5) define the amplification of the rubber matrix, and thus stress—softening effects of the model. Solving this set of equations for $\sigma$ requires iterative methods, e.g., Newton iteration.

By referring to the stress equilibrium between the strain amplified rubber matrix and the clusters, $\hat{\sigma}_{R,\mu}(\varepsilon_\mu) = G_A \varepsilon_{A,\mu}(\varepsilon_\mu)$, the cluster stress $\sigma_{A,\mu}$ responsible for filler-induced hysteresis is obtained by differentiation of Equation (7) with respect to cluster strain:

$$\sigma_{A,\nu} \equiv \frac{\partial W_A}{\partial \varepsilon_{A,\nu}} = \sum_\mu^{\frac{\partial \varepsilon}{\partial t}>0} \frac{1}{d} \int_{d(\frac{s_d}{\hat{\sigma}_{R,\mu}(\varepsilon_{\mu,max})})}^{d(\frac{s_d}{\hat{\sigma}_{R,\mu}(\varepsilon_\mu)})^\alpha} G_A(\xi_\mu) \varepsilon_{A,\mu}(\xi_\mu, \varepsilon_\mu) \frac{\partial \varepsilon_{A,\mu}(\xi_\mu)}{\partial \varepsilon_{A,\nu}(\xi_\mu)} \varphi(\xi_\mu) d\xi_\mu = \qquad (22)$$
$$\sum_\mu^{\frac{\partial \varepsilon}{\partial t}>0} \hat{\sigma}_{R,\mu}(\varepsilon_\mu) \langle \frac{\partial \varepsilon_{A,\mu}}{\partial \varepsilon_{A,\nu}} \rangle \frac{1}{d} \int_{d(\frac{s_d}{\hat{\sigma}_{R,\mu}(\varepsilon_{\mu,max})})}^{d(\frac{s_d}{\hat{\sigma}_{R,\mu}(\varepsilon_\mu)})^\alpha} \varphi(\xi_\mu) d\xi_\mu$$

The exponent $\alpha$ takes the two fracture criteria into account, i.e., $\alpha = 1/2$ for the "monodisperse" model and $\alpha = 1$ for the "hierarchical" model. In addition, we assume that the clusters, on average, deform like the sample:

$$\langle \frac{\partial \varepsilon_{A,\mu}}{\partial \varepsilon_{A,\nu}} \rangle = \frac{\partial \varepsilon_\mu}{\partial \varepsilon_\nu} \qquad (23)$$

The sum in Equation (22) runs over stretching directions, only, implying that the up- and down cycles are different. The cluster stress of the upcycle is positive while the downcycle gives a negative contribution, producing the filler-induced hysteresis.

For uniaxial deformations, realized on microscales $\lambda_2 = \lambda_3 = \lambda_1^{-1/2}; \partial \lambda_2 = \partial \lambda_3 = -1/2\, \lambda_1^{-3/2} \partial \lambda_1$) and macroscales $1 + \varepsilon_2 = 1 + \varepsilon_3 = (1 + \varepsilon_1)^{-1/2}; \partial \varepsilon_2 = \partial \varepsilon_3 = -1/2\,(1 + \varepsilon_1)^{-3/2} \partial \varepsilon_1$, the cluster stress in stretching direction for the upcycle ($\partial \varepsilon_1/\partial t > 0$) is obtained as:

$$\sigma_{A,1}^{up} = \hat{\sigma}_{R,1}(\varepsilon_1) \frac{1}{d} \int_{d(\frac{s_d}{\hat{\sigma}_{R,1}(\varepsilon_{1,max})})}^{d(\frac{s_d}{\hat{\sigma}_{R,1}(\varepsilon_1)})^\alpha} \varphi(\xi_1) d\xi_1 \qquad (24)$$

For the down cycle, the lateral directions contribute to the cluster stress ($\partial \varepsilon_2/\partial t > 0; \partial \varepsilon_3/\partial t > 0$):

$$\sigma_{A,1}^{down} = 2 \hat{\sigma}_{R,2}(\varepsilon_1) \left( -\frac{1}{2}(1 + \varepsilon_1)^{-\frac{3}{2}} \right) \frac{1}{d} \int_{d(\frac{s_d}{\hat{\sigma}_{R,2}(\varepsilon_{1,max})})}^{d(\frac{s_d}{\hat{\sigma}_{R,2}(\varepsilon_1)})^\alpha} \varphi(\xi_2) d\xi_2 \qquad (25)$$

With $\varphi(\xi_1) = \varphi(\xi_2) = \varphi(\xi_3) \equiv \varphi(\xi)$. This gives a negative stress contribution, which must be subtracted from the rubber stress. It can also be expressed by the rubber stress $\sigma_{R,1}$ in stretching direction. By assuming that the same energy is needed for stretching in 1-direction and compressing in 2- and 3-direction to obtain a final deformed state, the following relation is derived:

$$\hat{\sigma}_{R,2}(\varepsilon_2) \equiv \sigma_{R,2}(\varepsilon_2) - \sigma_{R,2}(\varepsilon_{2,min}) = -\lambda_1^{3/2} \sigma_{R,1}(\varepsilon_1) - \lambda_{1,max}^{3/2} \sigma_{R,1}(\varepsilon_{1,max}) \qquad (26)$$

Finally, for the evaluation of the (measured) total stress we have to consider an additional set stress $\sigma_{set}$ that appears as a remaining stress in the undeformed state after stretching and retraction. Note that this is also found for unfilled rubbers and depends on temperature and stretching rate. It probably results from long time relaxation effects of the polymer network. We introduce it in a purely empirical manner for the case of uniaxial deformations:

$$\sigma_{set,1} = s_{set,0} \left( \sqrt{\varepsilon_{1,max}} - \sqrt{\varepsilon_{1,min}} \right) \qquad (27)$$

Then for uniaxial deformations the total stress reads:

$$\sigma_{tot,1}(\varepsilon_1) = \sigma_{R,1}(\varepsilon_1) + \sigma_{A,1}^{up/down}(\varepsilon_1) + \sigma_{set,1}(\varepsilon_{1,max}, \varepsilon_{1,min}) \tag{28}$$

The stress of the rubber matrix $\sigma_{R,1}$ is given by Equation (4) with $\lambda_\mu = 1 + X(\varepsilon_{\mu,max})\varepsilon_\mu$ and strain amplification factor $X(\varepsilon_{\mu,max})$ specified by Equation (5). The cluster stress $\sigma_{A,1}$ depends on the direction of straining and is determined by Equations (24) and (25) for up and down, respectively. The theory presented here describes the complex quasi-static deformation behavior of filler reinforced elastomers for repeated stretching with increasing amplitude. A more general formulation of the DFM that applies for arbitrary deformation histories requires an additional term for the free energy density of soft filler clusters Equation (7), which considers the relaxation of cluster stress upon retraction [15,16]. For a test of the theory the present formulation for repeated stretching with increasing amplitude is sufficient because all open parameters are entering already and the extension to arbitrary deformation histories requires no additional fitting parameters.

### 2.4. Frame-Independent Formulation of Stress-Softening for Fast FEM Simulations

The implementation of the DFM into a FEM algorithm faces several problems. First, the DFM is a microscopic theory, where stresses are calculated by differentiation of the free energy density with respect to the internal strain variables $\lambda_\mu$ and $\lambda_{A,\mu}$, respectively. This is in discrepancy to continuum mechanical considerations, where corresponding differentiations have to be performed with respect to external strain variables. Second, the DFM is formulated in the main axis system and requires stress contributions of different directions, especially for description of filler-induced hysteresis, which all sum up to produce close cycles. This can hardly be transferred to a pure tensorial formulation. Nevertheless, a FEM implementation of the DFM was obtained by referring to the concept of representative directions, which considers uniaxial deformations along fibers in different spatial directions [24,25]. However, the computational cost of this workaround is very large, and the efficiency is low. Therefore, we want to focus here on a frame-independent tensor formulation of the stress-softening part of the DFM for fast and efficient FEM simulations.

In a first step we put the strain amplification factor $X_{max} \equiv X(\varepsilon_{\mu,max})$ in front of the deformation invariants, appearing in the free energy density of the extended non-affine tube model Equation (2) and replace the internal strain $\lambda_\mu = 1 + X(\varepsilon_{\mu,max})\varepsilon_\mu$ by the external strain $\lambda_\mu = 1 + \varepsilon_\mu$. This follows the ideas of Einstein [37] and Domurath et al. [38] and is done in close correlation to the evaluations in [22]. The free energy density reads:

$$W_R = \frac{G_c}{2}\left\{\frac{X_{max}\bar{I}_1\left(1 - \frac{T_e}{n_e}\right)}{1 - \frac{T_e}{n_e}X_{max}\bar{I}_1} + \ln\left[1 - \frac{T_e}{n_e}X_{max}\bar{I}_1\right]\right\} + 2G_eX_{max}\vec{I} \tag{29}$$

with the (frame-independent) first invariant of the left Cauchy–Green tensor:

$$\bar{I}_1 \equiv \lambda_1^2 + \lambda_2^2 + \lambda_3^2 - 3 \tag{30}$$

and the (frame-independent) generalized invariant:

$$\vec{I} \equiv \lambda_1^{-1} + \lambda_2^{-1} + \lambda_3^{-3} - 3 \tag{31}$$

Here, $\lambda_\mu = 1 + \varepsilon_\mu$ is the external strain of the sample. A frame-independent formulation of the strain amplification factor $X_{max}$ is obtained similar to Equation (5) by replacing the relative stress $\hat{\sigma}_{R,\mu}(\varepsilon_{\mu,\max})$ used for the calculation of the boundary cluster size $x_{\mu,\min}$ by the Frobenius norm $\|\sigma_R(\varepsilon_{\max})\|$ of the engineering stress.

$$X(\varepsilon_{\max}) = 1 + c\Phi_{\text{eff}}^{\frac{2}{3-d_f}} \frac{1}{d}\left[\int_0^{d(s_v/\sigma_R(\varepsilon_{\max}))^\alpha} \left(\frac{\xi}{d}\right)^{d_w-d_f} \varphi(\xi)d\xi + \int_{d(s_v/\sigma_R(\varepsilon_{\max}))^\alpha}^{\infty} \varphi(\xi)d\xi\right] \quad (32)$$

The iteration procedure for the evaluation of stresses (compare Equation (21a) or (21b) together with Equation (5)) is then replaced by its tensorial analog:

$$\sigma_{R,\mu}(\varepsilon_{\mu,\max}) = f(X_{\max}(\sigma_{R,\mu}(\varepsilon_{\mu,\max}))) \rightarrow \sigma_R(\varepsilon_{\max}) = f(X_{\max}(\sigma_R(\varepsilon_{\max}))) \quad (33)$$

The free energy density Equation (29) can be used in a standard continuum mechanical sense for the evaluation of stresses and tangent vectors. It can be further simplified by omitting the logarithmic term, which gives a minor contribution to the stress upturn. In addition. The generalized invariant can be approximated by the square root of the second invariant, which avoids the calculation of eigenvalues [39].

## 3. Experimental and Fitting Procedure

### 3.1. Materials

EPDM/A ("aging") samples were prepared by using an amorphous-type ethylene–propylene-diene rubber (EPDM, Keltan 4450) filled with 50-phr (parts per hundred mass parts rubber) carbon black (N339). The rubber has a Mooney viscosity of 46 MU (at 125 °C) and consists of 52% ethylene and 4.3% ethylidene norbornene. Moreover, 3 phr zinc oxide, 1 phr stearic acid and 1.5 phr N-isopropyl-N'-phenyl-1,4-phenylenediamine (IPPD, aging protection) were added. The curing system consists of 1.8-PHR sulfur, 1.5 phr 1,3-diphenylguanidine (DPG) and 2.4 phr N-cyclohexyl-2-benzothiazolylsulfenamide (CBS, accelerator). EPDM/CB ("carbon black") consists of the same polymer, additives and curing system, but has varying amounts of carbon black (N339): 20, 40 and 60 phr.

### 3.2. Mixing and Sample Preparation

All ingredients except of the vulcanization system were mixed in an internal mixer of type GK 1,5E (Werner & Pfleiderer Gummitechnik GmbH, Freudenberg, Germany) at a loading of 75% and a temperature less than 140 °C. The vulcanization system was added in a second step on a 150*350RR roller mill (KraussMaffei Berstorff GmbH, Hannover, Germany).

All samples were cured in a heated press of type WLP 63/3,5/3 (Wickert Maschinenbau GmbH, Landau, Germany) at 160 °C up to the $t_{90\%}$ time, where 90% of the torque obtained from a vulcameter measurement is reached (17:23 min). One minute curing time was added per millimeter sample thickness to account for heat diffusion.

From EPDM/A dumbbell test specimen were prepared. These were stored under thermo-oxidative aging conditions at 130 °C in an air-ventilated oven for 0, 1, 3, 7 and 14 days.

### 3.3. Test Methods and Fitting

Quasi-static multi-hysteresis experiments (multiple deformation cycles up to different strain levels) were performed in a Zwick 1445 (Zwick Roell, Ulm, Germany) universal testing machine using a crosshead speed of 20 mm/min. Every strain level was repeated five times. From EPDM/CB tensile test specimen of S2 type were prepared to achieve strains, which are not accessible using dumbbell samples. Multi-hysteresis experiments at 100 mm/min crosshead speed were performed in the same stretching machine. In both cases, the 5th cycles were separated for fittings with the DFM, which can be considered as equilibrium cycles.

The adaptation of stress-strain cycles to Equation (28) was performed by minimization of the error functional $\chi^2 = \sum_{\text{cycles}} \sum_n \left( y_{\text{mod},n} - y_{\text{exp},n} \right)^2$, where $y_{\text{mod},n}$ and $y_{\text{exp},n}$ represent the $n$th model and experimentally obtained 1. Piola–Kirchhoff ("engineering") stress data point. The set stress parameter $s_{\text{set},0}$ was included in the fitting procedure as defined by referring to Equation (27). The remaining two stress contributions of Equation (28), which involve seven fitting parameters, were obtained iteratively by using Equations (4)–(6) for the intrinsic stress of the rubber matrix $\sigma_{R,1}$ and Equations (24) and (25) for the evaluation of cluster stress $\sigma_{A,1}$. Note that both stress contributions involve the same cluster size distribution, which stabilizes the fitting procedure, significantly. The front factor of Equation (5) was fixed as $c = 2.5$ according to the Einstein equation [37] and the exponent was approximated as $d_w - d_f \approx 1$ to allow for an analytical solution of the integral. The three fitting parameters describing the rubber elastic network are the crosslink modulus $G_c$, the topological constraint modulus $G_e$ and the effective chain length for finite extensibility $n_{\text{eff}} \equiv n_e / T_e$. The filler clusters are described by four fitting parameters, i.e., the effective filler volume fraction $\Phi_{\text{eff}}$, the average cluster size $x_0 \equiv \langle x_\mu \rangle / d$, the tensile strength of virgin filler–filler bonds $s_v$ and the tensile strength of damaged filler–filler bonds $s_d$.

## 4. Results and Discussion

### 4.1. Micromechanical Investigations of Thermo-Oxidative Aging

The thermo-oxidative aging of elastomers is of high technological interest for the rubber industry, because it mostly increases the hardness of the samples and has a negative effect on the fracture toughness or crack resistance. This limits the life time of rubber goods, significantly. The reason for this property losses are mainly seen in a change of the rubber–elastic network due to post-curing, but the often-accompanied increase in electrical conductivity indicates that also the carbon black network is altering during thermo-oxidative aging of elastomers. This is hardly to distinguish by standard measurement techniques since changes of the polymer- or filler network structure are difficult to detect. We therefore refer here to the evaluation of micromechanical material parameters, which are obtained by fitting the stress–strain response of the aged samples to the DFM.

Figure 4 shows a series of fittings of multi-hysteresis stress–strain cycles of EPDM/A samples stored under thermo-oxidative aging conditions at 130 °C for 0, 1, 3, 7 and 14 days. In Figure 4a, the "monodisperse" bond fracture model ($\alpha = 1/2$) is used while Figure 4b refers to the "hierarchical" bond fracture model ($\alpha = 1$). Before we discuss the effect of thermomechanical aging on material parameters, we will first consider the quality of the fits for the two different bond fracture criteria. The "monodisperse" model depicted in Figure 4a delivers fair agreement between fits and experimental data with correlation coefficients between $R^2 = 0.994$ and $0.995$, but systematic deviations are seen, e.g., for the peak stresses in the medium strain regime. For the "hierarchical" model shown in Figure 4b, the fits are significantly better with correlation coefficients between $R^2 = 0.995$ and $R^2 = 0.998$. This indicates that the "hierarchical" model is more suited for describing the stress–strain cycles of filler reinforced elastomers. Even in the case of aged samples we get excellent adaptations in the small and medium strain regime up to 100% strain. Therefore, we will mainly focus on the "hierarchical" model with bond fracture exponent $\alpha = 1$ for the discussion of thermo-oxidative aging effects on a microscopic level. Nevertheless, we will see that the "monodisperse" model delivers similar values and trends of the fitting parameters.

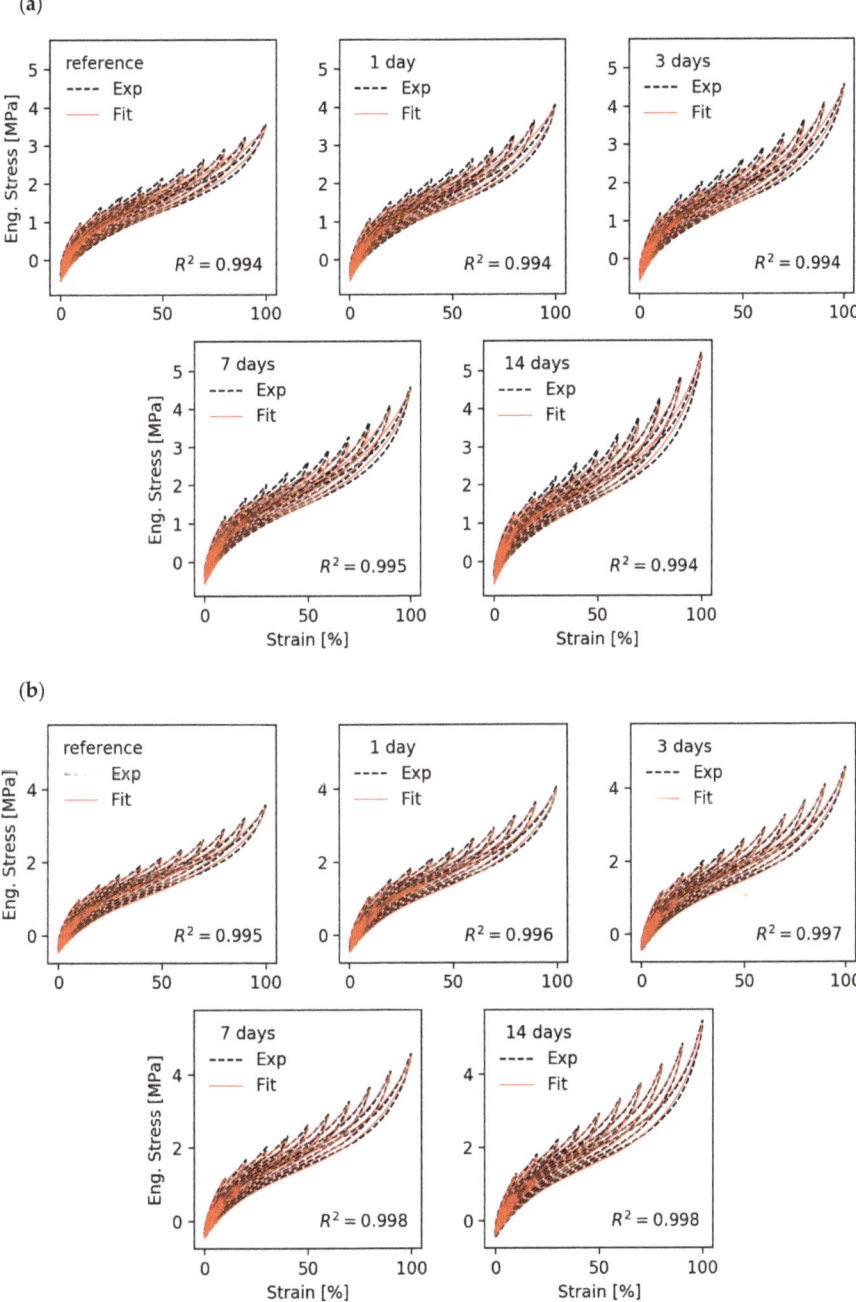

**Figure 4.** Fit of stress–strain cycles of ethylene–propylene-diene rubber EPDM/A samples for various aging times (**a**) with the "monodisperse" bond fracture model ($\alpha = 1/2$) and (**b**) with the "hierarchical" bond fracture model ($\alpha = 1$).

The stress–strain data in Figure 4 show that the average stress level increases with increasing aging time. The reason for this hardening of the samples shall be analyzed by referring to the evolution of fitting parameters that are depicted in Figure 5. The fitting parameters obtained with the "monodisperse" bond fracture model ($\alpha = 1/2$) are shown in Figure 5a and those from the "hierarchical" bond fracture model ($\alpha = 1$) in Figure 5b. Obviously, the "hierarchical" bond fracture model in Figure 5b delivers a smoother evolution of fitting parameters, which correlates with the higher correlation coefficients of the fits in Figure 4b. This confirms our view that the "hierarchical" bond fracture model is more suited for the discussion of aging effects. Looking first at the crosslink and topological constraint moduli of the polymer network, $G_c$ and $G_e$, we see that the former remains almost constant while the later increases successively with aging time. This indicates that the post-curing effect is not pronounced for the EPDM samples used in this study, but the topological constraints of the chains increase with aging time possibly due to an increasing number of entanglements close to the carbon black particles (surface-induced entanglements). This correlates with the observed decrease of the effective chain length, $n_\text{eff} \equiv n_e/T_e$, since the segment number $n_e$ between successive entanglements decreases with aging time if $G_e$ increases ($G_e \sim 1/n_e$). However, it must be noted that it is difficult to distinguish between the two parameters $G_c$ and $G_e$ in the frame of the DFM, since both act in a similar way.

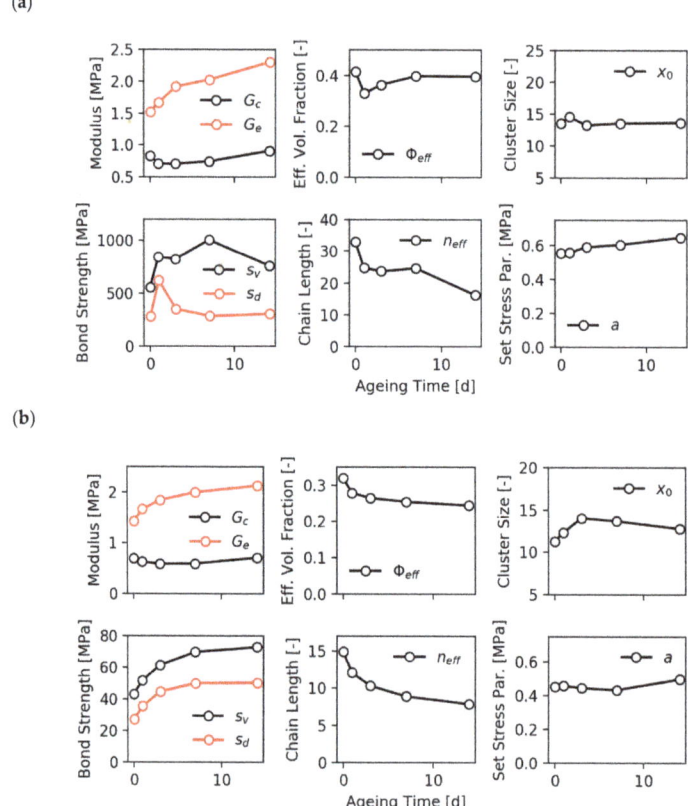

**Figure 5.** Evolution of fitting parameters obtained with (**a**) the "monodisperse" bond fracture model ($\alpha = 1/2$) and (**b**) the "hierarchical" bond fracture model ($\alpha = 1$) of the EPDM/A samples from Figure 4 for various aging times.

$n_{\text{eff}} \equiv n_e/T_e$. The filler clusters are described by four fitting parameters, i.e., the effective filler volume fraction $\Phi_{\text{eff}}$, the average cluster size $x_0 \equiv \langle x_\mu \rangle/d$, the tensile strength of virgin filler–filler bonds $s_v$ and the tensile strength of damaged filler–filler bonds $s_d$.

An additional significant effect of thermo-oxidative aging is observed for the strength of virgin and damaged filler–filler bonds, $s_v$ and $s_d$, which both increase systematically with aging time. This is clearly seen in the case of the "hierarchical" bond fracture model ($\alpha = 1$). It indicates that a relaxation of bonds takes place during heating of the samples at 130 °C, leading to more stable filler–filler joints. Note that this is related to confined polymer between the filler particles, which is assumed to be in a glassy-like state at room temperature [7] but can relax at elevated temperature. This relaxation process has been investigated recently by online dielectric spectroscopy during heat treatment of carbon black filled EPDM [40]. The increase of $s_v$ and $s_d$ with aging time observed in Figure 5b results in stronger stress-softening and hysteresis effects of the aged samples. Two further parameters describing the aging of the filler network are the effective filler volume fraction $\Phi_{\text{eff}}$ and the mean cluster size $x_0$. The former decreases slightly with aging time, but remains larger than the real filler volume fraction, as expected ($\Phi_{\text{eff}} > \Phi \approx 0.2$). This indicates a slight decrease of the occluded rubber during aging, which is hidden in the voids of the filler particles and acts like additional filler. The mean cluster size lies in a reasonable range of about 10 to 15 particle diameters and goes through a weak maximum with increasing aging time. This can be related to flocculation effects and restructuring of the filler network

### 4.2. Frame-Independent Model of Stress Softening

For the discussion of the frame-independent model of stress softening we will focus on the "hierarchical" bond fracture model with $\alpha = 1$, only. For the analysis of the stress-softening effect, described here, we refer to the EPDM/CB samples with varying amount of carbon black. Note that the hysteresis is not included in this model.

Figure 6a,b shows fits of stress–strain cycles of the EPDM/CB samples with the frame-independent model ($\alpha = 1$) up to 100% and 200%, respectively. In both cases the stress-softening effect is reproduced fairly well, though for the 200% fit systematic deviations are seen for the sample with 60-PHR N339 in the small strain regime. The effect of filler concentration on the fitting parameters is depicted for both cases in Figure 7a,b, respectively. All values of the parameters are found in a reasonable range. However, they strongly depend on the range of the fitted stress–strain data up to 100% and 200%, respectively. This indicates that the DFM cannot simply be extended to strains up to 200% because additional mechanisms of stress softening may appear at larger strains, e.g., detachment of the polymer from the filler surface. Nevertheless, the quite reasonable fits in Figure 6b show that the simplified DFM can be used as an empirical model also for larger strains up to rupture of the samples. Due to the simplifications of the frame independent model compared to the original DFM it makes no sense to discuss the dependence of fitting parameters on filler concentration in detail. The more interesting point is the high stability of the fitting procedure with parameters that all are positive and can easily be reproduced. Most important is the ability to implement the simplified model into a finite element code for fast FEM simulations by using standard methods. This is of major interest for the rubber industry and will be a task of future work.

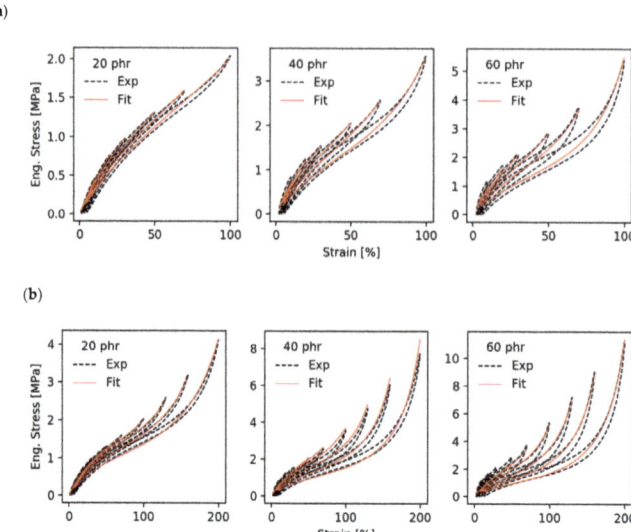

**Figure 6.** Fit of stress–strain cycles of EPDM/CB samples filled with various amounts of carbon black (N339) with the frame-independent model of stress softening ($\alpha = 1$) data up to 100% strain and (**b**) up to 200% strain.

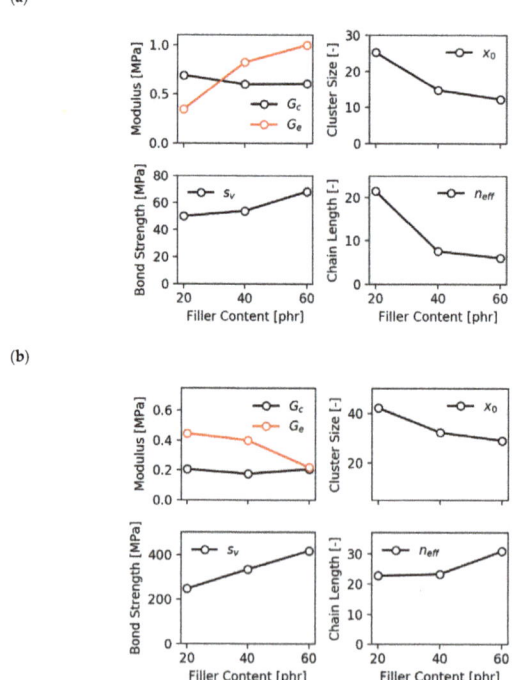

**Figure 7.** Fitting parameters obtained with the frame independent model of stress softening of the EPDM/2.4-PHR *N*-cyclohexyl-2-benzothiazolylsulfenamide (CBS) samples as shown in Figure 6, using (**a**) data up to 100% strain and (**b**) up to 200% strain.

## 5. Conclusions

A micromechanical model of stress-softening and hysteresis of filler reinforced elastomers was presented, which is based on a non-affine tube model of rubber elasticity and a generalized three-dimensional Kantor–Webman model of flexible chain aggregates, describing the deformation and fracture of filler clusters in the stress field of the rubber matrix. This dynamic flocculation model (DFM) has been shown to reproduce the complex quasi-static deformation behavior of filler reinforced elastomers upon repeated stretching with increasing amplitude fairly well. It is described in some detail in the theoretical section, whereby two different fracture mechanisms of filler–filler bonds, denoted "monodisperse" and "hierarchical" bond fracture mechanism, are considered. In the first approach all bonds are considered to be equal (monodisperse) having the same strength. In the second approach a hierarchy of bond strengths is realized, because during cluster–cluster aggregation the mobility of the clusters decreases with cluster size.

In the experimental section the DFM is adapted to a series of aged EPDM samples which were treated in an oven at 130 °C for different thermo-oxidative aging times. The fitting parameters indicate that the crosslinking density remains almost constant while the entanglement density increases slightly. This indicates that the sulfur bridges in EPDM networks are quite stable even at 130 °C aging temperature, which can be related to the mono-sulfidic nature of the crosslinks. The observed hardening of the composites with increasing aging time is mainly attributed to the relaxation of filler–filler bonds. This results in an increased strength of the bonds, which produces a larger stiffness and filler-induced hysteresis of the composites.

The two different bond fracture mechanisms are investigated by separate adaptations to the full series of aged EPDM composites. They show that the "hierarchical" bond fracture mechanism delivers better fits and more stable fitting parameters, though the evolution of fitting parameters with aging time is similar for both models. Therefore, it is concluded that the "hierarchical" bond fracture mechanism, which takes into account that the mobility of clusters decreases with cluster size, appears to be realized in filler reinforced elastomers.

In the last section a frame-independent simplified version of the DFM is proposed that focuses on an easy implementation of the stress-softening effect of filled rubbers into a finite element codes by using standard methods. The model is shown to reproduce the stress-softening effect of EPDM samples with varying amount of carbon black fairly well. Therefore, it appears to be well suited for performing fast FEM simulations of highly filled rubber goods, where stress-softening cannot be neglected.

**Author Contributions:** J.P. developed the curve fitting algorithm, made the graphs and provided the ideas for the frame-independent DFM. M.K. wrote the major part of the introduction, theory and results section. All authors have read and agreed to the published version of the manuscript.

**Funding:** The work received no special funding.

**Conflicts of Interest:** The authors declare no conflict of interest.

## References

1. Kraus, G. *Reinforcement of Elastomers Wiley*; Interscience Publishing: London Sydney, NY, USA, 1965.
2. Vilgis, T.A.; Heinrich, G.; Klüppel, M. *Reinforcement of Polymer Nano-Composites*; Cambridge University Press: Cambridge, UK, 2009.
3. Lorenz, H.; Steinhauser, D.; Klüppel, M. Morphology and Micro-Mechanics of Filled Elastomer Blends: Impact on Dynamic Crack Propagation. In *Fracture Mechanics and Statistical Mechanics of Reinforced Elastomeric Blends*; Lecture Notes in Applied and Computational, Mechanics; Grellmann, W., Heinrich, G., Kaliske, M., Klüppel, M., Schneider, K., Vilgis, T.A., Eds.; Springer: Berlin/Heidelberg, Germany, 2013; Volume 70.
4. Payne, A.R. Strainwork dependence of filler-loaded vulcanizates. *J. Appl. Polym. Sci.* **1964**, *8*, 2661. [CrossRef]
5. Medalia, A.I. Elastic Modulus of Vulcanizates as related to Carbon Black Structure. *Rubber Chem. Technol.* **1973**, *46*, 877. [CrossRef]

6. Mullins, L.; Tobin, N.R. Stress softening in rubber vulcanizates. Part I. Use of a strain amplification factor to describe the elastic behavior of filler-reinforced vulcanized rubber. *J. Appl. Polym. Sci.* **1965**, *9*, 2993. [CrossRef]
7. Klüppel, M. The Role of Disorder in Filler Reinforcement of Elastomers on Various Length Scales. *Adv. Polym. Sci.* **2003**, *164*, 1–86.
8. Klüppel, M.; Heinrich, G. Fractal structures in carbon black reinforced rubbers. *Rubber Chem. Technol.* **1995**, *68*, 623–651. [CrossRef]
9. Klüppel, M.; Schuster, R.H.; Heinrich, G. Structure and properties of reinforcing fractal filler networks in elastomers. *Rubber Chem. Technol.* **1997**, *70*, 243. [CrossRef]
10. Heinrich, G.; Klüppel, M. Recent Advances in the Theory of Filler Networking in Elastomers. *Adv. Polym. Sci.* **2002**, *160*, 1–44.
11. Kantor, Y.; Webman, I. Elastic Properties of Random Percolation Systems. *Phys. Rev. Lett.* **1984**, *52*, 1891. [CrossRef]
12. Klüppel, M.; Schramm, J. A generalized tube model of rubber elasticity and stress softening of filler reinforced elastomer systems. *Macromol. Theory Simul.* **2000**, *9*, 742–754. [CrossRef]
13. Klüppel, M.; Meier, J.; Dämgen, M. Modelling of stress softening and filler induced hysteresis of elastomer materials. In *Constitutive Models for Rubber IV*; Austrell, K., Ed.; Taylor & Francis: London, UK, 2005; Volume 171.
14. Lorenz, H.; Meier, J.; Klüppel, M. Micromechanics of Internal Friction of Filler Reinforced Elastomers. In *Elastomere Friction: Theory, Experiment and Simulation*; Lecture Notes in Applied and Computational Mechanics; Besdo, D., Heimann, B., Klüppel, M., Kröger, M., Wriggers, P., Nackenhorst, U., Eds.; Springer: Berlin/Heidelberg, Germany, 2010; Volume 51, ISBN 978-3-642-10656-9.
15. Lorenz, H.; Klüppel, M. Microstructure-based modeling of arbitrary deformation histories of filler-reinforced elastomers. *J. Mech. Phys. Solids* **2012**, *60*, 1842–1861. [CrossRef]
16. Lorenz, H.; Klüppel, M.; Heinrich, G. Micro-structure based modeling and FE-implementation of filler-induced stress softening and hysteresis of reinforced rubbers. *ZAMM J. Appl. Math. Mech. Z. Angew. Math. Mech.* **2012**, *92*, 608–631. [CrossRef]
17. Montes, H.; Lequeux, F.; Berriot, J. Influence of the glass transition temperature gradient on the nonlinear viscoelastic behavior in reinforced elastomers. *Macromolecules* **2003**, *36*, 8107. [CrossRef]
18. Berriot, J.; Lequeux, F.; Monnerie, L.; Montes, H.; Long, D.; Sotta, P.; Non-Cryst, J. Evidence for the shift of the glass transition near the particles in silica-filled elastomers. *Macromolecules* **2002**, *35*, 9756. [CrossRef]
19. Berriot, J.; Montes, H.; Lequeux, F.; Long, D.; Sotta, P. Gradient of glass transition temperature in filled elastomers. *Europhys. Lett.* **2003**, *64*, 50. [CrossRef]
20. Lin, C.-R.; Lee, Y.-D. Strain-dependent dynamic properties of filled rubber network systems. *Macromol. Theory Simul.* **1996**, *5*, 1075–1104. [CrossRef]
21. Witten, T.A.; Rubinstein, M.; Colby, R.H. Reinforcement of rubber by fractal aggregates. *J. Phys.* **1993**, *3*, 367–383. [CrossRef]
22. Plagge, J.; Klüppel, M. A physically based model of stress softening and hysteresis of filled rubber including rate- and temperature dependency. *Int. J. Plast.* **2017**, *89*, 173–196. [CrossRef]
23. Plagge, J.; Klüppel, M. Mullins effect revisited: Relaxation, recovery and high-strain damage. *Mater. Today Commun.* **2019**, *20*, 100588. [CrossRef]
24. Freund, M.; Lorenz, H.; Juhre, D.; Ihlemann, J.; Klüppel, M. Finite element implementation of a micro-structure based model for filled elastomers. *Int. J. Plast.* **2011**, *27*, 902–919. [CrossRef]
25. Ihlemann, J. Kontinuumsmechanische Nachbildung Hochbelasteter Technischer Gummiwerk-Stoffe. Ph.D. Thesis, Leibniz University Hannover, Hannover, Germany, 2003.
26. Ragunath, R.; Juhre, D.; Klüppel, M. A physically motivated model for filled elastomers including strain rate and amplitude dependency in finite viscoelasticity. *Int. J. Plast.* **2016**, *78*, 223–241. [CrossRef]
27. Lockhorn, D.; Klüppel, M. Structure-Property Relationships of Silica/Silane Formulations in Natural Rubber, Isoprene Rubber and Styrene-Butadiene Rubber Composites. *J. Appl. Polym. Sci.* **2019**, *48435*, 12. [CrossRef]
28. Kaliske, M.; Heinrich, G. An extended tube-model for rubber elasticity: Statistical-mechanical theory and finite element implementation. *Rubber Chem. Technol.* **1999**, *72*, 602. [CrossRef]
29. Edwards, S.F.; Vilgis, T.A. The tube model theory of rubber elasticity. *Rep. Prog. Phys.* **1988**, *51*, 243. [CrossRef]
30. Heinrich, G.; Straube, E.; Helmis, G. Rubber elasticity of polymer networks: Theories. *Adv. Polym. Sci.* **1988**, *85*, 33–87.
31. Huber, G.; Vilgis, T.A. Universal Properties of Filled Rubbers: Mechanisms for Reinforcement on Different Length Scales. *Kautsch. Gummi Kunstst.* **1998**, *3*, 217.

32. Medalia, A. Effect of carbon black on dynamic properties of rubber vulcanizates. *Rubber Chem. Technol.* **1978**, *51*, 437. [CrossRef]
33. van Dongen, P.G.J.; Ernst, M.H. Dynamic scaling in the kinetics of clustering. *Phys. Ref. Lett.* **1985**, *54*, 1396. [CrossRef]
34. Ziff, R.M.; McGrady, E.D.; Meakin, P. On the validity of Smoluchowski's equation for cluster–cluster aggregation kinetics. *J. Chem Phys.* **1985**, *82*, 5269. [CrossRef]
35. Jullien, R. The application of fractals to investigations of colloidal aggregation and random deposition. *New J. Chem (1987)* **1990**, *14*, 239–253.
36. Rubinstein, M.; Colby, R.H. *Polymer Physics*; Oxford University Press: New York, NY, USA, 2003; Volume 23.
37. Einstein, A. Eine neue Bestimmung der Moleküldimensionen. *Ann. Phys.* **1906**, *324*, 289–306. [CrossRef]
38. Domurath, J.; Saphiannikova, M.; Gilles, A.; Heinrich, G. Modelling of stress and strain amplification effects in filled polymer melts. *J. Newt. Fluid Mech.* **2012**, *171*, 8–16. [CrossRef]
39. Plagge, J.; Ricker, A.; Kröger, N.H.; Wriggers, P.; Klüppel, M. Efficient modeling of filled rubber assuming stress-induced microscopic restructurization. *Int. J. Eng. Sci.* **2020**, *151*, 103291. [CrossRef]
40. Steinhauser, D.; Möwes, M.; Klüppel, M. Carbon Black Networking in Elastomers Monitored by Simultaneous Rheological and Dielectric Investigations. *J. Phys. Condens. Matter* **2016**, *28*, 495103. [CrossRef] [PubMed]

© 2020 by the authors. Licensee MDPI, Basel, Switzerland. This article is an open access article distributed under the terms and conditions of the Creative Commons Attribution (CC BY) license (http://creativecommons.org/licenses/by/4.0/).

Article

# Graphene Layers Functionalized with A *Janus* Pyrrole-Based Compound in Natural Rubber Nanocomposites with Improved Ultimate and Fracture Properties

Gea Prioglio [1], Silvia Agnelli [2], Lucia Conzatti [3], Winoj Balasooriya [4], Bernd Schrittesser [4] and Maurizio Galimberti [1,*]

1. Department of Chemistry, Materials and Chemical Engineering Giulio Natta, Politecnico di Milano, Via Mancinelli 7, 20131 Milano (I), Italy; gea.prioglio@polimi.it
2. Department of Mechanical and Industrial Engineering, University of Brescia, Via Branze 38, 25123 Brescia (I), Italy; silvia.agnelli@unibs.it
3. Istituto di Scienze e Tecnologie Chimiche "Giulio Natta" (SCITEC), CNR, Via De Marini 6, 16149 Genova (I), Italy; lucia.conzatti@cnr.it
4. Polymer Competence Center Leoben GmbH Roseggerstrasse 12, A-8700 Leoben (A), Austria; Winoj.Balasooriya@pccl.at (W.B.); Bernd.Schrittesser@pccl.at (B.S.)
* Correspondence: maurizio.galimberti@polimi.it

Received: 18 March 2020; Accepted: 14 April 2020; Published: 18 April 2020

**Abstract:** The ultimate properties and resistance to fracture of nanocomposites based on poly(1,4-*cis*-isoprene) from *Hevea Brasiliensis* (natural rubber, NR) and a high surface area nanosized graphite (HSAG) were improved by using HSAG functionalized with 2-(2,5-dimethyl-1H-pyrrol-1-yl)propane-1,3-diol (serinol pyrrole) (HSAG-SP). The functionalization reaction occurred through a domino process, by simply mixing HSAG and serinol pyrrole and heating at 180 °C. The polarity of HSAG-SP allowed its dispersion in NR latex and the isolation of NR/HSAG-SP masterbatches via coagulation. Nanocomposites, based either on pristine HSAG or on HSAG-SP, were prepared through traditional melt blending and cured with a sulphur-based system. The samples containing HSAG-SP revealed ultimate dispersion of the graphitic filler with smaller aggregates and higher amounts of few layers stacks and isolated layers, as revealed by transmission electron microscopy. With HSAG-SP, better stress and elongation at break and higher fracture resistance were obtained. Indeed, in the case of HSAG-SP-based composites, fracture occurred at larger deformation and with higher values of load and, at the highest filler content (24 phr), deviation of fracture propagation was observed. These results have been obtained with a moderate functionalization of the graphene layers (about 5%) and normal lab facilities. This work reveals a simple and scalable way to prepare tougher NR-based nanocomposites and indicates that the dispersion of a graphitic material in a rubber matrix can be improved without using an extra-amount of mechanical energy, just by modifying the chemical nature of the graphitic material through a sustainable process, avoiding the traditional complex approach, which implies oxidation to graphite oxide and subsequent partial reduction.

**Keywords:** natural rubber; high surface area graphite; functionalization; serinol pyrrole; ultimate properties; fracture resistance

---

## 1. Introduction

Poly(1,4-*cis*-isoprene) from *Hevea Brasiliensis*, known as natural rubber (NR) [1,2], is the most important rubber, with a worldwide production of almost 14 million tons in 2018; it represents about 40% of the total rubber consumption [3,4]. Such a great success is based on the outstanding properties of NR: tack [5,6] and strength [7] in the uncured state and, upon vulcanizing, tensile strength [8,9] and resistance to fatigue and crack growth [10–13]. In particular, the latter properties are due to a peculiar property of NR, strain induced crystallization [14–17]. Indeed, NR is the selected rubber for demanding applications, for instance those in tire compounds (for cars, heavy trucks, and airplanes), anti-seismic base isolators, and anti-vibrating mounting pads. Moreover, NR is produced with a far lower energy input with respect to synthetic rubber: 15-16 MJ/kg versus a typical value of 100 MJ/kg [4]. It is also worth mentioning the efficient carbon sequestration performed by the *Hevea* tree: the photosynthetic rate of *Hevea* leaves is of about 11 µmol/m$^2$·s versus a value of 5–13 µmol/m$^2$·s in other trees [18]. Thus, the large amount of research performed on NR and NR-based composites can be easily understood.

In spite of its remarkable properties, NR is unable to meet the requirements of the above-mentioned applications. Better mechanical properties are required and achieved with the addition of reinforcing fillers [19,20]. Carbon black (CB) has been used for over a century [21]. After the discovery of fullerenes [22], an impressive amount of nanosized sp$^2$ carbon allotropes has been prepared: single [23,24] or multi-walled [25,26] carbon nanotubes, graphene [27–30], and graphene-related materials [31–34]. Graphene is of particular interest because of its outstanding properties: an elastic modulus larger than 1 TPa, exceptional thermal and electrical conductivity [35,36].

Papers and reviews on rubber composites with graphene and related materials (GRM), mainly dedicated to isoprene rubber, are available in the scientific literature [37–58]. The basic objective of the reported researches is to obtain an ultimate distribution and dispersion of the graphene layers. Indeed, it is widely acknowledged that, by increasing filler distribution and dispersion, superior mechanical and ultimate properties and abrasion resistance can be achieved [59]. Moreover, the amount of filler can be reduced, thus preparing lighter materials. In the field of rubber composites, the filler dispersion can be improved by applying mechanical energy. However, such an approach can alter the structure of the rubber and even of the filler. Melt blending is used to promote the dispersion of nanographite, even with a very high surface area [39,44,50]. The solution mixing has been documented [58], but this approach is troublesome, particularly in view of an industrial scale up. The favored technology reported in the literature is the dispersion of graphene layers in latexes of NR [37,38,41,45,46,48,50–52,55] or of poly(styrene-*co*-butadiene) (SBR) [40], combined with the rational design of the surface chemistry of the filler [40]. In fact, to prepare latex dispersions, polar graphitic materials, in most cases graphene/graphite oxide (GO), are used [23–26,37,38,40,41,44–46,48]. GO, which is typically prepared by means of the Hummers and Offeman method [60–63] is then reduced, typically with chemical methods [64–66]. In such an oxidation-reduction method, the main disadvantage is the use of dangerous and even toxic chemicals (such as potassium permanganate, sulphuric acid, hydrazine), and harsh reaction conditions. Moreover, it is acknowledged that the bulk graphene structure is only partially restored: an appreciable amount of sp$^3$ defects is present in the final product.

In this work, nanocomposites based on NR and graphene layers functionalized by means of a simple and sustainable method were prepared. The main objective of the research was to investigate the effect of an enhanced dispersion of the graphene layers on the dynamic-mechanical, tensile, and fracture properties of natural rubber nanocomposites. As the starting graphitic material, a high surface area graphite (HSAG) was selected, with a limited number of stacked layers (about 35) and a high order inside the basal plane, hence with a high shape anisotropy [66–68]. In order to achieve ultimate distribution and dispersion in the NR matrix, the nanocomposite was prepared via latex blending and, in order to obtain a higher compatibility with the latex, functionalization of HSAG was performed. A key objective of this work was to perform the functionalization with a sustainable and simple method, avoiding the above-mentioned oxidation-reduction cycle and harsh reaction conditions and preserving the structure of the graphene layers. Functionalization has been performed with a *Janus* molecule such as 2-(2,5-dimethyl-1*H*-pyrrol-1-yl)propane-1,3-diol (serinol pyrrole, SP), by simply mixing and heating HSAG and SP [69,70]. It has been shown [71] that a domino reaction occurs: a carbocatalyzed oxidation of the pyrrole compound occurs and then the pyrrole ring gives rise to a cycloaddition reaction with the graphitic substrate. The layers, whose bulk structure is substantially unaltered, are functionalized at the edges with OH groups. Nanocomposites prepared via latex blending have been compared with those obtained via melt blending. The dispersion of the graphitic filler was investigated with transmission electron microscopy (TEM). Dynamic-mechanical properties were studied through the application of stresses in the shear and axial modes, and tensile and fracture properties were also investigated.

## 2. Materials and Methods

### 2.1. Materials

High surface area graphite (HSAG) was Nano 27 from Asbury Graphite Mills, Inc. (Asbury, NJ, USA). According to the technical data sheet: carbon content is not lower than 99 mass%, surface area is 250 m$^2$/g, and the chemical composition from elemental analysis (U.S. Standard Test Sieves) is carbon 99.82%, ash 0.18%, and moisture 0.97%. The number of stacked layers was estimated, as already reported [66–68], to be about 35.

The NR latex was a medium ammonia grade from Centex FA, with a 60 wt % solid content, pH (at 20 °C) = 9–11, a density of 0.95 g/cm$^3$, and partial miscibility with water. 2,5-hexanedione ($M_W$ = 114.12 g/mol) from Sigma-Aldrich (purity ≥97%, St. Louis, MO, USA), 2-aminopropane-1,3-diol (serinol, purity ≥98%) was kindly provided by Bracco (Milan (MI), Italy). Acetone was from Sigma-Aldrich (St. Louis, MO, USA), purity ≥97%. All the chemicals were used without further purification.

The following chemicals were used for the preparation of the elastomeric compounds: Zinc Oxide (from Zincol Ossidi, Bellusco (MI), Italy), stearic acid (from Sogis, Sospiro (CR), Italy), 6PPD (N-(1,3-dimethylbutyl)-N'-phenyl-p-phenylenediamine, from Crompton, Middlebury, CT, USA), sulphur (from Solfotecnica, Cotignola (RA), Italy), and TBBS (N-tert-butyl-2-benzothiazyl) sulfonamide, from Eastman (Sauget, IL, USA).

### 2.2. Synthesis of 2-(2,5-Dimethyl-1H-Pyrrol-1-yl)Propane-1,3-Diol or Serinol Pyrrole (SP)

A mixture of 2,5-hexanedione (HD, 12.15 g, 0.107 mol) and 2-aminopropane-1,3-diol (S, 9.96 g, 0.107 mol) was poured into a 100 mL round-bottomed flask equipped with a magnetic stirrer and a condenser (single glass tube). The mixture was then stirred (300 rpm) at 155 °C for 3 h. Then the condenser was removed, and the mixture was left stirring at 155 °C for 30 min. Afterwards, the reaction mixture was cooled down to room temperature. Finally, 15.51 g of pure, dark amber, viscous product was obtained. 1H NMR (CDCl$_3$, 400 MHz); δ (ppm) = 2.27 (s, 6H); 3.99 (m, 4H); 4.42 (quintet, 1H); 5.79 (s, 2H). 13C NMR (DMSO-d6, 100 MHz); δ (ppm) = 127.7; 105.9; 71.6; 61.2; 13.9.

### 2.3. Functionalization of High Surface Area Graphite (HSAG) with Serinol Pyrrole (SP)

Two samples were prepared, with two different HSAG/SP mass ratios.

Sample 1. HSAG/SP mass ratio = 10/1. 10 g of HSAG and 50 mL of acetone were put in a 250 mL round-bottomed flask. A solution of 1 g of SP in 15 mL of acetone was then added. The suspension was sonicated for 10 min, using a 2 L ultrasonic bath 260 W (Sonica, Soltec Srl, Milan, Italy). The solvent was removed under reduced pressure using a rotary evaporator. The 250 mL round-bottomed flask was then equipped with a magnetic stirrer and a condenser, heated up to 180 °C in an oil bath, and left under stirring (300 rpm) for 3 h. One hundred mL of acetone were then added to the free-flowing powder and the suspension was stirred overnight at room temperature and then filtered on a Büchner funnel with a sintered glass disc. The free-flowing powder was recovered and dried in an oven at 95 °C for 4 h.

Sample 2. HSAG/SP mass ratio = 10/0.6. The same experimental procedure as above was used. Instead of 1 g of SP, 0.6 g was used.

## 2.4. Characterization of the HSAG-SP Adduct

Functionalization Yield was determined using Equation (1):

$$\text{Functionalization Yield (\%)} = \frac{\text{SP mass \% in (HSAG} - \text{SP adduct) after acetone wash}}{\text{SP mass \% in (HSAG} - \text{SP adduct) before acetone wash}} \quad (1)$$

The mass percentages of Serinol Pyrrole (SP) in the HSAG-SP adduct (before and after acetone washing) were obtained from TGA analysis. This technique is frequently used to check the presence of organic compounds on the surface of carbonaceous fillers [72].

The TGA instrument used to perform thermogravimetric analyses was a Mettler TGA SDTA/851 (Mettler Toledo, Columbus, OH, USA). The standard method ISO9924-1 was followed. The method used for the analysis of adducts (5–10 mg) consists in a heating ramp (10 °C/min) from 30 up to 300 °C, followed by a 10 min isotherm at 300 °C, then another heating ramp (20 °C/min) up to 550 °C and another isotherm at 550 °C (15 min); this isotherm is followed by a final heating ramp (10 °C/min) up to 900 °C after which the temperature is kept constant until the end of the experiment. At 102.5 min from the beginning of the test, the gas in the chamber is switched from $N_2$ to air. The whole experiment lasts 120 min.

## 2.5. Rubber Composites Preparation

Composites formulations, expressed in parts per hundred rubber (phr), are reported in Table 1.

**Table 1.** Recipes of NR-based composites with HSAG or HSAG-SP as reinforcing fillers [a,b].

| Composite | 0 | 1 | 2 | 3 | 4 | 5 | 6 |
|---|---|---|---|---|---|---|---|
| NR | 100 | 100 | 100 | 100 | 100 | 100 | 100 |
| HSAG | / | 5 | 15 | 24 | / | / | / |
| HSAG-SP | / | / | / | / | 5 [c] | 15 [c] | 24 [c] |

[a] Amount of ingredients in phr [b] Other ingredients: Stearic Acid 2, Zinc oxide 4, 6PPD 2, TBBS 1.7, sulphur 1.2
[c] Theoretical values. For the experimental values, see below in the text.

Composites were prepared by using the same NR grade, in two steps. In the first step, either NR or NR/HSAG-SP were coagulated from the latex. In the second step, further ingredients were added through melt blending. The amount of HSAG and HSAG/SP was: 5, 15, and 24 phr. The intention was to use 25 phr of HSAG and HSAG/SP. In the HSAG/SP masterbatch coagulated from the latex, 24 phr were experimentally found (by means of TGA) and it was decided to use the same amount of HSAG in the corresponding composite.

### 2.5.1. Coagulation of NR

Dilution of NR latex. A typical procedure is as follows: 83.33 g of NR latex were poured in a 500-mL beaker, equipped with a magnetic stirrer; 100 mL of distilled water was added. The mixture was left under stirring (300 rpm) for 10 min at room temperature.

Coagulation of NR from the latex. Rubber was then coagulated by adding 100 mL of a 1M sulphuric acid solution. The solid rubber was then squeezed, immersed in distilled water overnight, rinsed with water up to neutral pH, reduced to small pieces, and left to dry at room temperature.

### 2.5.2. Coagulation of NR/HSAG-SP

The same procedure, reported as follows, was adopted for the preparation of three NR/HSAG-SP masterbatches (containing 5, 15, and 24 phr of HSAG-SP).

Preparation of HSAG-SP dispersion in water. HSAG-SP was weighed and introduced in a 500-mL beaker, then distilled water was poured in the beaker, specifically 100 mL for each gram of adduct.

The dispersion was sonicated in a 2 L ultrasonic bath with power of 260 W for 15 min to produce a homogeneous dispersion of the adduct in water. Then a magnetic stirrer was introduced in the beaker.

*Dilution of NR latex.* A dispersion of 83.33 g of latex in 100 mL of distilled water was prepared as reported above.

*Coagulation of NR/HSAG-SP from the latex.* The diluted NR latex was added to the HSAG-SP dispersion in water. The mixture was left stirring (500 rpm) at room temperature for 1 h. Rubber was then precipitated by adding 100 mL of a 1M sulphuric acid solution. The precipitated rubber was washed and dried following the same procedure described above.

2.5.3. Melt Blending

*Composites with HSAG.* NR, coagulated from the latex, was fed into a 50 cc Brabender® internal mixer and masticated for 1 min at 80 °C and 60 rpm, with 85% as the fill factor. HSAG was then added and mixing was performed for 3 min. Stearic acid, 6PPD, and zinc oxide were added, and mixing was carried out for a further 3 min. The composite was then discharged and fed again into the mixer kept at 45 °C. After 1 min mixing, TBBS and sulphur were added. After 3 min mixing, the composite was discharged.

*Composites with HSAG-SP.* The same procedure was followed, except for filler addition. The total mixing time was kept equivalent.

*NR composite.* The same procedure reported above was adopted, except for HSAG, which was not added. The total mixing time was kept constant.

2.6. Curing

The crosslinking reaction was performed and monitored with a rubber process analyzer (RPA, Alpha Technologies, Hudson, OH, USA) at 170 °C for 10 min. 5 g of crude rubber compound were introduced in the rheometer. Before the crosslinking step, a strain sweep was performed at low deformations (0.1–25% strain), then the sample was kept at 50 °C for ten minutes and subjected to another strain sweep at 50 °C. Subsequently vulcanization was performed during which the torque-time curve, the minimum achievable torque ($M_L$), the maximum achievable torque ($M_H$), the time needed to have a torque equal to $M_L + 1$ ($T_{S1}$), and the time needed to reach 90% of the maximum torque ($T_{90}$) were measured. Acquisitions were performed with an oscillation angle of 6.98% and a frequency of 1.7 Hz.

2.7. Strain Sweep Tests

Shear dynamic-mechanical properties were measured using a rubber process analyser (RPA). A strain sweep was performed on the crude sample at low deformations (0.1–25% strain), then the sample was kept at 50 °C for ten minutes and subjected to another strain sweep at 50 °C before being vulcanized as described in the previous paragraph. After 20 min at 50 °C, shear dynamic-mechanical properties were measured, applying a 0.1–25% strain sweep at a frequency of 1 Hz. The measured properties were shear storage and loss moduli ($G'$, $G''$).

2.8. Morphological Analysis

The dispersion of the carbon allotrope in the NR matrix was investigated by transmission electron microscopy (TEM, Carl Zeiss AG, Oberkochen, Germany) with an 80 kV Zeiss EM900 microscope. Ultrathin cryosections of the cured composites were prepared at −130 °C by using a Leica EM FCS cryo-ultramicrotome (Leica Microsystems, Wetzlar, Germany).

2.9. Dynamic Mechanical Analysis (DMA)

The experiments were conducted with a DMA 861/40N testing device (Mettler Toledo GmbH, Schwerzenbach, Switzerland) in tension mode. The parallel parts between the shoulders of the dumbbell specimens (S2) (thickness of ~2 mm and width of 4 mm) were utilized as the specimens with a clamping distance of 19.5 mm. In a first step, amplitude tests were conducted at room temperature within 1–100 µm identifying the linear viscoelastic range of the material grades. Based on the results, temperature sweep tests were carried out with a dynamic amplitude of 8 µm, a static amplitude of 103%, a frequency of 2 Hz, and a heating rate of 3 K/min within the temperature range of −80 to 50 °C using liquid nitrogen as a cooling agent. Out of the measurements, the storage modulus ($E'$), loss modulus ($E''$), and loss factor (tan$\delta$) were plotted and compared in the investigated temperature range.

### 2.10. Quasi-Static Tensile Tests

The quasi-static tensile experiments were conducted based on DIN 53504 and the dumbbell-shaped test specimens (S2) were punched from the received ~2 mm thick plates. These specimens possess a width of 4 mm and were clamped at a distance of 43 mm. The tests were conducted with a Zwick universal testing machine (Zwick Roell Z001, Test expert, Ulm, Germany) utilizing a 1 kN load cell, at a constant crosshead speed of 200 mm/min according to the testing standard. The non-contact strain measurements based on the digital image correlation (DIC) technique were implemented using two cameras and the Mercury RT software (version 2.5, 2017) (Sobriety s.r.o., Kuřim, Czech Republic). Each material grade was characterized implementing five specimens to ensure reproducibility. Based on the results, stresses at 50 and 100% of strain ($\sigma_{50}$ and $\sigma_{100}$), as well as stress at break ($\sigma_B$) and elongation at break ($\varepsilon_B$) were determined in average values along with standard deviation.

### 2.11. Fracture Tests

The experimental procedure, described in detail in [73], is a single specimen procedure developed to obtain the fracture resistance of rubber mode I condition, and is based on the J-integral parameter [74]. The procedure is summarized here: Single Edge Notched in Tension (SENT) specimens, shown in Figure 1, are used, with the following dimensions: width $W$ = 10 mm, clamps distance $L$ = 30 mm, thickness $B$ = 2 mm, notch length $a_0$ = 3 mm. SENT specimens were tested at 10 mm/min of crosshead rate. During loading, the load- crosshead displacement curve was monitored. J-integral values at the fracture initiation point, $J_c$ (kJ/m$^2$) values, were calculated by Equation (2):

$$J_c = \frac{\eta \cdot U_c}{B \cdot (W - a_0)} \quad (2)$$

where $U_c$ is the energy, the area under the stress-strain curve, up to the initiation point, and $\eta$ is a geometry factor, previously calibrated for the dimensions used in this work (30 × 10 mm$^2$) following the multispecimen procedure described in [73,75]. The $\eta$ values are obtained from the following equation after measuring the actual $a_0$ values from the fracture surfaces:

$$\eta = -1.155 \frac{a_0^2}{W} + 1.8938 \frac{a_0}{W} - 0.0011 \quad (3)$$

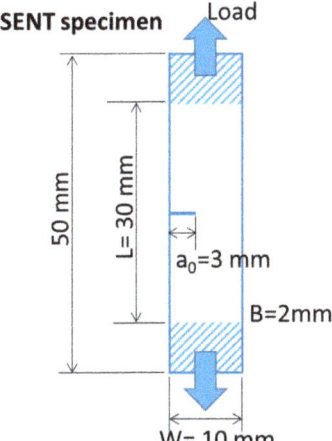

**Figure 1.** Sketch of the specimen used for fracture tests, with notch length ($a_0$), thickness of the specimen (B), and width (W) and length ($L_0$) of the specimen.

The point of fracture initiation was optically monitored by a camera placed in front of the notch taking an image every 2 s, as shown in Figure 2. The black notch surfaces were sputtered before testing by a white talc powder. In this way the new fracture surface can be easily distinguished as a black line at the notch root. Due to the progressive evolution of the fracture process, initiation is conventionally taken when the new fracture surface spans over all the specimen thickness and is 0.1 mm high (see ref. [73]). Five test repetitions were performed for each material.

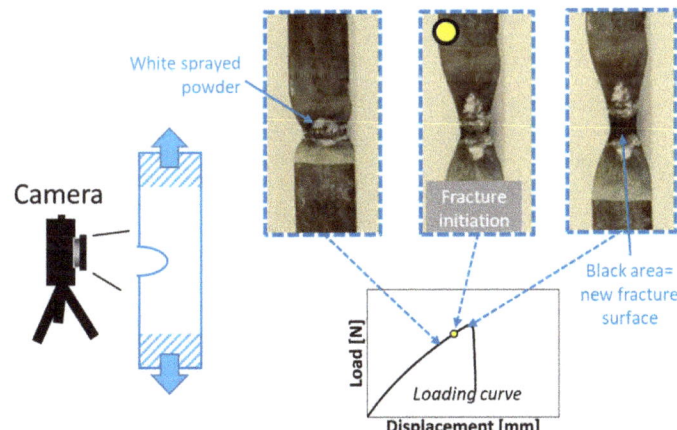

**Figure 2.** Experimental setup for fracture test: camera in front of the notch root, example of frames taken during a fracture test with the notch surfaces of a specimen. In particular, fracture initiation is indicated.

## 3. Results and Discussion

### 3.1. Functionalization of HSAG

The functionalization of HSAG was performed through the reaction with serinol pyrrole. As mentioned in the Introduction, the functionalization process is indeed simple and occurs through the domino reaction shown in Figure 3.

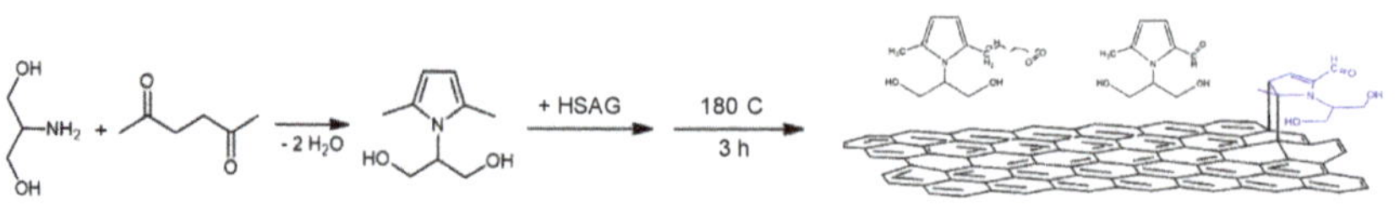

**Figure 3.** Scheme of the domino reaction for the functionalization of HSAG with serinol pyrrole.

Details about the reaction are provided in the Materials and Methods section. In brief, serinol pyrrole is obtained by simply mixing and heating the primary amine (serinol) and the dicarbonyl compound and functionalization occurs by simply mixing the pyrrole compound and the graphitic substrate, then heating the mixture at 180 °C for 3 h. The domino functionalization process occurs via the following steps: adsorption of the pyrrole compound onto the graphitic material, carbocatalyzed oxidation of SP, and cycloaddition reaction with the graphene layers. In this work, SP was obtained by performing the reaction between the diketone and serinol in a flask, isolating the product. However, as already reported [76], the reactants can be mixed on top of the $sp^2$ carbon allotrope. In this case, the first step of the domino process becomes the synthesis of the pyrrole compound. Such functionalization method avoids the above-mentioned harsh reaction conditions typically used for the traditional oxidation-reduction approach. In this work, two functionalization reactions were performed, with two HSAG/SP mass ratios: 10:1 and 10:0.6. The objective was to check the efficiency of the reaction by increasing the amount of the pyrrole compound.

The efficiency of the functionalization reaction was evaluated by means of thermogravimetric analysis (TGA), as described in Materials and Methods, on the HSAG/SP adduct isolated after the reaction and an exhaustive acetone extraction, curves are reported in Figure 4. The detected mass losses are in Table 2.

**Table 2.** Mass losses for HSAG and HSAG-SP adduct, as detected from TGA analysis.

| Sample | | Mass Loss (%) | | | |
|---|---|---|---|---|---|
| | | T < 150 °C | 150 °C < T < 400 °C | 400 °C < T < 700 °C | T > 700 °C |
| HSAG | (a) | 1.3 | 1.7 | 4.6 | 92.4 |
| | (c) | 0.0 | 0.1 | 0.4 | 99.5 |
| HSAG-SP Sample 1 | (b) | 0.9 | 6 | 9.4 | 83.6 |
| | (c) | 1.0 | 5 | 8.5 | 85.5 |
| HSAG-SP Sample 2 | (b) | 0.5 | 2.1 | 4.1 | 93 |
| | (c) | 0.6 | 1.4 | 4 | 94 |

(a) Pristine (b) after the functionalization reaction (c) after washing with acetone (see Materials and Methods section).

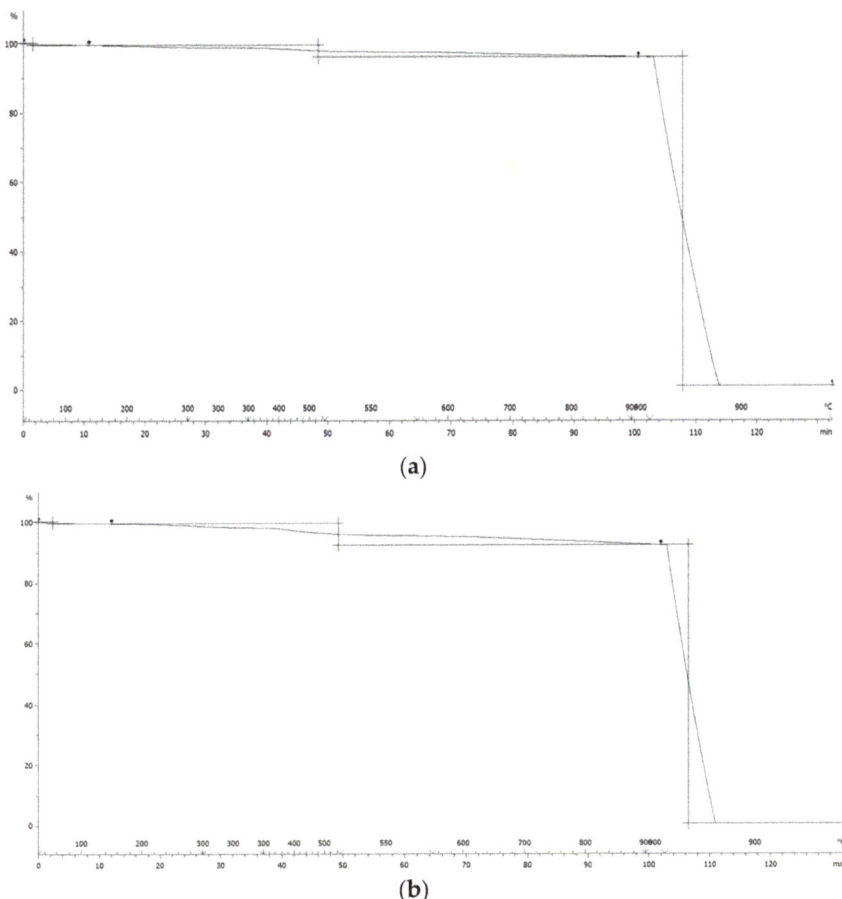

**Figure 4.** Thermogravimetric analysis of: (**a**) Pristine HSAG and, (**b**) HSAG-SP after acetone extraction (Sample 2 in Table 2 below).

Mass losses below 150 °C can be attributed to the removal of the adsorbed water. It is not revealed by washed pristine HSAG and increases with the amount of serinol pyrrole in the adduct. It is worth repeating (see Materials and Methods) that all the functionalization reactions were performed on washed HSAG samples. The mass loss at temperatures above 150 °C, for the washed sample of pristine HSAG can be attributed to alkenylic defects. In the case of HSAG/PyC adducts as well, mass loss is due to the presence of the organic modifier introduced with the functionalization reaction [69]. Such mass loss is indeed appreciable in the samples exhaustively extracted with acetone. Stable adducts were thus formed. The equation and procedure for estimating the functionalization yield was reported above in the Materials and Methods section. High functionalization yields were obtained by using the two levels of modifier, about 10% and 6% by mass: they were 88% and 87% for Sample 1 and Sample 2, respectively. For the preparation of the nanocomposites, Sample 2 (lower amount of SP) was used. This choice was motivated by two main objectives: to verify if such a low level of functionalization was enough to promote the dispersion of HSAG in the NR latex and then in the NR-based composite and then to reduce the interaction of the polar groups with the polymer chains. Indeed, the objective of this work was to investigate the effect of graphene layers on an NR-based composite.

The infrared analysis of HSAG-SP adducts has been discussed elsewhere [70]. In the case of the samples prepared in this work, the typical spectral features of the pyrrole compound (PyC) derivative have been identified.

### 3.2. Preparation and Characterization of Rubber Composites

The same NR grade, coagulated from a latex, was used for the preparation of all the composites. Coagulation was performed of either pristine NR latex or NR/HSAG-SP dispersion. Melt blending was then carried out, adding the ingredients reported in Table 1 above. Three different amounts of HSAG or HSAG-SP were used: 5, 15, and 24 parts per hundred rubber (phr). The presence of the functional group in HSAG-SP was neglected and the same amount of HSAG and HSAG-SP were used. Hence, when discussing the results reported in the following paragraphs, it is important to bear in mind that in the case of composites based on HSAG-SP, graphitic content is slightly lower.

#### 3.2.1. Curing

Sulphur-based crosslinking was performed with a typical recipe, based on sulphur and a sulphenamide. Data from rheometric tests, collected as described in Materials and Methods, are shown in Table 3.

Table 3. Curing data of NR–HSAG and NR–HSAG-SP compounds.

| Composite n. | 1 | 2 | 3 | 4 | 5 | 6 |
|---|---|---|---|---|---|---|
| Filler Type | HSAG | HSAG | HSAG | HSAG-SP | HSAG-SP | HSAG-SP |
| Filler Amount | 5 phr | 15 phr | 24 phr | 5phr | 15 phr | 24 phr |
| $M_L$ (dNm) | 0.3 | 0.8 | 1.3 | 0.6 | 0.9 | 1 |
| $M_H$ (dNm) | 6.1 | 8.4 | 10.2 | 6.4 | 8.2 | 9.1 |
| $M_H - M_L$ (dNm) | 5.8 | 7.6 | 8.9 | 5.8 | 7.3 | 8 |
| $t_{90}$ (min) | 3.9 | 4.0 | 3.8 | 4.3 | 4.2 | 3.8 |
| $t_{S1}$ (dNm) | 2.5 | 2.5 | 2.3 | 2.7 | 2.5 | 2.3 |
| $(M_H - M_L)/(t_{90} - t_{s1})$ | 4.1 | 5.1 | 5.9 | 3.6 | 4.3 | 5.3 |

The values of $M_L$ and of $M_H$ increase, as expected, with the amount of graphite. With pristine HSAG, higher $M_L$ and $M_H$ values were obtained for the largest content of the graphitic filler. The $M_L$ values are usually correlated with the viscosity of the sample. It is also worth commenting that the $M_H$ values are affected by the presence of a filler network when the considered filler content is above its percolation threshold, as the strain amplitude sweep is not large enough to completely disrupt the network. In a previous study [39], the percolation threshold of HSAG in a poly(isoprene) matrix was reported to occur at about 21 phr. The lower viscosity and lower $M_H$ of the composite with HSAG-SP could be attributed to a better dispersion of the filler in the matrix. Differences, if any, are small amongst the values of $t_{s1}$, $t_{90}$ and $(M_H - M_L)/(t_{90} - t_{s1})$, which indicate, respectively, the induction and optimum vulcanization times and the vulcanization rate. It seems thus possible to comment that the functionalizing agent does not appreciably affect the curing kinetics. However, it can be observed that values of $t_{s1}$ and $t_{90}$ remain almost constant for the composites based on HSAG, whereas they decrease, though to a minor extent, in the case of composites based on HSAG-SP. In the field of elastomer composites, it is acknowledged that a sp$^2$ carbon allotrope promotes faster sulphur-based vulcanization [77]. The decrease of $t_{s1}$ and $t_{90}$ for the HSAG-SP composites is thus in line with prior work and could indicate a better interfacial area between the filler and the composite ingredients, and hence a better filler dispersion.

#### 3.2.2. Strain Sweep Tests

Strain sweep experiments were performed, using a rubber process analyzer (RPA), as described in the Materials and Methods section. Storage ($G'$) and loss ($G''$) moduli and Tan Delta$_{max}$ ($G''/G'$)

were measured. The obtained results are reported in Table 4. Low levels of moduli were obtained, as expected, in consideration of the low amount of the graphitic filler. It should be also considered that the nanographite is made of graphene nanoplatelets, disposed parallel to the rotors of the RPA. Composites with lamellar nanographite have been reported to have mechanical anisotropy, revealing an orthotropic and transversally isotropic response: modulus values were observed to be very similar in all directions in the sheet plane and much larger (almost double) in the orthogonal direction [66].

**Table 4.** Storage modulus $G'$ at minimum (0.1%) shear strain amplitude, $\Delta G'(G'_{0.1\%} - G'_{25\%})$, maximum loss modulus $G''$, and maximum Tan Delta$_{max}$ ($G''/G'$) from strain sweep tests.

| Composite n. | 1 | 2 | 3 | 4 | 5 | 6 |
|---|---|---|---|---|---|---|
| Filler Type | HSAG | HSAG | HSAG | HSAG-SP | HSAG-SP | HSAG-SP |
| Filler Amount (phr) | 5 | 15 | 24 | 5 | 15 | 24 |
| $G'_{0.1\%}$ | 0.3 | 0.5 | 0.9 | 0.4 | 0.6 | 0.7 |
| $\Delta G' = (G'_{0.1\%} - G'_{25\%})$ | ~0 | 0.1 | 0.3 | ~0 | 0.1 | 0.2 |
| $G''_{max}$ | 0.02 | 0.03 | 0.06 | 0.02 | 0.03 | 0.05 |
| Tan Delta$_{max}$ | 0.06 | 0.06 | 0.08 | 0.05 | 0.06 | 0.08 |

Values of $\Delta G'$ and Tan Delta$_{max}$ increase with the HSAG content, as commonly observed in filled rubbers. The $\Delta G'(G'_{0.1\%} - G'_{25\%})$ value is taken as an indicator of the so-called Payne Effect, which expresses the non-linearity of the viscoelastic modulus due to the disruption of the filler network formed by the interaction of the filler particles, either directly or mediated by polymer layers [78,79]. Hence, the behavior at low strain of the nanocomposites, based on either HSAG or HSAG-SP, is similar at the same filler content.

3.2.3. Dynamic-Mechanical Tests

The mechanical properties of rapidly varying stress conditions over a broad range of temperatures of the material grades were studied using the DMA method. The storage modulus ($E'$) and loss factor along the temperature range are shown in Figure 5a,b, respectively. The storage modulus reveals the elastic component of material behavior, which is related to the stiffness of the material. The peak of the loss factor corresponds to the glass transition temperature ($T_g$), and the peak intensity, as well as the width, revealing the damping properties of the material [80,81].

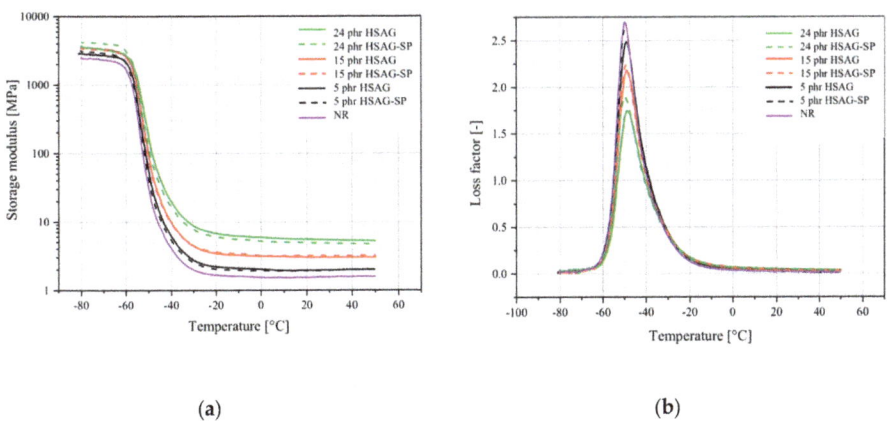

**Figure 5.** Dynamic mechanical analysis of vulcanized NR-based composites containing either HSAG or HSAG-SP (**a**) Storage modulus deviation versus temperature; (**b**) Loss factor deviation versus temperature.

$E'$ is affected by the presence of nanographite, independent of the functionalization: the storage modulus increases with the nanofiller amount, to a larger extent above the glass transition temperature. In the case of 5 phr- and 15 phr-filled nanocomposites, the functionalization with serinol pyrrole only slightly influenced the $E'$ values, whereas a lower $E'$ was obtained at 24 phr with HSAG-SP, even though the difference is still pretty low.

The $T_g$ appears to be substantially unaffected by the nanofiller type and content, except for the nanocomposite containing 24 phr of filler: in the presence of HSAG-SP, the $T_g$ value was about 1.5 °C lower than that obtained with HSAG. The intensity of the loss factor peak clearly depends on nanographite type and amount. Such intensity decreased by increasing the HSAG amount. As to the filler type, larger damping was obtained with HSAG-SP, at all the filler contents.

These findings are in line with what is reported in the literature for composites based on graphene related materials (GRM) and NR. Enhancement of $E'$ with GRM content, to a different extent, was found by using NR latexes; and thermally reduced GO, up to 4 phr (in prevulcanized natural rubber) [45], chemically reduced GO, up to 2 phr [43] or 5 phr [46], GO, up to 0.5 phr [54]. Particular enhancement of $E'$, of two orders of magnitude, was found with GO up to 5% in the composite. In this case, it was commented that GO acted as a physical crosslinker [51]. Indeed, the interpretation for the $E'$ enhancement is based on the hydrodynamic effect promoted by rigid inextensible filler particles and on their interaction with the polymer chains. The larger $E'$ increase, with respect to NR, observed in this work, in comparison with those commented above for composites based on reduced graphene oxides, has to be attributed to the larger content of HSAG (up to 24 phr). To explain the larger $E'$ observed in Figure 6 for the composite with 24 phr HSAG, a lower mobility could be hypothesized for the polymer chains trapped in the network of HSAG aggregates. In fact, the HSAG percolation threshold in a poly(isoprene) matrix was reported to occur at about 21 phr [39]. Less aggregates and more exfoliated layers are present in HSAG-SP based composites, as is shown by TEM pictures (discussed in the text below). It is also worth adding that HSAG-SP was reported to exfoliate in a water suspension [68]. Further explanation for the larger $E'$ of HSAG-based composites could be based on the consideration that graphene layers stacked in a crystalline aggregate are estimated to have larger volume than exfoliated layers. Hence, HSAG aggregates could be expected to give more mechanical reinforcement than exfoliated HSAG.

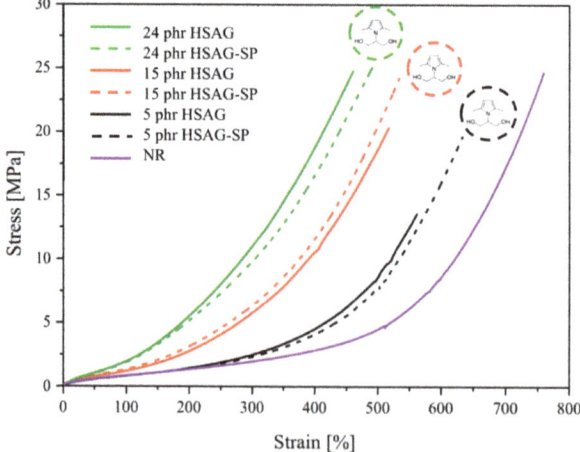

**Figure 6.** Stress-strain curves of vulcanized NR-based composites containing either HSAG or HSAG-SP. The number of each composite is close to the corresponding curve. Moreover, the chemical structure of serinol pyrrole is shown close to the curves of composites based on HSAG-SP.

In almost all the mentioned papers [43,51,54], $T_g$ was found to remain at the same temperature. A shift to slightly higher temperature was found in the presence of a surfactant in Ref. [45], which was commented to improve the interaction between the filler and the polymer chains. The slight reduction of $T_g$ commented above for the HSAG-SP based composite could be interpreted by the larger mobility of the polymer chains, which are in the presence of a larger amount of exfoliated graphene layers.

The reduction of tan delta peak value is a common feature in all the examined papers and is explained by the good interaction between the filler and the polymer chains. According to the literature [80,81], the source of damping is the reversible interaction of polymer chains with the filler surface. A more stable polymer-filler interaction is expected to reduce the loss factor intensity. Hence, it could be commented that a better HSAG-SP dispersion and exfoliation could favor the chain segment motion, thus reducing $T_g$. At the same time, the larger extent of reversible interactions between the HSAG-SP layers and the polymer chains leads to larger damping.

The above findings present the peculiar behaviour of HSAG-SP: functionalization with the polar pyrrole compound allows to improve the dispersion of the filler without establishing a strong interaction with the NR chains; in the case of GO, it leads to higher $T_g$ values.

### 3.2.4. Quasi-Static Tensile Tests

Quasi-static tensile measurements were performed on vulcanized composites, based on HSAG and HSAG-SP. A gum stock, with vulcanized NR and without any filler, was tested for comparison. Data from the measurements, with standard deviation, are given in Table 5 and representative stress-strain curves in Figure 6. Upon adding nanographite (with or without SP) to NR, higher stresses at every strain were obtained. HSAG and HSAG-SP based composites show comparable values of stress for a given value of deformation. As far as the ultimate properties are concerned, the strain at break decreased for all the composites loaded with the graphitic filler. The stress at break of the composites with 15 and 24 phr of graphite has substantially the same value as the NR gum stock, whereas it is lower for the composite with 5 phr, particularly with HSAG as the filler. As a matter of fact, at all the filler loadings, HSAG-SP led to higher values of elongation at break and stress at break, with respect to pristine HSAG.

**Table 5.** Stress at 50% ($\sigma_{50}$) and 100% ($\sigma_{100}$) strain, stress at break, and strain at break of NR/HSAG and NR/HSAG-SP vulcanized samples.

| Composite n. | 0 | 1 | 2 | 3 | 4 | 5 | 6 |
|---|---|---|---|---|---|---|---|
| Filler Type | NR | HSAG | HSAG | HSAG | HSAG-SP | HSAG-SP | HSAG-SP |
| Filler Amount (phr) | 0 | 5 | 15 | 24 | 5 | 15 | 24 |
| $\sigma_{50}$ (Mpa) | 0.57 ± 0.02 | 0.6 ± 0.01 | 0.7 ± 0.10 | 1.1 ± 0.06 | 0.5 ± 0.02 | 0.8 ± 0.03 | 1.0 ± 0.02 |
| $\sigma_{100}$ (Mpa) | 0.8 ± 0.02 | 0.8 ± 0.01 | 1.2 ± 0.06 | 1.9 ± 0.10 | 0.8 ± 0.01 | 1.3 ± 0.03 | 1.9 ± 0.04 |
| $\sigma_{300}$ (Mpa) | 1.93 ± 0.08 | 2.53 ± 0.04 | 6.04 ± 0.31 | 10.5 ± 0.6 | 2.31 ± 0.02 | 6.19 ± 0.22 | 9.91 ± 0.24 |
| $\sigma_B$ (Mpa) | 24.5 ± 3.4 | 13.6 ± 1.9 | 21.4 ± 1.1 | 24.3 ± 1.8 | 19.6 ± 0.8 | 23.1 ± 1.6 | 25.4 ± 0.6 |
| $\varepsilon_B$ (%) | 760 ± 37 | 561.2 ± 3 | 516.5 ± 10 | 468.7 ± 35 | 633.7 ± 18 | 536.1 ± 4 | 493 ± 1 |

As far as the strain at break is concerned, the results in Figure 5 and Table 5 are in line with those reported in the literature, although the level of filler content here explored is much larger. In most literature works based on composites from NR latexes, the strain at break was found to decrease, to a different extent, with the content of GRM, by using thermally [45] or chemically [46] reduced graphene oxide, mechanically exfoliated graphite [53] or GO [48,51]. In these works, the maximum amount of the graphitic material was 4–5 phr (or %). When the GRM content was lower than 1%, the strain at break was substantially unaltered [51,54,55]. An increase of $\varepsilon_B$ was also reported, however with graphene nanosheets prepared through the oxidation of graphite worms and dispersed with the help of ultrafine carbon black and silane as coupling agent and adopting 4 phr as the maximum content of the filler [57]. In the case of composites based on synthetic isoprene rubber and the same HSAG used in the present work and prepared via melt blending, a consistent reduction of $\varepsilon_B$ was observed with a filler content up to 60 phr [44].

The examination of literature stress at break data does not allow a straightforward summary and this hampers the comparison with the results reported in this manuscript. With reduced graphene oxide, the value of stress at break increases and then decreases, moving from 1 to 4 phr in ref [45]; an optimum level of 0.5 phr is reported in [46], where the strain induced crystallization of NR is supposed to play a main role. By using GO, with a filler concentration of 0.5%, an exceptional increase was found for the tensile strength, which decreased then to a value not far from that of NR, at 5% as GO content. The effect of GO on the tensile strength was the increase for a content of 1% and then a decrease to a value lower than that of NR at 5% [51]. The hybrid filler with a graphene layer, ultrafine carbon black, and the silane coupling agent led to a simultaneous and consistent increase of elongation and stress at break [57]. In the work with HSAG in synthetic poly(isoprene) a consistent decrease of both elongation and stress at break were found [44]. The present work shows that the stress at break can have substantially the same value of the NR gum stock even at 24 phr as filler content. The better ultimate properties obtained with HSAG-SP could be attributed to better filer dispersion. This could explain in particular the difference observed at 5 phr as the filler content. It could be hypothesized that, at this filler content, HSAG does not achieve a good dispersion in a composite prepared via melt blending, because of the low shear stress experienced during the mixing. The lower shear stress found also with HSAG-SP appears to be in line with some of the literature reports. It could be speculated that the increase of the tensile strength at larger filler contents could be mainly due to the increased interactions between the filler and the polymer chains. However, further experiments are needed, not to stretch too far inferences not supported by experimental evidences.

### 3.2.5. Mooney Rivlin Plot

The stress–strain curves were converted into the Mooney–Rivlin plots, shown in Figure 7, where the reduced stress given by Equation (4)

$$\sigma^* \; (\sigma^* = \sigma/(\alpha^2 - \alpha^{-1})) \tag{4}$$

was plotted versus the reciprocal of the extension ratio $\alpha$.

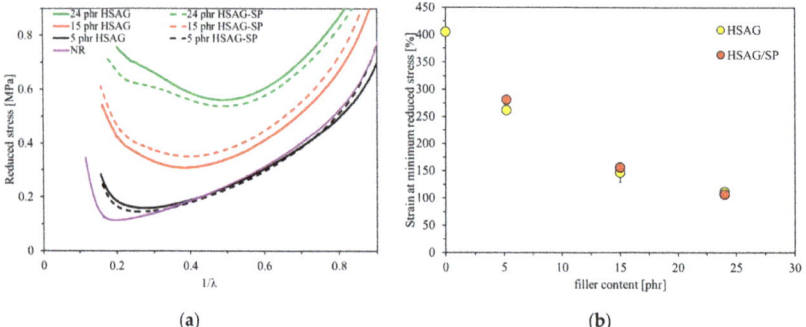

**Figure 7.** (a) Mooney–Rivlin plots of reduced stress versus reciprocal extension ratio of unfilled NR and nanocomposites based on either HSAG or HSAG-SP. Only representative curves are reported. (b) Strain at the minimum reduced stress. This parameter is conventionally taken as indication of the stress upturn point.

The Mooney Rivlin plot reveals the stress upturn, which is considered to be due either to the entanglements which behave as effective crosslinking points at higher elongations. [82] or to the finite extensibility of the network chain segments [83,84] or to the strain induced crystallization [82–88].

Curves in Figure 7 refer to the NR gum stock and the composites based on HSAG and HSAG-SP, with different filler contents. In the graph in Figure 7a, a remarkable decrease in stress at low strain levels can be observed, attributed to the Payne effect in the case of filled rubbers. Furthermore, NR and composites with low amount of filler (5 phr) show a nearly linear region, followed by a stress upturn. The higher the filler content, the less definite is the central linear region of the plot, which turns into a smooth curvature. It appears that the upturn occurs at a lower extension ratio for the HSAG-based nanocomposites: the larger the amount of filler, the lower is the extension ratio. At 5 phr of filler content, the upturn occurs at an appreciably lower strain in the case of HSAG-SP.

In the literature, the Mooney-Rivlin plot has been elaborated upon for NR/GRM composites from latex blending, containing a low amount of GRM. With 1 phr of either GO or chemically reduced GO [89], the upturn occurred at a lower strain than in the case of NR, in particular for the reduced GO. Decrease of stress upturn was found by increasing the amount of reduced GO from 1 to 2 phr [43]. Abrupt upturn was observed with 0.5 phr of reduced GO [46]. These findings were attributed to the strain-induced crystallization, a phenomenon that was documented to be favored by such a level of GRM content [90]. However, when the content of reduced GO was increased up to 5 phr [46], a gentle instead of an abrupt upturn was observed and attributed mainly to limited extensibility of the network chains. In this study, this latter interpretation could be adopted to explain the results of Figure 7. The difference observed between HSAG and HSAG-SP based composites, at 5 phr as the filler content, could be explained by the larger mobility of polymer chains in the latter.

3.2.6. Fracture Tests

In the rubber field, natural rubber is the material of choice to prevent crack propagation. The fracture behaviour of the NR–HSAG and NR–HSAG-SP composites was studied by applying a fracture mechanics approach based on J-testing methodology [73]. Fracture tests allow to evaluate the resistance of rubber in the presence of a defect. Due to the innovative character of the materials and their limited availability, fracture resistance was characterized by the single specimen method based on the J integral parameter, described in Ref. [73] and in Materials and Methods, on quite small specimens. The J integral parameter describes the energy required to produce a unitary fracture surface. Due to the viscoelastic nature of rubber and the size of specimens—smaller than usual—the results are also influenced by energy dissipation in the specimen bulk. However, since all the materials were tested with the same specimen dimensions, the results can be useful to rank the materials on the basis of the energy required to initiate fracture.

Loading curves obtained from fracture tests are reported in Figure 8: Figure 8a–c for composites with filler content, either HSAG (darker curves) or HSAG-SP (lighter curves), equal to 5, 15, and 24 phr, respectively. The yellow dots and red triangles on the curves indicate the points of fracture initiation.

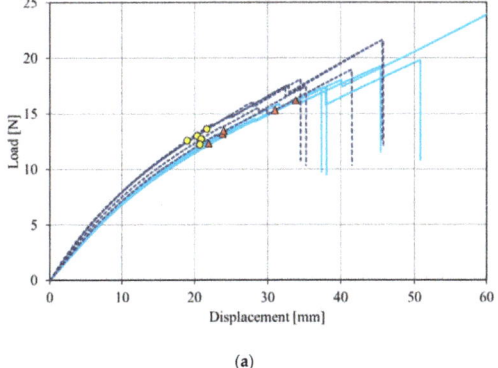

(a)

**Figure 8.** *Cont.*

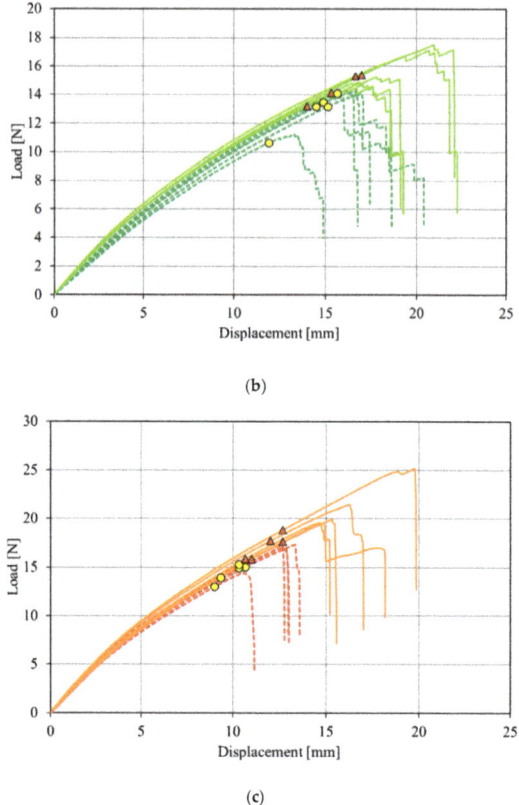

**Figure 8.** Fracture curves for NR-based composites containing 5 (**a**), 15 (**b**), and 24 phr (**c**) graphitic nanofiller. Lighter continuous lines and darker dashed lines refer to composites with HSAG-SP and HSAG, respectively. Dots and triangles on the curves indicate the points of fracture initiation for HSAG and HSAG-SP, respectively.

The elongation before fracture decreases by increasing the amount of the graphitic material for all the composites—the material becomes stiffer. It is worth pointing out that the functionalization of HSAG with SP appears to improve the fracture behaviour: curves with and without SP at the same filler amount are similar, but fracture occurs at larger deformation and with higher values of load in the case of HSAG-SP based composites. This causes slightly higher values of fracture resistance, $J_c$. Average fracture toughness ($J_c$) values obtained from Equation (2) are reported with the corresponding deviations in Table 6.

**Table 6.** $J_c$ values for NR–HSAG and NR–HSAG-SP composites.

| Composite n. | 1 | 2 | 3 | 4 | 5 | 6 |
|---|---|---|---|---|---|---|
| Filler Type | HSAG | HSAG | HSAG | HSAG-SP | HSAG-SP | HSAG-SP |
| Filler Amount (phr) | 5 | 15 | 24 | 5 | 15 | 24 |
| $J_c$ (kJ/m$^2$) | 5.2 ± 0.7 | 3.5 ± 0.3 | 2.6 ± 0.3 | 5.4 ± 0.6 | 4.0 ± 0.4 | 3.6 ± 0.3 |

Composites containing HSAG-SP are tougher; they show higher values of stored energy before fracture initiation with respect to composites containing pristine HSAG, and this difference is significant at 24 phr. It is acknowledged that in tensile tests, undispersed filler particles promote specimen failure. Hence, these experimental findings seem to suggest a better dispersion of HSAG-SP. The difference between HSAG and HSAG-SP based composites appears more relevant for 24 phr, as the filler content. As already reported, the percolation threshold of HSAG in a poly(isoprene) matrix occurs at about 21 phr [39]. Hence, the beneficial effect of a better filler dispersion seems to be more evident above the filler percolation. Similar conclusions were achieved in a recent study [54], which compared the fracture properties of NR/GO nanocomposites via melt and latex compounding. The same fracture method applied in the present work was used, but with different specimen dimensions and at very low filler content (< 0.5 phr). In spite of these differences, an improvement in fracture properties with latex blending due to better nanofiller dispersion was observed. It is reasonable that a polar filler such as GO benefit from the blending in rubber latex. This was also the hypothesis of the origin of this work, inspired also by the chance of using a polar graphite functionalized in a simpler and more sustainable way, with respect to GO.

Values of pure NR are not reported, although the experiments were performed under the same experimental conditions. As a matter of fact, NR specimens were elongated up to about 700% at complete break during fracture tests. At such a high deformation, the specimen completely lost its shape and stress state completely changed. This behaviour can be attributed to the extremely high deformability of NR combined with a high fracture toughness, and to the small size of the specimens, which does not provide enough constraint during the test. Therefore, although a value of fracture toughness equal to 18.1 ± 1.2 kJ/m$^2$ was calculated for NR, this figure was considered to be not reliable and thus not comparable with the other results. It is, however, clear that NR has a higher fracture toughness compared to the other composites, most likely as a consequence of its strain-induced crystallization.

The morphology of the fracture surface was investigated. For neat NR and composites with 5 and 15 phr of both fillers, and composite with 24 phr of HSAG, the fracture propagated straight and perpendicular to the applied load. Only the composite with 24 phr of HSAG-SP showed a different behaviour. Pictures of fractured samples of composites with 24 phr of filler are shown in Figure 8 (side view). The straight fracture propagation visible in the NR/HSAG composite indicates that the crack follows the shortest possible path (Figure 9 right), whereas the change of direction is evident in the specimen with HSAG-SP (Figure 9 left).

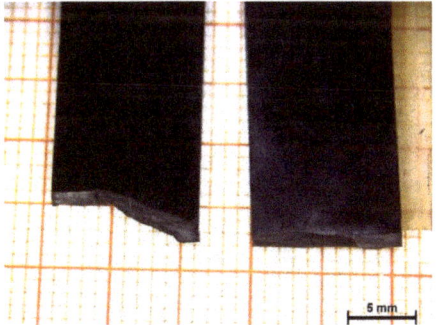

**Figure 9.** Side view of fractured samples containing 24 phr of HSAG-SP (**on the left**) and HSAG (**on the right**).

Fracture propagation deviates in HSAG-SP composites with 24 phr of filler, as if it had encountered obstacles that force the change of direction. This makes the material tougher: a higher amount of energy is required to break the material. To account for such a behaviour, it could be useful to remember that NR gives rise to strain-induced crystallization (SIC). HSAG-SP based samples achieve greater elongation; this could favour the crystallization. Moreover, the dispersion at the nano level of two-dimensional reinforcing fillers has been shown to favor SIC [46,48]. The crystallites could be responsible for the modification of the direction of propagation of the fracture found at 24 phr of filler content.

### 3.2.7. Structure of the Composite; TEM Analysis

Experimental findings from dynamic-mechanical, tensile, and fracture tests suggest that, thanks to functionalization with SP, a good dispersion of the graphitic filler was achieved. To confirm such hypothesis, the structure of cured composites based on HSAG or HSAG-SP was investigated by TEM analysis at different magnifications. Representative micrographs of the composites containing 24 phr of nanofiller are shown in Figure 10.

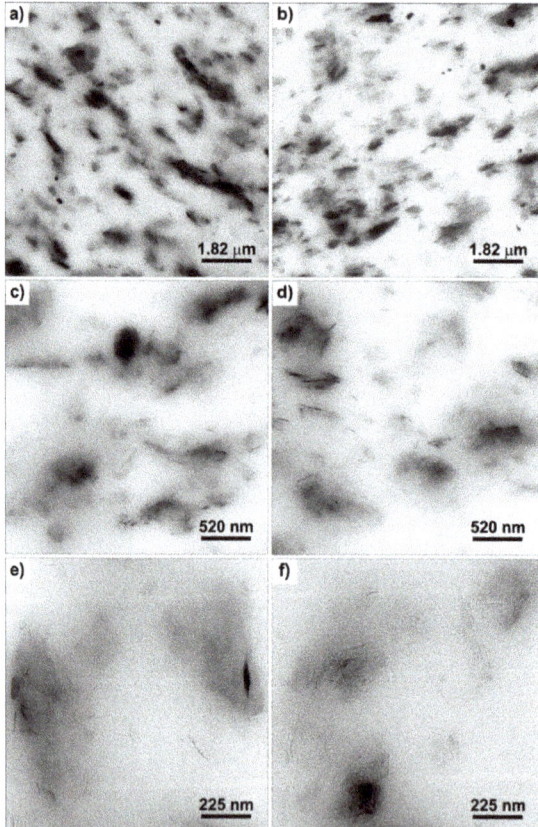

**Figure 10.** TEM micrograph taken at different magnifications of the samples containing 24 phr of HSAG (**a,c,e**) and HSAG-SP (**b,d,f**).

A relatively homogenous distribution of sub-micrometric aggregates was observed at low magnifications (Figure 10a) for both the samples, suggesting the presence of a well-structured nanofiller network. The morphological analysis carried out at different magnifications indicates a similar nanofiller dispersion in the two samples; however, in the sample containing HSAG-SP (Figure 10b,d,f) smaller aggregates and higher amounts of few layers stacks and isolated layers can be observed within the elastomer matrix.

## 4. Conclusions

This work reports a simple and sustainable way of improving the ultimate and fracture properties of a nanocomposite based on natural rubber and a graphitic material. The nanocomposite is characterized by the ultimate dispersion of a nanosized high surface area graphite, achieved by modifying the chemical nature of the carbon allotrope, thus without using an extra amount of mechanical energy. The polarity of HSAG was increased, thanks to functionalization with serinol pyrrole; a masterbatch of HSAG-SP in NR was prepared by coagulating the HSAG-SP dispersion in NR latex. The functionalization with the pyrrole compound, with respect to the traditional approach based on GO, has the following main features: it does not require a reduction step (either thermal or chemical) and the use of hazardous chemicals and harsh reaction conditions, and it leaves the bulk structure of the graphitic substrate substantially unaltered. The polar HSAG-SP does not act as GO, as a crosslinking agent for the polymer chains, but promotes the ultimate dispersion of the graphene layers. Composites, with either pristine HSAG or HSAG-SP, were then prepared with melt blending, using traditional equipment and energy. Good dispersion of the graphene layers appears to be responsible for better ultimate and fracture properties of the sulphur-crosslinked HSAG-SP composites. These results were obtained by using HSAG with a low amount of SP and by preparing the dispersion of HSAG-SP in water and then in NR latex by means of magnetic stirring. Recent studies [34,91] have demonstrated that the shear rate of order of magnitude of $10^4$ s$^{-1}$ can lead to extensive exfoliation of the graphitic aggregates and that the affinity of graphite for the solvent (obtained through functionalization) is highly beneficial. The perspective of the present research proceeds by obtaining a larger exfoliation of graphitic adducts with SP by using appropriate technology, also in view of a scale up, which appears feasible. Through tailor-made functionalization of the graphitic material, with the appropriate pyrrole compound, composites based on different elastomers will be prepared, with evenly dispersed graphene layers and hybrid filler systems, aimed at reproducing the improvements of the properties shown in the present work.

**Author Contributions:** Conceptualization, M.G., S.A. and G.P.; methodology, M.G., S.A.; formal analysis, G.P., S.A. and W.B.; investigation, G.P., S.A., W.B. and L.C.; resources, M.G., S.A., L.C. and B.S.; data curation, G.P., S.A. and W.B.; writing—original draft preparation, M.G., G.P., S.A., L.C., W.B. and B.S.; writing—review and editing, M.G., G.P.; supervision, M.G., B.S.; project administration, M.G. All authors have read and agreed to the published version of the manuscript.

**Funding:** This research was funded under two of the projects (grant number: 854178, grant number: 21647053) by the Austrian Research Promotion Agency (FFG).

**Acknowledgments:** Pirelli Tyre S.p.A. is gratefully acknowledged for supporting the PhD activity of Gea Prioglio. Part of this research work was performed at the Polymer Competence Center Leoben GmbH (PCCL, Austria) and within the COMET-modul "Polymers4Hydrogen" within the framework of the COMET-program of the Federal Ministry for Transport, Innovation and Technology and Federal Ministry for Economy, Family and Youth, with contributions by the Department of Polymer Engineering and Science (Montanuniversitaet Leoben). The PCCL is funded by the Austrian Government and the State Governments of Styria, Lower Austria and Upper Austria.

**Conflicts of Interest:** The authors declare no conflict of interest.

## References

1. Eng, A.H.; Ong, E.L. Hevea natural rubber. In *Handbook of Elastomers*; Bhowmick, A.K., Stephens, H., Eds.; CRC Press: Boca Raton, FL, USA, 2000; pp. 29–60.
2. Barlow, F.W. *Basic Elastomer Technology*; Baranwal, K.C., Stephens, H.L., Eds.; Rubber Division American Chemical Society: Akron, OH, USA, 2001; Chapter 9; pp. 235–258.
3. Natural Rubber Statistic 2018. Available online: http://www.lgm.gov.my/nrstat/Statistics%20Website%202018%20(Jan-Dec).pdf (accessed on 18 March 2020).
4. Chapman, A.V. Natural rubber and NR-based polymers: Renewable materials with unique properties. In Proceedings of the 24th International H.F. Mark-Symposium, 'Advances in the Field of Elastomers & Thermoplastic Elastomers, Vienna, Austria, 15–16 November 2007.

5. Hamed, G.R. Tack and green strength of elastomeric materials. *Rubber Chem. Technol.* **1981**, *54*, 576–595. [CrossRef]
6. Wool, R.P. Molecular Aspects of Tack Rubber. *Rubber Chem. Technol.* **1984**, *57*, 307–319. [CrossRef]
7. Amnuaypornsri, S.; Sakdapipanich, J.; Tanaka, Y. Green strength of natural rubber: The origin of the stress–strain behavior of natural rubber. *J. Appl. Polym. Sci.* **2009**, *111*, 2127–2133. [CrossRef]
8. Thomas, A.G.; Whittle, J.M. Tensile rupture of rubber. *Rubber Chem. Technol.* **1970**, *43*, 222–228. [CrossRef]
9. Gent, A.N.; Zhang, L.-Q. Strain-induced crystallization and strength of rubber. *Rubber Chem. Technol.* **2002**, *75*, 923–934. [CrossRef]
10. Mars, W.V.; Fatemi, A. Factors that Affect the Fatigue Life of Rubber: A Literature Survey. *Rubber Chem. Technol.* **2004**, *77*, 391–412. [CrossRef]
11. Saintier, N.; Cailletaud, G.; Piques, R. Cyclic loadings and crystallization of natural rubber: An explanation of fatigue crack propagation reinforcement under a positive loading ratio. *Mater. Sci. Eng.* **2011**, *528*, 1078–1086. [CrossRef]
12. Hamed, G.R. Molecular Aspects of the Fatigue and Fracture of Rubber. *Rubber Chem. Technol.* **1994**, *67*, 529–536. [CrossRef]
13. Lake, G.J. Fatigue and Fracture of Elastomers. *Rubber Chem. Technol.* **1995**, *68*, 435–460. [CrossRef]
14. Magill, J.H. Crystallization and Morphology of Rubber. *Rubber Chem. Technol.* **1995**, *68*, 507–539. [CrossRef]
15. Huneau, B. Strain-Induced Crystallization of natural rubber: A Review of X-ray diffraction investigations. *Rubber Chem. Technol.* **2011**, *84*, 425–452. [CrossRef]
16. Musto, S.; Barbera, V.; Maggio, M.; Mauro, M.; Guerra, G.; Galimberti, M. Crystallinity and Crystalline Phase Orientation of Poly(1,4-cis-isoprene) from Hevea brasilliensis and Taraxacum kok-saghyz. *Polym. Adv. Technol.* **2016**, *27*, 1082–1090. [CrossRef]
17. Musto, S.; Barbera, V.; Guerra, G.; Galimberti, M. Processing and strain induced crystallization and reinforcement under strain of poly(1,4-cis-isoprene) from Ziegler-Natta catalysis, hevea brasiliensis, taraxacum kok-saghyz and partenium argentatum. *Adv. Ind. Eng. Polym. Res.* **2019**, *2*, 1–12. [CrossRef]
18. Jones, K.P. The paradoxical nature of natural rubber. *Kautsch. Gummi Kunstst.* **2000**, *53*, 735–742.
19. Medalia, A.I.; Kraus, G. *The Science and Technology of Rubber*, 2nd ed.; Mark, J.E., Erman, B., Eirich, F.R., Eds.; Elsevier Academic Press: Berlin, Germany, 1994; Chapter 8; pp. 387–418.
20. Donnet, J.B.; Custodero, E. *The Science and Technology of Rubber*, 3rd ed.; Mark, J.E., Erman, B., Eirich, F.R., Eds.; Elsevier Academic Press: Berlin, Germany, 2005; Chapter 8; pp. 367–400.
21. Donnet, J.B. *Carbon Black: Science and Technology*; CRC Press: Boca Raton, FL, USA, 1993.
22. Kroto, H.W.; McKay, K. The formation of quasi-icosahedral spiral shell carbon particles. *Nature* **1988**, *331*, 328–331. [CrossRef]
23. Iijima, S.; Ichihashi, T. Single-shell carbon nanotubes of 1-nm diameter. *Nature* **1993**, *363*, 603–605. [CrossRef]
24. Bethune, D.S.; Kiang, C.H.; de Vries, M.S.; Gorman, G.; Savoy, R.; Vazquez, J.; Beyers, R. Cobalt-catalysed growth of carbon nanotubes with single-atomic-layer walls. *Nature* **1993**, *363*, 605–607. [CrossRef]
25. Iijima, S. Helical microtubules of graphitic carbon. *Nature* **1991**, *354*, 56–58. [CrossRef]
26. Monthioux, M.; Kuznetsov, V.L. Who should be given the credit for the discovery of carbon nanotubes? *Carbon* **2006**, *44*, 1621–1623. [CrossRef]
27. Novoselov, K.S.; Geim, A.K.; Morozov, S.V.; Jiang, D.; Zhang, Y.; Dubonos, S.V.; Grigorieva, I.V.; Firsov, A.A. Electric Field Effect in Atomically Thin Carbon Films. *Science* **2004**, *306*, 666–669. [CrossRef]
28. Geim, A.K.; Novoselov, K.S. The rise of graphene. *Nat. Mater.* **2007**, *6*, 183–191. [CrossRef] [PubMed]
29. Allen, M.J.; Tung, V.C.; Kaner, R.B. Honeycomb carbon: A review of graphene. *Chem. Rev.* **2010**, *110*, 132–145. [CrossRef] [PubMed]
30. Zhu, Y.W.; Murali, S.; Cai, W.; Li, X.; Suk, J.W.; Potts, J.R.; Ruoff, R.S. Graphene and Graphene Oxide: Synthesis, Properties, and Applications. *Adv. Mater.* **2010**, *22*, 3906–3924. [CrossRef] [PubMed]
31. Geng, Y.; Wang, S.J.; Kim, J.-K. Preparation of graphite nanoplatelets and graphene sheets. *J. Colloid Interface Sci.* **2009**, *336*, 592–598. [CrossRef] [PubMed]
32. Kavan, L.; Yum, J.H.; Gratzel, M. Optically Transparent Cathode for Dye-Sensitized Solar Cells Based on Graphene Nanoplatelets. *ACS Nano* **2011**, *5*, 165–172. [CrossRef]
33. Nieto, A.; Lahiri, D.; Agarwal, A. Synthesis and properties of bulk graphene nanoplatelets consolidated by spark plasma sintering. *Carbon* **2012**, *50*, 4068–4070. [CrossRef]

34. Paton, K.R.; Varrla, E.; Backes, C.; Smith, R.J.; Khan, U.; O'Neill, A.; Boland, C.; Lotya, M.; Istrate, O.M.; King, P.; et al. Scalable production of large quantities of defect-free few-layer graphene by shear exfoliation in liquids. *Nat. Mater.* **2014**, *13*, 624–630. [CrossRef]
35. Soldano, C.; Mahmood, A.; Dujardin, E. Production, properties and potential of graphene. *Carbon* **2010**, *48*, 2127–2150. [CrossRef]
36. Kauling, A.P.; Seefeldt, A.T.; Pisoni, D.P.; Pradeep, R.C.; Bentini, R.; Oliveira, R.V.; Castro Neto, A.H. The Worldwide Graphene Flake Production. *Adv. Mater.* **2018**, *30*, 1803784. [CrossRef]
37. Potts, J.R.; Shankar, O.; Du, L.; Ruoff, R.S. Processing–morphology–property relationships and composite theory analysis of reduced graphene oxide/natural rubber nanocomposites. *Macromolecules* **2012**, *45*, 6045–6055. [CrossRef]
38. Potts, J.R.; Shankar, O.; Murali, S.; Du, L.; Ruoff, R.S. Latex and two-roll mill processing of thermally-exfoliated graphite oxide/natural rubber nanocomposites. *Compos. Sci. Technol.* **2013**, *74*, 166–172. [CrossRef]
39. Galimberti, M.; Kumar, V.; Coombs, M.; Cipolletti, V.; Agnelli, S.; Pandini, S.; Conzatti, L. Filler networking of a nanographite with a high shape anisotropy and synergism with carbon balck in poly(1,4-cis-isoprene)–based nanocomposites. *Rubber Chem. Technol.* **2014**, *87*, 197–218. [CrossRef]
40. Tang, Z.; Zhang, L.; Feng, W.; Guo, B.; Liu, F.; Jia, D. Rational design of graphene surface chemistry for high-performance rubber/graphene composites. *Macromolecules* **2014**, *47*, 8663–8673. [CrossRef]
41. Luo, Y.; Zhao, P.; Yang, Q.; He, D.; Kong, L.; Peng, Z. Fabrication of conductive elastic nanocomposites via framing intact interconnected graphene networks. *Compos. Sci. Technol.* **2014**, *100*, 143–151. [CrossRef]
42. Araby, S.; Meng, Q.; Zhang, L.; Kang, H.; Majewski, P.; Tang, Y.; Ma, J. Electrically and thermally conductive elastomer/graphene nanocomposites by solution mixing. *Polymer* **2014**, *55*, 201–210. [CrossRef]
43. Zhan, Y.; Wu, J.; Xia, H.; Yan, N.; Fei, G.; Yuan, G. Dispersion and exfoliation of graphene in rubber by an ultrasonically-assisted latex mixing and in situ reduction process. *Macromol. Mater. Eng.* **2011**, *296*, 590–602. [CrossRef]
44. Kumar, V.; Hanel, T.; Giannini, L.; Galimberti, M.; Giese, U. Graphene reinforced synthetic isoprene rubber nanocomposites. *KGK Kautsch. Gummi Kunstst.* **2014**, *67*, 38–46.
45. Aguilar-Bolados, H.; Brasero, J.; López-Manchado, M.A.; Yazdani-Pedram, M. High performance natural rubber/thermally reduced graphite oxide nanocomposites by latex technology. *Compos. Part B Eng.* **2014**, *67*, 449–454. [CrossRef]
46. Xing, W.; Wu, J.; Huang, G.; Li, H.; Tang, M.; Fu, X. Enhanced mechanical properties of graphene/natural rubber nanocomposites at low content. *Polym. Int.* **2014**, *63*, 1674–1681. [CrossRef]
47. Papageorgiou, D.G.; Kinloch, I.A.; Young, R.J. Graphene/elastomer nanocomposites. *Carbon* **2015**, *95*, 460–484. [CrossRef]
48. Dong, B.; Liu, C.; Zhang, L.; Wu, Y. Preparation, fracture, and fatigue of exfoliated graphene oxide/natural rubber composites. *Rsc. Adv.* **2015**, *5*, 17140–17148. [CrossRef]
49. Kumar, V.; Scotti, R.; Hanel, H.; Giannini, L.; Galimberti, M.; Giese, U. Influence of Nanographite Surface Area on mechanical Reinforcement of Nanocomposites based on Poly (styrene-co-butadiene). *KGK-Kautsch. Gummi Kunstst.* **2016**, *69*, 33–39.
50. Dong, B.; Zhang, L.; Wu, Y. Highly conductive natural rubber–graphene hybrid films prepared by solution casting and in situ reduction for solvent-sensing application. *J. Mater. Sci.* **2016**, *51*, 10561–10573. [CrossRef]
51. Zhang, C.; Zhai, T.; Dan, Y.; Turng, L.S. Reinforced natural rubber nanocomposites using graphene oxide as a reinforcing agent and their in situ reduction into highly conductive materials. *Polym. Compos.* **2017**, *38*, E199–E207. [CrossRef]
52. Zhan, Y.; Meng, Y.; Li, Y. Electric heating behavior of flexible graphene/natural rubber conductor with self-healing conductive network. *Mater. Lett.* **2017**, *192*, 115–118. [CrossRef]
53. George, G.; Sisupal, S.B.; Tomy, T.; Pottammal, B.A.; Kumaran, A.; Suvekbala, V.; Ragupathy, L. Thermally conductive thin films derived from defect free graphene-natural rubber latex nanocomposite: Preparation and properties. *Carbon* **2017**, *119*, 527–534. [CrossRef]
54. Berki, P.; László, K.; Tung, N.T.; Karger-Kocsis, J. Natural rubber/graphene oxide nanocomposites via melt and latex compounding: Comparison at very low graphene oxide content. *J. Reinf. Plast. Compos.* **2017**, *36*, 808–817. [CrossRef]

55. Lim, L.P.; Juan, J.C.; Huang, N.M.; Goh, L.K.; Leng, F.P.; Loh, Y.Y. Enhanced tensile strength and thermal conductivity of natural rubber graphene composite properties via rubber-graphene interaction. *Mater. Sci. Eng. B* **2019**, *246*, 112–119. [CrossRef]
56. Bokobza, L. Natural Rubber Nanocomposites: A Review. *Nanomaterials* **2019**, *9*, 12. [CrossRef]
57. Qin, H.; Deng, C.; Lu, S.; Yang, Y.; Guan, G.; Liu, Z.; Yu, Q. Enhanced mechanical property, thermal and electrical conductivity of natural rubber/graphene nanosheets nanocomposites. *Polym. Compos.* **2020**, *41*, 1299–1309. [CrossRef]
58. Hao, S.; Wang, J.; Lavorgna, M.; Fei, G.; Wang, Z.; Xia, H. Constructing 3D Graphene Network in Rubber Nanocomposite via Liquid-Phase Redispersion and Self-Assembly. *ACS Appl. Mater. Interfaces* **2020**, *12*, 9682–9692. [CrossRef] [PubMed]
59. Grunert, F.; Wehmeier, A.; Blume, A. New Insights into the Morphology of Silica and Carbon Black Based on Their Different Dispersion Behavior. *Polymers* **2020**, *12*, 567. [CrossRef] [PubMed]
60. Hummers, W.S.; Offeman, R.E. Preparation of graphitic oxide. *J. Am. Chem. Soc.* **1958**, *80*, 1339. [CrossRef]
61. He, H.; Klinowski, J.; Forster, M.; Lerf, A. Density functional theory study of atomic oxygen, $O_2$ and $O_3$ adsorptions on the H-capped (5, 0) single-walled carbon nano-tube. *Chem. Phys. Lett.* **1998**, *287*, 53–56. [CrossRef]
62. Lerf, A.; He, H.; Forster, M.; Klinowski, J.J. Structure of graphite oxide revisited. *Phys. Chem. B* **1998**, *102*, 4477. [CrossRef]
63. Stankovich, S.; Piner, R.D.; Nguyen, S.T.; Ruoff, R.S. Synthesis and exfoliation of isocyanate-treated graphene oxide nanoplatelets. *Carbon* **2006**, *44*, 3342. [CrossRef]
64. Stankovich, S.; Dikin, D.A.; Piner, R.D.; Kohlhaas, K.A.; Kleinhammes, A.; Jia, Y.; Wu, Y.; Nguyen, S.T.; Ruoff, R.S. Synthesis of graphene-based nanosheets via chemical reduction of exfoliated graphite oxide. *Carbon* **2007**, *45*, 1558. [CrossRef]
65. Pei, S.; Cheng, H.-M. The reduction of graphene oxide. *Carbon* **2012**, *50*, 3210–3228. [CrossRef]
66. Mauro, M.; Cipolletti, V.; Galimberti, M.; Longo, P.; Guerra, G. Chemically reduced graphite oxide with improved shape anisotropy. *J. Phys. Chem. C* **2012**, *116*, 24809–24813. [CrossRef]
67. Agnelli, S.; Pandini, S.; Torricelli, F.; Romele, P.; Serafini, A.; Barbera, V.; Galimberti, M. Anisotropic properties of elastomeric nanocomposites based on natural rubber and sp2 carbon allotropes. *eXPRESS Polym. Lett.* **2018**, *12*, 713–730. [CrossRef]
68. Barbera, V.; Guerra, S.; Brambilla, L.; Maggio, M.; Serafini, A.; Conzatti, L.; Vitale, A.; Galimberti, M. Carbon papers and aerogels based on graphene layers and chitosan: Direct preparation from high surface area graphite. *Biomacromolecules* **2017**, *18*, 3978–3991. [CrossRef] [PubMed]
69. Barbera, V.; Bernardi, A.; Palazzolo, A.; Rosengart, A.; Brambilla, L.; Galimberti, M. Facile and sustainable functionalization of graphene layers with pyrrole compounds. *Pure Appl. Chem.* **2018**, *90*, 253–270. [CrossRef]
70. Galimberti, M.; Barbera, V.; Guerra, S.; Bernardi, A. Facile functionalization of $sp^2$ carbon allotropes with a biobased Janus molecule. *Rubber Chem. Technol.* **2017**, *90*, 285–307. [CrossRef]
71. Barbera, V.; Brambilla, L.; Milani, A.; Palazzolo, A.; Castiglioni, C.; Vitale, A.; Bongiovanni, R.; Galimberti, M. Domino reaction for the sustainable functionalization of few-layer graphene. *Nanomaterials* **2019**, *9*, 44. [CrossRef]
72. Yang, D.; Kong, X.; Ni, Y.; Ruan, M.; Huang, S.; Shao, P.; Zhang, L. Improved mechanical and electrochemical properties of XNBR dielectric elastomer actuator by poly (dopamine) functionalized graphene nano-sheets. *Polymers* **2019**, *11*, 218. [CrossRef]
73. Ramorino, G.; Agnelli, S.; De Santis, R.; Riccò, T. Investigation of fracture resistance of natural rubber/clay nanocomposites by J.-testing. *Eng. Fract. Mech.* **2010**, *77*, 1527–1536. [CrossRef]
74. Begley, J.A.; Landes, J.D. The J integral as a fracture criterion. In Fracture Toughness: Part II. *ASTM Int.* **1972**. [CrossRef]
75. Agnelli, S.; Baldi, F.; Bignotti, F.; Salvadori, A.; Peroni, I. Fracture characterization of hyperelastic polyacrylamide hydrogels. *Eng. Fract. Mech.* **2018**, *203*, 54–65. [CrossRef]
76. Barbera, V.; Citterio, A.; Leonardi, G.; Sebastiano, R.; Shisodia, S.U.; Valerio, A.M. Process for the Synthesis of 2-(2,5-Dimethyl-1H-Pyrrol-1-yl)-1,3-Propanediol and its Substituted Derivatives. U.S. Patent EP3154939B1, 25 June 2019.

77. Musto, S.; Barbera, V.; Cipolletti, V.; Citterio, A.; Galimberti, M. Master curves for the sulphur assisted crosslinking reaction of natural rubber in the presence of nano- and nano-structured sp2 carbon allotropes. *eXPRESS Polym. Lett.* **2017**, *11*, 435–448. [CrossRef]
78. Payne, A.R. The dynamic properties of carbon black-loaded natural rubber vulcanizates. Part I. *J. Appl. Polym. Sci.* **1962**, *6*, 57. [CrossRef]
79. Robertson, C.G. Dynamic Mechanical Properties, Section 3: Rubbers and Elastomers. In *Encyclopedia of Polymeric Nanomaterials*; Kobayashi, S., Müllen, K., Eds.; Springer-Verlag: Berlin/Heidelberg, Germany, 2015.
80. Kohls, D.J.; Beaucage, G. Rational design of reinforced rubber. *Curr. Opin. Solid State Mater. Sci.* **2002**, *6*, 183–194. [CrossRef]
81. Balasooriya, W.; Schrittesser, B.; Pinter, G.; Schwarz, T.; Conzatti, L. The Effect of the Surface Area of Carbon Black Grades on HNBR in Harsh Environments. *Polymers* **2019**, *11*, 61. [CrossRef] [PubMed]
82. López-Manchado, M.A.; Valentín, J.L.; Carretero, J.; Barroso, F.; Arroyo, M. Rubber network in elastomer nanocomposites. *Eur. Polym. J.* **2007**, *43*, 4143–4150. [CrossRef]
83. Treloar, L.R.G. The photoelastic properties of short-chain molecular networks. *Trans. Faraday Soc.* **1954**, *50*, 881–896. [CrossRef]
84. Isihara, A.; Hashitsume, N.; Tatibana, M. Statistical Theory of Rubber-Like Elasticity. IV. (Two-Dimensional Stretching). *J. Chem. Phys.* **1951**, *19*, 1508–1512. [CrossRef]
85. Flory, P.J. Effects of Molecular Structure on Physical Properties of Butyl Rubber. *Ind. Eng. Chem.* **1946**, *38*, 417–436. [CrossRef]
86. Toki, S.; Fujimaki, T.; Okuyama, M. Strain-induced crystallization of natural rubber as detected real-time by wide-angle X-ray diffraction technique. *Polymer* **2000**, *41*, 5423–5429. [CrossRef]
87. Nie, Y.; Huang, G.; Qu, L.; Wang, X.; Weng, G.; Wu, J. New insights into thermodynamic description of strain-induced crystallization of peroxide cross-linked natural rubber filled with clay by tube model. *Polymer* **2011**, *52*, 3234–3242. [CrossRef]
88. Marano, C.; Boggio, M.; Cazzoni, E.; Rink, M. Fracture phenomenology and toughness of filled natural rubber compounds via the pure shear test specimen. *Rubber Chem. Technol.* **2014**, *87*, 501–515. [CrossRef]
89. Li, F.; Yan, N.; Zhan, Y.; Fei, G.; Xia, H. Probing the reinforcing mechanism of graphene and graphene oxide in natural rubber. *J. Appl. Polym. Sci.* **2013**, *129*, 2342–2351. [CrossRef]
90. Ozbas, B.; Toki, S.; Hsiao, B.S.; Chu, B.; Register, R.A.; Aksay, I.A.; Adamson, D.H. Strain-induced crystallization and mechanical properties of functionalized graphene sheet-filled natural rubber. *J. Polym Sci. Part B Polym. Phys.* **2012**, *50*, 718–723. [CrossRef]
91. Karagiannidis, P.G.; Hodge, S.A.; Lombardi, L.; Tomarchio, F.; Decorde, N.; Milana, S.; Leary, R.K. Microfluidization of graphite and formulation of graphene-based conductive inks. *ACS Nano* **2017**, *11*, 2742–2755. [CrossRef] [PubMed]

© 2020 by the authors. Licensee MDPI, Basel, Switzerland. This article is an open access article distributed under the terms and conditions of the Creative Commons Attribution (CC BY) license (http://creativecommons.org/licenses/by/4.0/).

Article

# Fatigue Life Assessment of Filled Rubber by Hysteresis Induced Self-Heating Temperature

Wenbo Luo [1,2,*], Youjian Huang [2,3], Boyuan Yin [4], Xia Jiang [2] and Xiaoling Hu [1,2,*]

1. Hunan Key Laboratory of Geomechanics and Engineering Safety, Xiangtan University, Xiangtan 411105, China
2. College of Civil Engineering and Mechanics, Xiangtan University, Xiangtan 411105, China; byhyj@21cn.com (Y.H.); jiangxia127@163.com (X.J.)
3. Zhuzhou Times New Materials Technology Co. Ltd., Zhuzhou 412001, China
4. School of Civil Engineering, Hunan University of Science and Technology, Xiangtan 411201, China; yinboyuanxtu@163.com
* Correspondence: luowenbo@xtu.edu.cn (W.L.); huxiaoling@xtu.edu.cn (X.H.); Tel.: +86-731-58298659 (W.L.); +86-731-58293084 (X.H.)

Received: 18 February 2020; Accepted: 5 April 2020; Published: 7 April 2020

**Abstract:** As a viscohyperelastic material, filled rubber is widely used as a damping element in mechanical engineering and vehicle engineering. Academic and industrial researchers commonly need to evaluate the fatigue life of these rubber components under cyclic load, quickly and efficiently. The currently used method for fatigue life evaluation is based on the S–N curve, which requires very long and costly fatigue tests. In this paper, fatigue-to-failure experiments were conducted using an hourglass rubber specimen; during testing, the surface temperature of the specimen was measured with a thermal imaging camera. Due to the hysteresis loss during cyclic deformation, the temperature of the material was found to first rise and then level off to a steady state temperature, and then it rose sharply again as failure approached. The S–N curve in the traditional sense was experimentally determined using the maximum principal strain as the fatigue parameter, and a relationship between the steady state temperature increase and the maximum principal strain was then established. Consequently, the steady state temperature increase was connected with the fatigue life. A couple of thousand cycles was sufficient for the temperature to reach its steady state value during fatigue testing, which was less than one tenth of the fatigue life, so the fatigue life of the rubber component could be efficiently assessed by the steady state temperature increase.

**Keywords:** fatigue life; filled rubber; hysteresis loss; temperature increase; S–N curve

## 1. Introduction

Because of their hyperelasticity and the energy damping behavior of elastomeric materials, rubber damping elements, such as V-springs, tapered springs, air springs, track dampers, engine mounting pads, joint bushings, etc., are widely used in mechanical engineering, aerospace engineering and vehicle engineering [1]. Due to the diversity of their structural forms, complex viscohyperelasticity of the rubber material and significant physical and geometric nonlinear behaviors under external forces, it is very difficult to predict the fatigue life of these rubber dampers.

Currently, for practical evaluation, fatigue failure tests are carried out on structurally similar prototypes or real structures by applying the designated load spectrum to obtain the S–N curve, which relates the fatigue life with a certain mechanical quantity. The mechanical quantities concerned are usually based on stress (e.g., the maximum stress or stress amplitude) [2,3]; strain (e.g., the maximum strain or strain amplitude) [4,5]; or energy (e.g., the strain energy release rate and the cracking energy density, etc.) [6–8]. For testing convenience, the S–N curve is usually established by mechanical parameters based on stress or strain. Obviously, the fatigue life data are only valid for the specific structures tested and are not applicable to other structures. As a consequence, testing requires a considerable time and number of specimens. To reduce this cost, it is necessary to find alternative fatigue control parameters rather than stress- or strain-based quantities. In recent years, rapid estimation methods for mean fatigue limits of metallic materials have been developed based on temperature measurements [9]. Some recent publications [10–13] proved that the characterization of the fatigue properties of fiber-reinforced thermoplastics can be very much accelerated by the use of the "heat build-up" approach. Le Saux and Marco, et al. [14] tried to use this method for rubber-like materials. They linked the temperature rise to the principal maximum strain, and discussed the relationship between the thermal measurements and the fatigue properties of 15 industrial materials. However, there are few models that bridge the heat build-up with the fatigue life of rubber materials.

The stress–strain curve of rubber materials under cyclic loading shows a hysteresis loop due to viscoelasticity, and the area of the loop represents the energy loss per unit volume in a deformation cycle. The loss of energy eventually dissipates into heat. When the heat is not allowed to flow out to the environment in time, the temperature of the material rises [15–18], showing a sharp increase as failure approaches. Such a sudden rise in temperature can be regarded as the precursor to fatigue failure. Therefore, the objective of this paper is to correlate the fatigue life with the self-heating temperature increase, and to develop an efficient method for evaluating the fatigue life of rubber structures.

## 2. Experiments

### 2.1. Materials and Specimens

The rubber material used for the tests was provided by Zhuzhou Times New Material Technology Co., Ltd. in Zhuzhou, China. The formulation of the rubber compounds was as follows: 100 phr Thailand RSS3 natural rubber, 20 phr N550 carbon black, 10 phr zinc oxide, 5 phr antioxidant, 2.5 phr sulfur, 2 phr stearic acid, 2 phr wax, 2 phr solid coumarone resin, 1.4 phr vulcanization activator. Hourglass rubber specimens, 24 mm long and with a minimum diameter of 14.6 mm, were used in the fatigue failure tests.

### 2.2. Fatigue Tests and Temperature Measurements

The fatigue-to-failure experiments were conducted on the hourglass specimens using an electromagnetic dynamic testing machine (CARE M-3000, CARE Measurement & Control Co. Ltd., Tianjin, China) in force control mode at room temperature. The sinusoidally varying loads were applied to the specimens at a frequency of 5 Hz and with a load ratio of 0; that is, the minimum load was fixed to be 0N and the maximum load varied from 250 N to 400 N in the four independent fatigue tests. The cycles to failure were recorded for each specimen. Duplicate tests were conducted for each loading case. In order to quantify the hysteresis dissipation, the surface temperature of the specimen was measured with a FLIR ThermaCAM SC3000 thermal imaging camera (FLIR Systems Inc., Orlando, FL, USA) during the testing. The test setup is shown in Figure 1.

The maximum temperature values on the surface of the hourglass specimens were recorded during the fatigue-to-failure experiments by the thermal imaging camera. The discrepancies between duplicate measurements were small in all cases. Figure 2 shows the measured temperature evolution curves for four loading cases. It is obvious that the self-heating temperature increase was dependent on the loading conditions. The surface temperature rose rapidly during the first 3000–4000 cycles, and subsequently leveled off at a steady state temperature $T_\infty$, until it rose again sharply as failure approached. Fatigue failure of rubber material often occurs via crack propagation. At the moment of material fatigue failure, the crack growth rate in the material suddenly increases to infinity, that is to say, in a very short period of time, a large amount of energy is released due to the crack growth, which results in a sharp increase in the material temperature. Thus, the sharp increase in temperature can be regarded as a precursor to fatigue failure. It is also clear that the steady state temperature increases $\theta_\infty$

($=T_\infty - T_0$, where $T_0$ is the initial temperature of the specimen) are different under various loading conditions, therefore $\theta_\infty$ reflects the fatigue life in another manner, and can be considered a promising and alternative fatigue parameter.

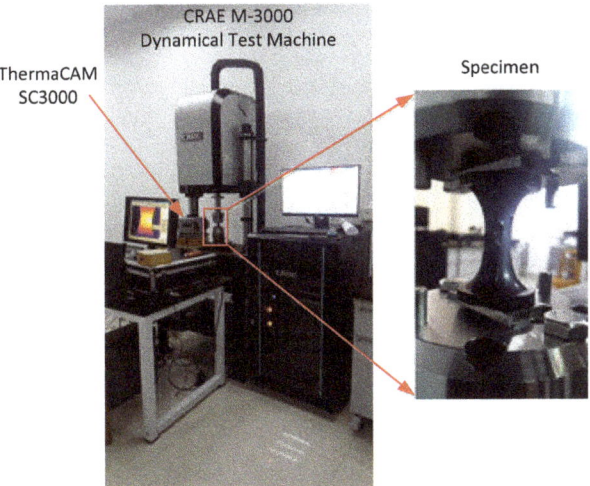

**Figure 1.** Test setup for fatigue and infrared thermal imaging measurement.

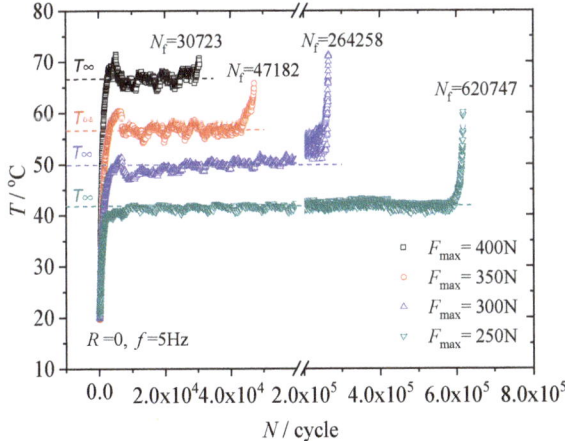

**Figure 2.** Surface temperature evolution of the specimen.

## 3. Modeling and Discussions

### 3.1. Fatigue Damage Parameter Determination

The maximum principal strain $\varepsilon_{\max}$ was selected to be the fatigue parameter for constructing the S–N curve, as done in most studies in the literature [4,5,19]. To obtain the $\varepsilon_{\max}$ of the hourglass specimen under different force control fatigue loadings, a finite element simulation was employed, in which the material constitutive model was essential. As the material constitutive model determined from a single deformation mode experiment may not describe the mechanical response in other deformation modes, tests under three basic deformation modes are usually required in order to accurately establish the true constitutive model of rubber materials: simple tension (ST), equal-biaxial tension (ET) and planar tension (PT). Due to the incompressibility of the material, the ET of a rubber specimen creates a state of strain equivalent to pure compression, which can be accomplished by stretching the circumference of a circular specimen in 16 directions in a plane [20]. The PT test provides a state of pure shear in the specimen at a 45° angle to the stretching direction because of the perfectly lateral constrain, which can be easily performed on a universal tensile testing machine using a special fixture.

The stress–stretch data of the filled rubber material in ST, ET and PT tests at 23 °C are shown in Figure 3. Such behavior is often modeled via hyperelastic idealization. By fitting a hyperelastic model to the test data, the constitutive model of the material can be determined. Many hyperelastic models have been developed. The relatively simple neo-Hookean and Mooney–Rivlin solids were the first hyperelastic models developed [21]; currently, the Arruda–Boyce model [22] and the Ogden model [23] are widely used to describe the strain state-dependent hyperelastic behavior, as depicted in Figure 3.

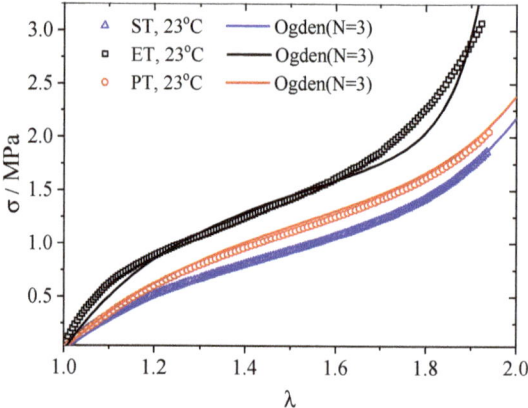

**Figure 3.** Stress–stretch curves for simple tension (ST), planar tension (PT) and equal-biaxial tension (ET) tests of rubber and their Ogden model fits.

The Ogden strain density function [23], defined as

$$W_{\text{Ogden}} = \sum_{n=1}^{N} \frac{\mu_n}{\alpha_n} \left( \lambda_1^{\alpha_n} + \lambda_2^{\alpha_n} + \lambda_3^{\alpha_n} - 3 \right) \tag{1}$$

where $\lambda_i$ are the principal stretches, $\mu_i$ and $\alpha_i$ are experimentally determined material constants, and $N$ is the number of terms in the function, is considered one of the most successful functions in describing the hyperelasticity of rubber-like materials. The three-term model ($N = 3$) is used in this work to fit the experimental data, and the model parameters are identified as $\mu_1 = 1.9042$; $\mu_2 = -1.924 \times 10^{-10}$; $\mu_3 = 3.1850 \times 10^{-4}$; $\alpha_1 = 1.0625$; $\alpha_2 = -17.7$; $\alpha_3 = 12.3795$. The model fits are also shown by lines in Figure 3, indicating a satisfactory agreement with the tests.

A finite element model of the hourglass specimen used to calculate the $\varepsilon_{\max}$ is shown in Figure 4. The rubber part of the model was constructed by using four-node axisymmetric quadrilateral hybrid elements (CAX4H). The metal parts for clamping and loading at the top and bottom of the specimen used four-node axisymmetric quadrilateral elements (CAX4). In order to verify the three-term Ogden model described above, the uniaxial tensile behavior of the hourglass rubber specimen when stretched up to about 800N was numerically analyzed using the finite element model. The obtained force versus displacement relation of the specimen was compared with the lab test data, as shown in Figure 5. The good agreement between the numerical results and the experiment further indicates that the three-term Ogden model is an appropriate constitutive model for describing the hyperelastic behavior of the filled rubber material investigated in this study.

**Figure 4.** Axisymmetric geometric model of the hourglass rubber specimen (left: before rotation, right: after rotation).

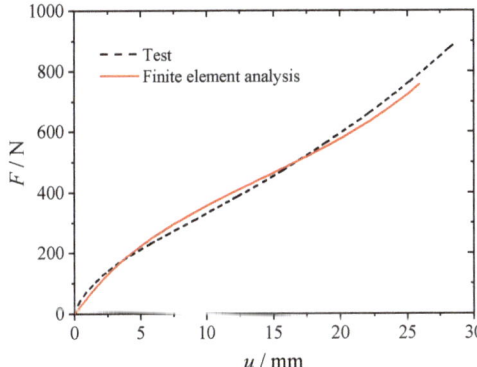

**Figure 5.** Tensile force-displacement curves of the hourglass specimen obtained from numerical analysis and lab tests.

To get the maximum principal strain corresponding to the four loading cases in fatigue-to-failure experiments, finite element analyses with the three-term Ogden hyperelastic constitutive model were conducted. The bottom surface of the specimen was fixed, and four constant tensile forces (250 N, 300 N, 350 N, and 400 N) were applied to the top surface, respectively. Figure 6 shows the maximum principal strain contour plots of the specimen under the four designated tensile loads. The calculated maximum principal strains are listed in Table 1.

**Table 1.** The maximum principal strains of the hourglass specimen loaded with different maximum forces.

| $F_{max}$/N | 250 | 300 | 350 | 400 |
|---|---|---|---|---|
| $\varepsilon_{max}$ | 0.4133 | 0.5487 | 0.7136 | 0.9072 |

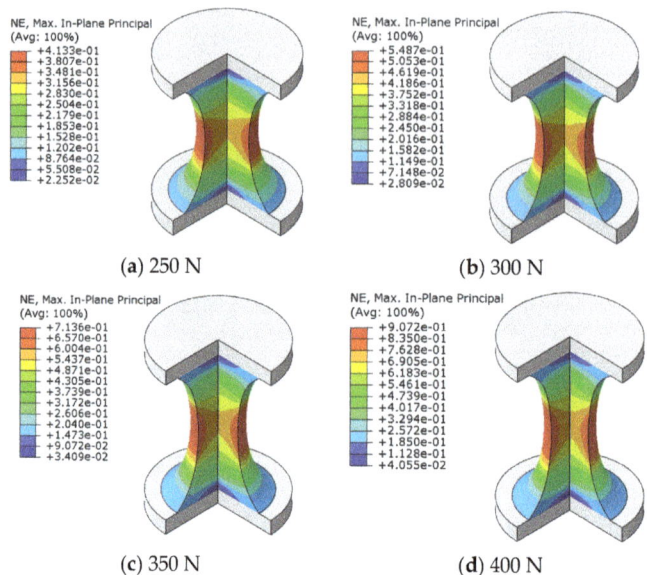

**Figure 6.** Maximum principal strain contours of the hourglass rubber specimen. (**a**) 250 N; (**b**) 300 N; (**c**) 350 N; (**d**) 400 N.

*3.2. Fatigue Life Assessment*

In the fatigue-to-failure experiments, the fatigue lives for each loading case were recorded as seen in Figure 2. The maximum principal strains were obtained via finite element analysis, as described in the previous subsection. Considering maximum principal strain as the fatigue parameter, S–N curves can be built, as shown in Figure 7.

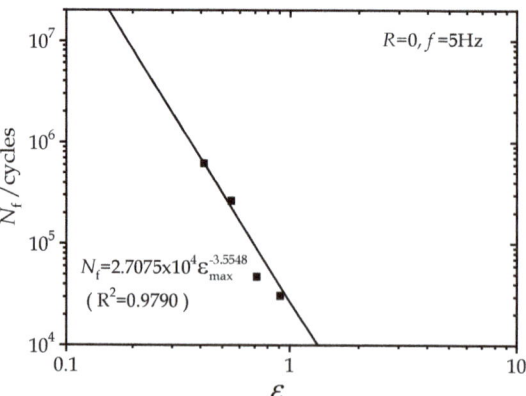

**Figure 7.** S–N curve of the hourglass rubber specimen.

Numerous studies demonstrate that the power law is an excellent model for relating the fatigue life and the maximum strain [4,5,19]. As depicted in Figure 7, the power law model describes the experimental S–N curve very well with Equation (2).

$$N_f = k\varepsilon_{max}^n = 2.7075 \times 10^4 \varepsilon_{max}^{-3.5548} \qquad (2)$$

As mentioned in Section 2, the steady state temperature increase $\theta_\infty$ can be considered as a promising and alternative fatigue parameter. We plotted the observed $\theta_\infty$ with the corresponding maximum principal strain obtained by finite element simulation in Figure 8. Le Saux et al. [14] conducted similar heat build-up tests on several industrial rubber materials and discussed the correlation between the temperature rise and the principal maximum strain. In Figure 9, we redrew the $\theta_\infty$ vs. $\varepsilon_{max}$ data within the strain range from 0.3 to 1.5, approximately the same range as used in Figure 8, for carbon black filled natural rubber with different carbon black contents (22 phr, 39 phr and 43 phr). It is clear from Figures 8 and 9 that the steady state temperature increase is approximately linearly proportional to the maximum strain in the considered range; therefore, the fatigue life would also be related to the steady temperature increase by a power law. We suggest a power law relation in the form of

$$N_f = A\left(\frac{\theta_\infty}{T_0}\right)^n \tag{3}$$

where $T_0$ is the initial temperature at the beginning of the fatigue test. Fitting the data in Figure 10 to the above equation with $T_0 = 20\ °C$ yields the model parameters $A$ and $n$ in Equation (3); $A = 1.06 \times 10^6$, and $n = -4.46$. Equation (3) provides a promising criterion for the fatigue life prediction. As long as the steady state temperature increase at the point of the maximum strain is determined, the fatigue life of the structure can be predicted by the model.

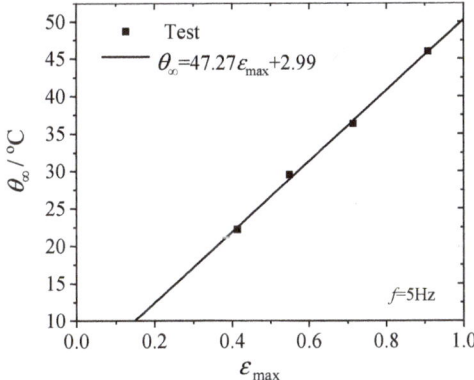

**Figure 8.** Steady state temperature increase of the hourglass specimen under fatigue with different $\varepsilon_{max}$.

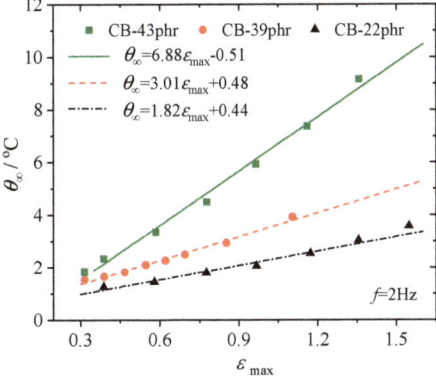

**Figure 9.** $\theta_\infty$ vs. $\varepsilon_{max}$ data for cyclic loaded filled rubber with various carbon black contents [14].

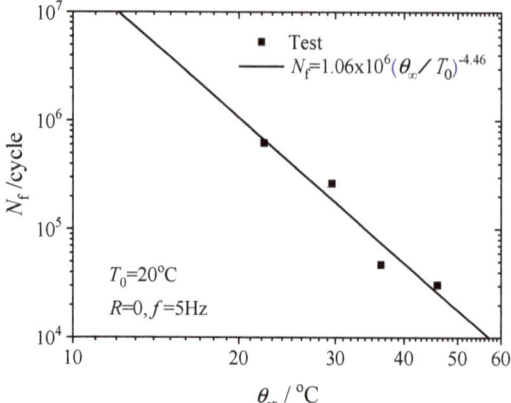

**Figure 10.** Fatigue lives vs. steady state temperature increases.

Equation (3) bridges the fatigue life with the steady state temperature increase in the rubber material, rather than the maximum strain as expressed in Equation (2). The advantage is that the steady state temperature increase can be measured by thermal imaging in a few thousand cycles in fatigue tests, much fewer than the number of cycles to fatigue failure. Thus, the time cost for fatigue testing can be reduced and the characterization of the fatigue properties can be considerably accelerated by the method provided in this work.

Based on the fatigue criterion suggested by Equation (3), thermomechanical coupling finite element simulation is expected to be the most promising method for fatigue life assessment. In such simulations, the loss energy density due to hysteresis in the dynamic viscoelasticity is considered as the heat resource in the deformed body. The steady temperature increase can be obtained by solving the heat equation with given initial boundary conditions [24], and the fatigue life of the rubber component can be subsequently determined.

## 4. Conclusions

Fatigue tests and infrared thermal imaging measurements were carried out to correlate the fatigue life with the temperature increase induced by the hysteresis loss. Experiments show that the surface temperature of the specimen keeps its steady value for a prolonged period, which accounts for the majority of the fatigue life, and then sharply increases until the specimen ruptures. The sharp rise in temperature can be regarded as a precursor to fatigue failure. Moreover, the steady state temperature increase is linearly proportional to the maximum principal strain. By replacing the maximum principal strain, which is used as the fatigue life predictor in classic fatigue models, with the steady state temperature increase, a promising method for quickly evaluating the fatigue life of rubber structures is developed based on the power law model. Since only a couple of thousand cycles are required for the temperature to reach its steady state value during fatigue testing, which is less than one tenth of the fatigue life, the fatigue life of rubber components can be efficiently assessed by the steady state temperature increase.

**Author Contributions:** Conceptualization, W.L.; Data curation, Y.H., B.Y. and X.J.; Formal analysis, W.L. and Y.H.; Funding acquisition, W.L.; Methodology, W.L. and X.H.; Writing—original draft, W.L., Y.H. and B.Y.; Writing—review and editing, W.L. and X.H. All authors have read and agreed to the published version of the manuscript.

**Funding:** This research was funded by National Natural Science Foundation of China, Grant No. 11572275.

**Conflicts of Interest:** The authors declare no conflict of interest.

## References

1. Fan, R.P.; Meng, G.; Yang, J.; He, C. Experimental study of the effect of viscoelastic damping materials on noise and vibration reduction within railway vehicles. *J. Sound Vib.* **2009**, *319*, 58–76. [CrossRef]
2. Brunac, J.B.; Gérardin, O.; Leblond, J.B. On the heuristic extension of Haighs diagram for the fatigue of elastomers to arbitrary loadings. *Int. J. Fatigue* **2009**, *31*, 859–867. [CrossRef]
3. Ayoub, G.; Nait-Abdelaziz, M.; Zairi, F.; Gloaguen, J.M.; Charrier, P. Fatigue life prediction of rubber-like materials under multiaxial loading using a continuum damage mechanics approach: Effects of two-blocks loading and R ratio. *Mech. Mater.* **2012**, *52*, 87–102. [CrossRef]
4. Li, Q.; Zhao, J.; Zhao, B. Fatigue life prediction of a rubber mount based on test of material properties and finite element analysis. *Eng. Fail. Anal.* **2009**, *16*, 2304–2310. [CrossRef]
5. Kim, W.; Lee, H.; Kim, J.; Koh, S.K. Fatigue life estimation of an engine rubber mount. *Int. J. Fatigue* **2004**, *26*, 553–560. [CrossRef]
6. Gent, A.N.; Lindley, P.B.; Thomas, A.G. Cut growth and fatigue of rubbers. I. the relationship between cut growth and fatigue. *J. Appl. Polym. Sci.* **1964**, *8*, 455–466. [CrossRef]
7. Verron, E.; Andriyana, A. Definition of a new predictor for multiaxial fatigue crack nucleation in rubber. *J. Mech. Phys. Solids* **2008**, *56*, 417–443. [CrossRef]
8. Mars, W.V. Cracking energy density as a predictor of fatigue life under multiaxial conditions. *Rubber Chem. Technol.* **2002**, *75*, 1–17. [CrossRef]
9. Doudard, C.; Calloch, S.; Cugy, P.; Galtier, A.; Hild, F. A probabilistic two-scale model for high-cycle fatigue life predictions. *Fatigue Fract. Eng. Mater. Struct.* **2005**, *28*, 279–288. [CrossRef]
10. Rosa, G.L.; Risitano, A. Thermographic methodology for rapid determination of the fatigue limit of materials and mechanical components. *Int. J. Fatigue* **2000**, *22*, 65–73. [CrossRef]
11. Meneghetti, G.; Quaresimin, M. Fatigue strength assessment of a short fiber composite based on the specific heat dissipation. *Compos. Part B Eng.* **2011**, *42*, 217–225. [CrossRef]
12. Jegou, L.; Marco, Y.; Le Saux, V.; Calloch, S. Fast prediction of the Wöhler curve from heat build-up measurements on short fiber reinforced plastic. *Int. J. Fatigue* **2013**, *47*, 259–267. [CrossRef]
13. Serrano, L.; Marco, Y.; Le Saux, V.; Robert, G.; Charrier, P. Fast prediction of the fatigue behavior of short-fiber-reinforced thermoplastics based on heat build-up measurements: Application to heterogeneous cases. *Contin. Mech. Thermodyn.* **2017**, *29*, 1–21. [CrossRef]
14. Le Saux, V.; Marco, Y.; Calloch, S.; Doudard, C.; Charrier, P. Fast evaluation of the fatigue lifetime of rubber-like materials based on a heat build-up protocol and micro-tomography measurements. *Int. J. Fatigue* **2010**, *32*, 1582–1590. [CrossRef]
15. Luo, W.B.; Yin, B.Y.; Hu, X.L.; Zhou, Z.; Deng, Y.; Song, K. Modeling of the heat build-up of carbon black filled rubber. *Polym. Test.* **2018**, *69*, 116–124. [CrossRef]
16. Zhi, J.; Wang, S.; Zhang, M.; Wang, H.; Lu, H.; Lin, W.; Qiao, C.; Hu, C.; Jia, Y. Numerical analysis of the dependence of rubber hysteresis loss and heat generation on temperature and frequency. *Mech. Time-Depend. Mat.* **2019**, *23*, 427–442. [CrossRef]
17. Le Saux, V.; Marco, Y.; Calloch, S.; Charrier, P.; Taveau, D. Heat build-up of rubber under cyclic loadings: Validation of an efficient demarch to predict the temperature fields. *Rubber Chem. Technol.* **2013**, *86*, 38–56. [CrossRef]
18. Katunin, A. Criticality of the self-heating effect in polymers and polymer matrix composites during fatigue, and their application in non-destructive testing. *Polymers* **2019**, *11*, 19. [CrossRef]
19. Mars, W.V.; Fatemi, A. A literature survey on fatigue analysis approaches for rubber. *Int. J. Fatigue* **2002**, *24*, 949–961. [CrossRef]
20. Li, X.B.; Wei, Y.T. Classic strain energy functions and constitutive tests of rubber-like material. *Rubber Chem. Technol.* **2015**, *88*, 604–627. [CrossRef]
21. Mooney, M. A theory of large elastic deformation. *J. Appl. Phys.* **1940**, *11*, 582–592. [CrossRef]
22. Arruda, E.M.; Boyce, M.C. Three-dimensional constitutive model for the large stretch behavior of rubber elastic materials. *J. Mech. Phys. Solids* **1993**, *41*, 289–412. [CrossRef]

23. Ogden, R.W. Large deformation isotropic elasticity—On the correlation of theory and experiment for incompressible rubberlike solids. *Proc. R. Soc. Lond. A Math. Phys. Sci.* **1972**, *326*, 565–584. [CrossRef]
24. Li, F.Z.; Liu, J.; Yang, H.B.; Lu, Y.L.; Zhang, L.Q. Numerical simulation and experimental verification of heat build-up for rubber compounds. *Polymer* **2016**, *101*, 199–207. [CrossRef]

© 2020 by the authors. Licensee MDPI, Basel, Switzerland. This article is an open access article distributed under the terms and conditions of the Creative Commons Attribution (CC BY) license (http://creativecommons.org/licenses/by/4.0/).

Article

# Effect of Stabilizer States (Solid Vs Liquid) on Properties of Stabilized Natural Rubber s

Khwanchat Promhuad [1] and Wirasak Smitthipong [1,2,*]

[1] Specialized center of Rubber and Polymer Materials in agriculture and industry (RPM), Department of Materials Science, Faculty of Science, Kasetsart University, Chatuchak, Bangkok 10900, Thailand; pond_kh@hotmail.com

[2] Office of research integration on target-based natural rubber, National Research Council of Thailand (NRCT), Chatuchak, Bangkok 10900, Thailand

* Correspondence: fsciwssm@ku.ac.th

Received: 29 February 2020; Accepted: 20 March 2020; Published: 27 March 2020

**Abstract:** The main objective of this work is to study the effect of hydroxylamine sulfate or stabilizer states (solid vs liquid) on the storage hardening of natural rubber (NR). Several types of natural rubber samples were prepared: unstabilized NR samples and stabilized NR samples: (i) dry NR with 0.2 and 2.0 parts per hundred rubber (phr) of dry hydroxylamine sulfate, and (ii) natural latex with 0.2 and 2.0 phr of liquid hydroxylamine sulfate. The samples were characterized immediately (time 0) and after 12 weeks of storage at room temperature, respectively. We found that the Mooney viscosity, gel content, and Wallace plasticity of NR without a stabilizer increases with storage hardening for 12 weeks. However, two types of stabilized NR samples represent constant values of those three parameters, because hydroxylamine sulfate inhibits network and gel formation in NR. Interestingly, the mixing states (solid vs liquid) between natural rubber and the stabilizer affect the properties of stabilized NR. This could be explained by the better dispersion and homogeneous nature of liquid stabilizers in natural latex (liquid state), and thus the higher loading of the stabilizer in the liquid state. This is important, as the stabilization of NR properties as a function of time is required by rubber industry. This study is a utilization model from theory to application.

**Keywords:** natural rubber; natural latex; viscosity stabilizer; hydroxylamine sulfate; storage hardening

## 1. Introduction

Natural rubber (NR) can be obtained from *Hevea brasiliensis* plant. NR has been widely used across industries for surgical gloves, pillows and mattresses, elastic bands, medical products, and tires, plus many more. NR consists of rubber and non-rubber components, the main component is *cis*-1,4-polyisoprene [1–5]. The non-rubber fraction consists of proteins ($\omega$-terminal) and phospholipids ($\alpha$-terminal), which are not only attached at each end chain of the polyisoprene, but also included in the serum [6]. When we stored the NR for a prolonged period of time, proteins and phospholipids at the end chain could promote formation of the NR network through protein-protein interactions at the $\omega$-terminal or phospholipid-phospholipid interactions at the $\alpha$-terminal. The increased of the branch chain or network of NR as a function of time can be determined by the Mooney viscosity via the storage hardening [7]. The storage hardening of NR can be inhibited by a stabilizer. A recent study [8] presented the effect of polar chemicals on the storage hardening of NR. Three types of polar chemicals were used: phenol, diethylene glycol, and hydroxylamine hydrochloride. They found that gel content and Mooney viscosity of NR samples with phenol and diethylene glycol are increased with storage hardening. However, NR samples with hydroxylamine hydrochloride (stabilizer) represent constant values of gel content and Mooney viscosity. Another example study proposed a model wherein the interaction between non-rubber components at the terminal ends of NR molecules and the viscosity stabilizer could be hydrogen bonding [9].

Most of the previous research has presented only the study of dry NR with dry stabilizers, and also fixed the concentration of the stabilizer at only 0.2 parts per hundred rubber (phr) [8–10]. It would be interesting to better understand the mixing state between NR and the stabilizer, as well as the concentration of the stabilizer. So, the objective of this present work is to investigate the effect of stabilizer states (dry stabilizer with dry NR and liquid stabilizer with natural latex) at different stabilizer concentrations on the properties of stabilized NR compared to unstabilized NR. This study could allow us to better control the properties of stabilized natural rubber.

## 2. Materials and Methods

*2.1. Materials*

Fresh natural latex was used, with a dry rubber content 29.6 wt.% (Num rubber and latex Co. Ltd, Trang, Thailand), hydroxylamine sulfate (HS) (Sigma-Aldrich, Bangkok, Thailand, preparing by Thai eastern group, Thailand), and toluene (AR grade, RCI Labscan Limited, Bangkok, Thailand).

*2.2. Preparation of NR Samples*

Five types of NR samples (400 g each) were prepared (Table 1): control NR or unstabilized NR, dry NR with 0.2 and 2.0 phr of dry hydroxylamine sulfate (NRD/HS), and natural latex with 0.2 and 2.0 phr of liquid hydroxylamine sulfate (NRL/HS). Then, all the rubber samples were individually masticated by a two-roll mill with a front-rotor speed of 18 rpm and a back-rotor speed of 20 rpm. The mastication for each sample was carried out at 70 °C for 10 min. Finally, the NR samples were characterized after preparation (time 0, or 0 weeks) and also after 12 weeks of storage at room temperature, respectively.

**Table 1.** List of natural rubber (NR) samples used in this study (phr means parts per hundred rubber).

| Name | Samples |
|---|---|
| Control NR | Unstabilized NR |
| NRD/0.2 HS | Dry NR with dry hydroxylamine sulfate 0.2 phr |
| NRD/2.0 HS | Dry NR with dry hydroxylamine sulfate 2.0 phr |
| NRL/0.2 HS | Natural latex with liquid hydroxylamine sulfate 0.2 phr |
| NRL/2.0 HS | Natural latex with liquid hydroxylamine sulfate 2.0 phr |

*2.3. Characterizations*

Gel content was measured by dissolving the NR samples in toluene. After that, the solution was kept in the dark at room temperature for one week. Finally, the solution was filtered, weighed, and then calculated with respect to the original weight sample as an equation seen below [1]:

$$\%\text{Gel content} = \frac{\text{Weigth of gel} \times 100}{\text{Weigth of sample}} \quad (1)$$

Mooney viscosity analysis was performed using a Mooney viscometer (viscTECH+, TECHPRO, Columbia City, IN, USA), within a sealed, pressurize, d and heated cavity. A rubber sample was heated at 100 °C for 1 min before analysis. After that, the rubber sample was continuously measured for 4 min by the torque required to keep the rotor rotating at a constant rate as a function of time for reading the Mooney viscosity, which was recorded as torque in newton metre (Nm) [9].

The functional group of rubber samples was determined by attenuated total reflection Fourier transform infrared (ATR-FTIR VERTEX 70, Bruker, Billerica, MA, USA). Rubber samples were cut into small pieces and put on a Ge crystal probe at 600–4000 cm$^{-1}$ [9].

The Wallace plasticity value was determined according to the International Organization for Standardization (ISO 2007). The small rubber disk was compressed between two platens at 100 °C for 15 s at a fixed thickness of 1 mm. After this preheating period, the rubber specimen was subjected to a constant compressive force of 100 N for 15 additional seconds. The thickness of the specimen at the end of this period was taken as the Wallace plasticity value. Plasticity retention index (PRI) is a measure of the resistance of raw natural rubber to oxidation. The oxidation effect was assessed by measuring the plasticity before aging (Po) and after aging for 30 min in the MonTech aging oven for plasticity testing at 140 °C ($P_{30}$) [11]:

$$\text{PRI} = \frac{P_{30} \times 100}{Po} \quad (2)$$

The processability under strain sweep modes of the rubber samples was investigated by a rubber processing analyzer (RPA 2000, Alpha Technologies, Hudson, OH, USA). The strain sweep mode was in the range of 0.5–100%, the process was carried out at 100 °C and 1 Hz [9].

The viscoelastic properties of uncrosslinked rubbers were determined by dynamic mechanical analysis (DMA1, Mettler Toledo, Columbus, OH, USA) based on Williams–Landel–Ferry (WLF) analysis. The time-temperature superposition principle was used to establish a master curve of the storage modulus (E') as a function of reduced frequency at a reference temperature $T_{ref}$ of 298 K. Shift factors ($a_T$) for establishment of master curves were determined according to the universal WLF equation, where $c_1$ and $c_2$ are constants depending on the nature of the elastomer and the reference temperature [12]:

$$\log a_T = \frac{\left(-c_1(T_{\exp} - T_{ref})\right)}{\left(c_2 + (T_{\exp} - T_{ref})\right)} \quad (3)$$

## 3. Results and Discussion

Concerning to the visual aspect of samples (Figure 1), we found that the control NR without any stabilizer had a dark brown color at 0 weeks. However, the dry rubber with dry HS samples had a light brown color. Moreover, the latex with liquid 2.0 phr of the HS sample seemed to be the lightest brown color. After 12 weeks (data not shown), the color of the control NR became darker. However, the color of all the rubbers with stabilizers were stable.

NR control     NRD/0.2HS     NRD/2.0HS     NRL/0.2HS     NRL/2.0HS

**Figure 1.** The natural rubber samples used in this study for 0 weeks: NRD/0.2HS means dry NR with dry hydroxylamine sulfate 0.2 phr, NRD/2.0HS means dry NR with dry hydroxylamine sulfate 2.0 phr, NRL/0.2HS means natural latex with liquid hydroxylamine sulfate 0.2 phr, NRL/2.0HS means natural latex with liquid hydroxylamine sulfate 2.0 phr.

We investigated the Mooney viscosity of the samples, which depends on the macrostructure of the rubber samples. The results of Mooney viscosity of the samples at various storage times are shown in Figure 2. At time 0, hydroxylamine sulfate additive stabilized both dry rubber and latex which, are NRD/0.2HS, NRD/2.0HS, NRL/0.2HS, and NRL/2.0HS, respectively. When the amount of hydroxylamine sulfate in rubber was increased, the Mooney viscosity of sample was decreased. However, the latex samples showed a Mooney viscosity lower than the samples from dry rubber. After 12 weeks, the Mooney viscosity increased for the control sample, which is a sign of the storage hardening phenomenon of natural rubber [13]. However, the NR samples supplemented with hydroxylamine sulfate were stable (NRD/0.2HS, NRD/2.0HS, NRL/0.2HS, and NRL/2.0HS). Considering the effect of the viscosity stabilizer, all NR samples with the viscosity stabilizer HS (0 and 12 weeks) had a more stable Mooney viscosity than those without. Interestingly, when the liquid hydroxylamine sulfate was mixed with natural latex (NRL/HS), its Mooney viscosity seemed to be lower than that of the NRD/HS sample at a given concentration. This could be explained by the better dispersion and homogeneous nature of liquid hydroxylamine sulfate in natural latex (liquid state) compared to the mixing of HS and NR in a dry state.

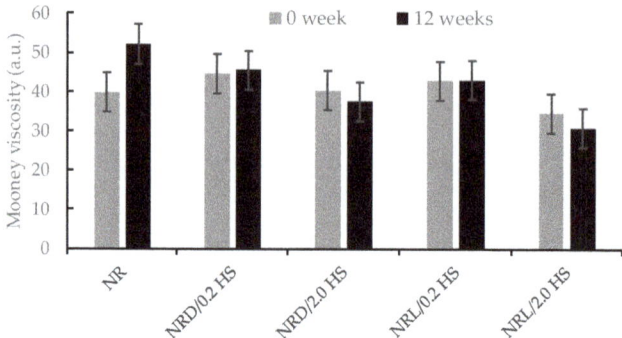

**Figure 2.** The Mooney viscosity of the samples at 0 and 12 weeks.

The gel content of the rubber samples was also interesting for investigating the effect of stabilizer states (solid vs liquid), and the amounts of gel content are shown in Figure 3. At time 0, the control NR sample has a gel content slightly higher than the samples with hydroxylamine sulfate (NRD/0.2HS, NRD/2.0HS, NRL/0.2HS, and NRL/2.0HS). After 12 weeks, the NR control had a higher gel content, resulting from gel formation during prolonged storage [1,14,15]. The gel content depends on proteins and lipids that are the major components in the formation of the gel fraction during NR storage [16]. In contrast, the gel content was stable for NRD/0.2HS, NRD/2.0HS, NRL/0.2HS, and NRL/2.0HS. The effect of proteins and lipids causes a gel network that is called storage hardening in NR. However, the viscosity stabilizer hydroxylamine sulfate inhibited gel formation in the NR samples. Again, the NRL/HS sample had a lower gel content than the NRD/HS sample at a given concentration. So, the effects of stabilizers in liquid state were more pronounced. This result is in good agreement with the results of the Mooney viscosity analysis.

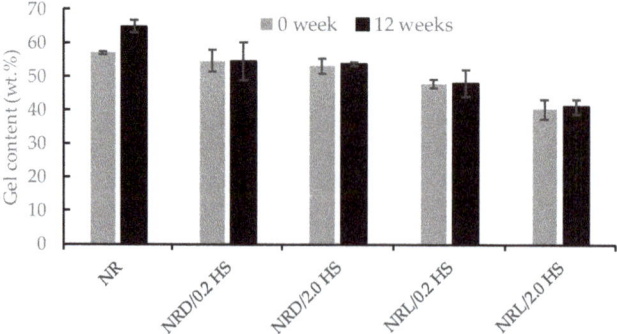

**Figure 3.** Gel content of NR samples for times 0 and 12 weeks.

The chemical compositions of the NR samples and non-rubber components were analyzed by attenuated total reflection Fourier transform infrared or ATR-FTIR (Figure 4). The FTIR spectra of the NR with hydroxylamine sulfate samples presented at 3600–2600 cm$^{-1}$ and 1800–1600 cm$^{-1}$. At time 0, samples showed peaks for the amine group (N–H) at 3280 cm$^{-1}$, the fatty acid ester group at 1740 cm$^{-1}$, the aldehyde group at 1710 cm$^{-1}$, and the amide I group at 1660–1630 cm$^{-1}$ [13].

After 12 weeks, all the peaks of the control NR sample were increased, and this result is in good agreement with a previous study [9]. For the stabilized NR samples, there was almost no change in the FTIR spectra, except NRL/0.2HS, which was close to the control NR sample. However, the ATR-FTIR technique focuses on a few microns from the sample surface, which is quite different from the bulk results of the Mooney viscosity and gel content analyses between control NR and NRL/0.2HS, in particular with this low concentration of liquid stabilizer (NRL/0.2HS), which probably would dilute the ATR-FTIR measurement.

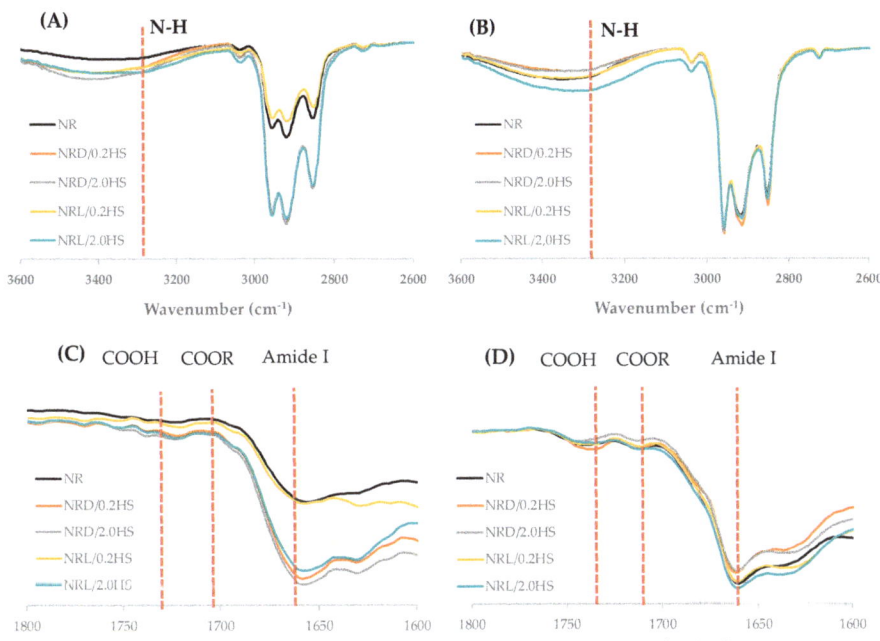

**Figure 4.** FTIR spectra of NR with hydroxylamine sulfate samples: (**A**) and (**C**) at 0 weeks, (**B**) and (**D**) at 12 weeks.

We were also interested in the elasticity of rubber samples, which could be determined by the original Wallace plasticity (Po). When the Po value of rubber is high, its elasticity is high [8]. The Po value of control NR sample increased after being stored for 12 weeks, whereas the other NR samples with hydroxylamine sulfate had a constant Po value (Table 2). However, the NRL/HS samples possessed lower Po values than NRD/HS, no matter the concentration of HS and the storage time. This result is in good agreement with the results of the Mooney viscosity and gel content analyses. The proteins and phospholipids in NR represent the network and the gel formation, which causes the increase in Po. This phenomenon is called storage hardening [11]. Unlike the NR with stabilizer samples [16], they represented almost stable Po values as a function of time within the uncertainty values (± 5 a.u.).

The results of Plasticity Retention Index (PRI) are also presented in Table 2. This test estimates the resistance to oxidation and breakage of rubber molecules at higher temperatures. Similar to the Po results, the PRI values of NR with stabilizer samples were stable, unlike the PRI value of the control NR sample, which decreased after 12 weeks of storage time [13].

**Table 2.** The original Wallace plasticity (Po) and Plasticity Retention Index (PRI) of the rubber samples at times 0 and 12 weeks.

| Sample Name | Po (± 5 a.u.) | | PRI (± 5 a.u.) | |
| --- | --- | --- | --- | --- |
| | 0 Weeks | 12 Weeks | 0 Weeks | 12 Weeks |
| NR | 20.0 | 28.0 | 85.0 | 67.9 |
| NRD/0.2 HS | 28.5 | 32.0 | 61.4 | 59.4 |
| NRD/2.0 HS | 22.0 | 21.5 | 54.6 | 51.2 |
| NRL/0.2 HS | 22.5 | 24.5 | 62.2 | 59.2 |
| NRL/2.0 HS | 18.5 | 18.0 | 56.8 | 54.2 |

Concerning the rigidity or storage modulus (G′) of NR samples as a function of strain, its values for 0 and 12 weeks can be seen in Figure 5. At time 0, the results show that all the NR samples with or without stabilizer possessed almost the same level of rigidity (G′), whereas NRL/2.0HS had the lowest G′. This result is in good agreement with the result of the Mooney viscosity analysis. After 12 weeks, the rigidity of the control NR increased slightly compared to that at 0 weeks. However, the rigidity of NR with HS almost decreased compared to that of 0 weeks. This may be explained by the increased Mooney viscosity and gel content of the control NR sample after 12 weeks.

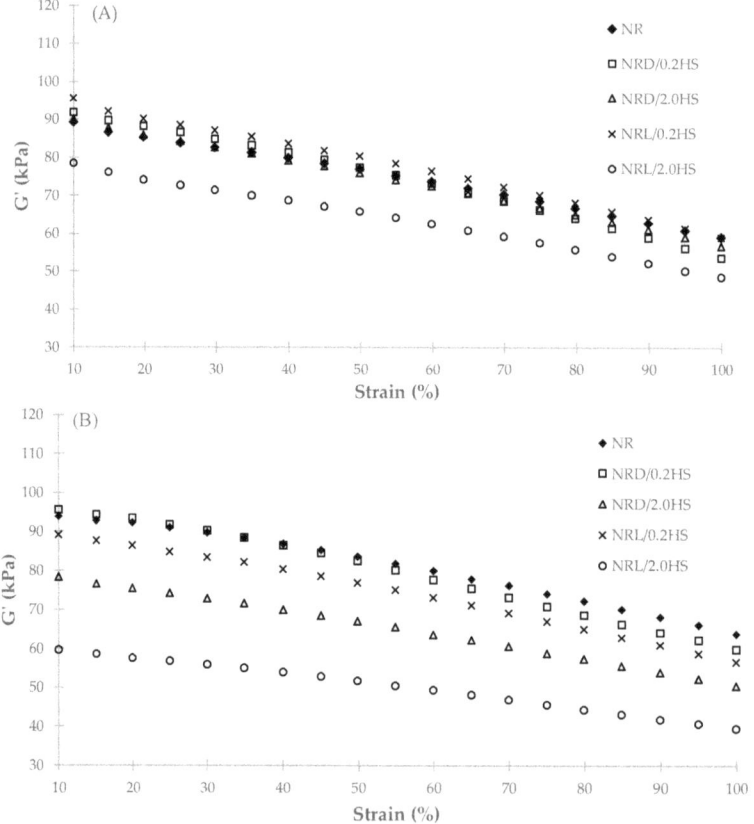

**Figure 5.** Relationship between storage modulus (G′) and the strain of NR with hydroxylamine sulfate samples for 0 weeks (**A**) and 12 weeks (**B**).

The tan delta is the ratio of the loss modulus (G") and storage modulus (G'), a high tan delta value means the rubber samples have lost more energy or have a higher heat build-up. At 0 weeks, the control NR sample had a tan delta level close to those of NR with higher amounts of stabilizer (2 phr), whereas the tan delta of NR with lower amounts of stabilizer (0.2 phr) was lower (Figure 6). At 12 weeks, all samples with hydroxylamine sulfate had an increasing tan delta. However, the control NR sample had the decreasing of tan delta because of the increasing network structure and longer molecular chains, which reduced the molecular friction or heat build-up [9].

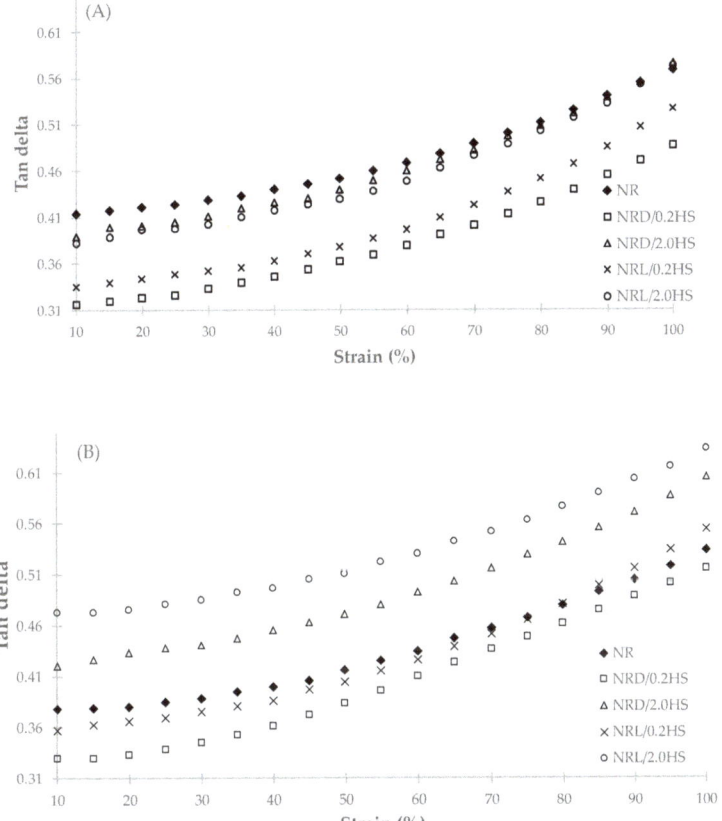

**Figure 6.** Relationship between tan delta and the strain of NR with hydroxylamine sulfate samples for 0 weeks (**A**) and 12 weeks (**B**).

One can study the viscoelastic properties of NR samples as master curves (time-temperature superposition) determined by dynamic mechanical analysis, which represents the storage modulus of rubber samples as a function of reduced frequency. The mean values of the constants $c_1$ and $c_2$ for all NR samples (8.50 and 186.50) were rather in good agreement with the Ferry (5.94 and 151.60) reference [12]. The shift factor ($a_T$) as a function of temperature for the control NR sample is presented as an insert in Figure 7, and the other samples apply the same shift factor. We found that all the NR samples with or without the stabilizer [16] possessed similar viscoelastic properties (Figure 7) at 12 weeks, which was also similar to the results for 0 weeks.

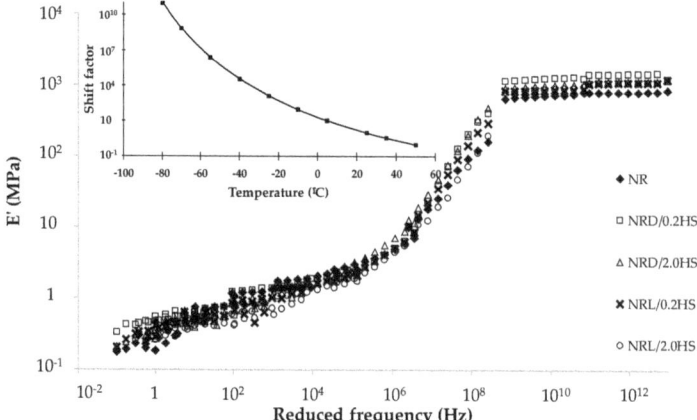

**Figure 7.** Storage modulus (E') versus reduced frequency master curves of NR with or without hydroxylamine sulfate for 12 weeks, the insert figure is the shift factor as a function of temperature for control NR sample.

Figure 7 presents the master curves in three zones: glassy plateau at high reduced frequency, transition state, and rubbery plateau at low reduced frequency. Rubber molecules are frozen in the glassy state below the glass transition temperature. The rotation around the molecular bonds increases with increasing the temperature to reach a rubbery state, which means that the rubber molecules become entangled.

## 4. Conclusions

A molecular chain of NR is mostly *cis*-1,4-polyisoprene, whereas the terminal chains are divided into proteins and phospholipids (non-rubber components), which causes storage hardening of NR as a function of time. We investigated the effects of stabilizer states (solid vs liquid), concentration of stabilizers (0.2 and 2.0 phr), and storage time (0 and 12 weeks) on the properties of stabilized NR compared to unstabilized NR. We found that NR with hydroxylamine sulfate (0.2 and 2.0 phr) had more constant values of gel content, Mooney viscosity, and Wallace plasticity compared to those of unstabilized NR. The result of processability under strain sweep is in good agreement with the result of the Mooney viscosity analysis. Interestingly, the mixing condition between NR and the stabilizer in solid vs liquid states affects the properties of stabilized NR. When we mixed NR with stabilizer in a liquid state, this type of sample obtained better performance to keep a lower and more stable Mooney viscosity compared to the sample mixed in a dry state. This opens up more applications of stabilized natural rubber in industry, in particular, it shows a better compromise between the processing and properties.

**Author Contributions:** K.P. carried out the preparation and characterization of NR samples; W.S. conceived the study, designed the study, contributed to the discussion and worked with the manuscript. All authors gave final approval for publication and agree to be held accountable for the work performed therein. All authors have read and agreed to the published version of the manuscript.

**Funding:** This research received no external funding.

**Acknowledgments:** We are thankful to Thai Eastern Rubber Co., Ltd. for the support of hydroxylamine sulfate and Num rubber and latex. Co. Ltd. for the support of fresh natural latex. This work was partly supported by National Research Council of Thailand (NRCT).

**Conflicts of Interest:** The authors declare no conflict of interest in reported research. The funders had no role in the design of the study; in the collection, analyses, or interpretation of data; in the writing of the manuscript, or in the decision to publish the results. All authors have read and agreed to the published version of the manuscript.

## References

1. Smitthipong, W.; Nardin, M.; Schultz, J.; Nipithakul, T.; Suchiva, K. Study of tack properties of uncrosslinked natural rubber. *J. Adhes. Sci. Technol.* **2004**, *18*, 1449–1463. [CrossRef]
2. Smitthipong, W.; Nardin, M.; Schultz, J.; Suchiva, K. Adhesion and self-adhesion of rubbers, crosslinked by electron beam irradiation. *Int. J. Adhes. Adhes.* **2007**, *27*, 352–357. [CrossRef]
3. Smitthipong, W.; Nardin, M.; Schultz, J.; Suchiva, K. Adhesion and self-adhesion of immiscible rubber blends. *Int. J. Adhes. Adhes.* **2009**, *29*, 253–258. [CrossRef]
4. Suksup, R.; Sun, Y.; Sukatta, U.; Smitthipong, W. Foam rubber from centrifuged and creamed latex. *J. Polym. Eng.* **2019**, *39*, 336–342. [CrossRef]
5. Backhaus, R.A. Rubber formation in plants—A mini-review. *Isr. J. Bot.* **1985**, *34*, 283–293.
6. Tarachiwin, L.; Tanaka, Y.; Sakdapipanich, J. Structure and origin of long-chain branching and gel in natural rubber. *Kautschuk Gummi Kunststoffe* **2005**, *58*, 115–122.
7. Tarachiwin, L.; Sakdapipanich, J.; Ute, K.; Kitayama, T.; Tanaka, Y. Structural Characterization of α-Terminal Group of Natural Rubber. 2. Decomposition of Branch-Points by Phospholipase and Chemical Treatments. *Biomacromolecules* **2005**, *6*, 1858–1863. [CrossRef] [PubMed]
8. Nimpaiboon, A.; Sriring, M.; Sakdapipanich, J. Molecular structure and storage hardening of natural rubber: Insight into the reactions between hydroxylamine and phospholipids linked to natural rubber molecule. *J. Appl. Polym. Sci.* **2016**, *133*. [CrossRef]
9. Chollakup, R.; Suwanruji, P.; Tantatherdtam, R.; Smitthipong, W. New approach on structure-property relationships of stabilized natural rubbers. *J. Polym. Res.* **2019**, *26*, 37. [CrossRef]
10. Wei, Y.; Ding, A.; Jin, L.; Zhang, H.; Liao, S. Quantitative Analysis of Abnormal Groups on Molecular Chain of Natural Rubber. *Polym. Sci. Ser. B* **2019**, *61*, 856–864.
11. Nimpaiboon, A.; Amnuaypornsri, S.; Sakdapipanich, J. Obstruction of storage hardening in nr by using polar chemicals. *Rubber Chem. Technol.* **2016**, *89*, 358–368. [CrossRef]
12. Ferry, J.D. *Viscoelastic Properties of Polymers*; John Wiley & Sons: Hoboken, NJ, USA, 1980.
13. Montha, S.; Suwandittakul, P.; Poonsrisawat, A.; Oungeun, P.; Kongkaew, C. Maillard Reaction in Natural Rubber Latex: Characterization and Physical Properties of Solid Natural Rubber. *Adv. Mater. Sci. Eng.* **2016**, *2016*, 1–6. [CrossRef]
14. Rungsanthie, K.; Suwanruji, P.; Tantatherdtam, R.; Chollakup, R. Effect of non-rubber components on viscosity stabilization of natural rubber. *Int. Conf. Polym. Process. Soc.* **2012**, *11*, 15.
15. Smitthipong, W.; Tantatherdtam, R.; Rungsanthien, K.; Suwanruji, P.; Klanarong, S.; Radabutra, S.; Thanawan, S.; Vallat, M.F.; Nardin, M.; Mougin, K.; et al. Effect of Non-Rubber Components on Properties of Sulphur Crosslinked Natural Rubbers. *Adv. Mater. Res.* **2013**, *844*, 345–348. [CrossRef]
16. Amnuaypornsri, S.; Sakdapipanich, J.; Toki, S.; Hsiao, B.; Ichikawa, N.; Tanaka, Y. Strain-Induced Crystallization of Natural Rubber: Effect of Proteins and Phospholipids. *Rubber Chem. Technol.* **2008**, *81*, 753–766. [CrossRef]

© 2020 by the authors. Licensee MDPI, Basel, Switzerland. This article is an open access article distributed under the terms and conditions of the Creative Commons Attribution (CC BY) license (http://creativecommons.org/licenses/by/4.0/).

Article

# Decoupling the Contributions of ZnO and Silica in the Characterization of Industrially-Mixed Filled Rubbers by Combining Small Angle Neutron and X-Ray Scattering

Mariapaola Staropoli [1,*], Dominik Gerstner [2], Aurel Radulescu [3], Michael Sztucki [4], Benoit Duez [2], Stephan Westermann [1], Damien Lenoble [1] and Wim Pyckhout-Hintzen [5]

1. Luxembourg Institute of Science and Technology, L-4422 Belvaux, Luxembourg; stephan.westermann@list.lu (S.W.); damien.lenoble@list.lu (D.L.)
2. GOODYEAR S.A., L-7750 Colmar-Berg, Luxembourg; dominik_gerstner@goodyear.com (D.G.); benoit_duez@goodyear.com (B.D.)
3. Forschungszentrum Jülich/MLZ Garching, D-85748 Garching, Germany; a.radulescu@fz-juelich.de
4. European Synchroton Radiation Facility, F-38000 Grenoble, France; sztucki@esrf.fr
5. Forschungszentrum Jülich, D-52425 Jülich, Germany; w.pyckhout@fz-juelich.de
* Correspondence: mariapaola.staropoli@list.lu

Received: 28 January 2020; Accepted: 22 February 2020; Published: 25 February 2020

**Abstract:** Scattering techniques with neutrons and X-rays are powerful methods for the investigation of the hierarchical structure of reinforcing fillers in rubber matrices. However, when using only X-ray scattering, the independent determination of the filler response itself sometimes remains an issue because of a strong parasitic contribution of the ZnO catalyst and activator in the vulcanization process. Microscopic characterization of filler-rubber mixtures even with only catalytic amounts of ZnO is, therefore, inevitably complex. Here, we present a study of silica aggregates dispersed in an SBR rubber in the presence of the catalyst and show that accurate partial structure factors of both components can be determined separately from the combination of the two scattering probes, neutrons, and X-rays. A unique separation of the silica filler scattering function devoid of parasitic catalyst scattering becomes possible. From the combined analysis, the catalyst contribution is determined as well and results to be prominent in the correction scheme. The experimental nano-structure of the ZnO after the mixing process as the by-product of the scattering decomposition was found also to be affected by the presence or absence of silica in the rubber mixture, correlated with the shear forces in the mixing and milling processes during sample preparation. The presented method is well suited for studies of novel dual filler systems.

**Keywords:** SANS; SAXS; silica; zinc oxide; partial structure factors; small angle scattering

## 1. Introduction

Nanofiller particles added to elastomers are known to form hierarchical structures varying from clusters of the order of a few nm up to agglomerates in the range of several μm. As a consequence of this, the resulting composite materials show strongly improved mechanical properties, which lead to important applications as, e.g., most prominently in tires [1–4]. The complex arrangement of the nanoparticles in the mixtures is thereby at the basis of the associated reinforcing mechanisms. The understanding of the correlation between the filler structure and the ultimate macroscopic properties of the filled rubbers widely employs Small Angle Scattering (SAS) methods [5–11].

These techniques exploit the different sensitivity of the respective radiation probe for the components in the mixture, allowing the identification of the structural distribution of the different species. Whereas for Small Angle X-ray Scattering (SAXS) the difference in the electron density between typical soft and hard constituents determines the contrast, in the case of Small Angle Neutron Scattering (SANS) different contrasts depending on nuclear properties rather than on electronic ones may exist. In the ideal case, these contrasts can be tuned in order to elaborate the scattering of specific components only [12–14]. For multicomponent systems, the combination of both SAXS and SANS can be therefore used to highlight the contribution of different components to the total scattering function, due to the different but complementary scattering contrasts obtained on the same sample. The underlying manuscript will deal with an application of these two probes to identify the ZnO structure for the first time in such industrial silica-containing composites, allowing its subtraction from the structural contribution of the reinforcing fillers.

ZnO is typically used as an activator in the vulcanization process and is known to improve the tire abrasion resistance and heat transfer at braking as well as to reduce the rubber shrinking during the molding process [15–17].

For conventional ZnO nanoparticles with a radius between 20 and 90 nm [18], however, the scattering contribution will occur in the same scattering vector range as the silica filler aggregates. This interference impairs, therefore, a direct evaluation of the scattering response of the reinforcing fillers, which is the only component determining the performance of the rubber. The problem of the interference and also additional inhomogeneities have been discussed in the literature [5,6,14,19–22]. Due to the difficulty in disentangling the contribution from the catalyst and the filler particles, the ZnO contribution was sometimes neglected or a simple weighted subtraction of a ZnO-containing background was applied as data correction. Recently, a more accurate correction was reported [22], based on different scattering techniques. The weighted subtraction assumed a priori an identical contribution of ZnO in filled and non-filled compounds. In addition, more recently, the use of contrast variation in such mixtures was discussed. With anomalous small angle X-ray scattering (ASAXS) at the Zn edge, which can be realized only at synchrotron sources varying the energy of the X-ray beam, a certain variation of the scattering length of the ZnO might be obtained, making it only partially contributing to the total intensity [19]. Alternatively, but for the case of neutron scattering another contrast variation procedure, in which the nuclear polarization of hydrogen atoms was modified, was applied for such compounds. However, the latter is technically difficult and is not compatible with standard neutron instrumentation [23]. In the underlying work using a combination of scattering techniques, we aim to extract the form and structure factors belonging to the silica filler complex in the presence of the full cure package and simultaneously of the necessary presence of ZnO.

Dealing with these problems is an important aspect for an enhanced understanding and the development of structure-property relationships. For the extraction of the single contributions to the total scattering function, we make use of a polymer-based approach in analogy to the Singular Value Decomposition (SVD) method widely applied to multi-component 'green' nanocomposites in the presence of solvents [14,24] but which relies on several samples with different labeling and contrasts. For our purpose, the unfilled rubber will be treated as a three-component mixture formed by the processing oil, matrix polymer, and ZnO [25], all leading to different scattering contrasts. The same approach is then extended to the case of silica-filled compounds in order to separate the contribution of silica and ZnO. This approximation is reasonable as these components have the strongest scattering contribution due to not only their concentration in the mixture. As we show, the approach relies on the combination of X-ray and neutron scattering intensities on identical samples and on the different contrast factors involved in the two methods to decouple the partial contributions to the total intensity. The ZnO correction is required, and its contribution appears in all samples analyzed.

## 2. Materials and Methods

### 2.1. Samples and Preparation

The SBR used in this work has high styrene (40%) and medium vinyl content (24%). The random copolymer is characterized by a glass transition temperature of −34.5 °C and is pre-mixed with 37.5 phr (parts in weight per 100 parts of rubber) of an aromatic distillate oil (TDAE). The rubber was mixed with amorphous precipitated silica (1165 MP, Rhodia/Solvay Group, Aubervilliers, France) of a specific surface of 165.8 m$^2$/g (N2 BET). Several samples with different silica volume fractions were prepared in a standard way [21] using an internal mixer of the Banbury type, followed by a two-roll milling process. Rubber and fillers were mixed in the non-productive stage using bis-3-triethoxysilylpropyldisulfide (Si266) as a coupling agent. In this stage N-(1,3)-dimethylbutyl-N-phenyl-p-phenylenediamine was used as an antioxidant while N -cyclohexyl-2-benzothiazolesulfeneamide and diphenylguanidine, respectively, as first and secondary accelerators. During the productive stage, the curing package was introduced, including ZnO with a specific surface of 5 m$^2$/g used as the catalyst. Samples with different silica mass fractions: 0 (A), 30 (B), 60 (C) phr, and an unfilled sample free of ZnO (E) were studied in this work. The sample list and the silica filling degree are specified, respectively, in Tables 1 and 2. After the milling process, the slabs were cured by compression-molding vulcanization at 170 °C for 10 min in a hydraulic press, and square sheet samples of thickness ~0.7mm were obtained.

**Table 1.** Composition of investigated samples. From left to right column the complexity is decreased. The compositions are given in parts per 100 g rubber (phr).

| Component | phr | | | |
|---|---|---|---|---|
| | Sample C (60 phr) | Sample B (30 phr) | Sample A (unfilled) | Sample E (unfilled, no ZnO) |
| SBR | 137.5 | 137.5 | 137.5 | 137.5 |
| Antioxidant | 0.75 | 0.75 | 0.75 | 0.75 |
| Oil | 2 | 2 | 2 | 2 |
| Activator | 3 | 3 | 3 | - |
| Silica 1165MP | 60 | 30 | - | - |
| Silane Si266 | 4.8 | 2.4 | - | - |
| Antioxidant | 1.75 | 1.75 | 1.75 | 1.75 |
| Antioxidant | 0.5 | 0.5 | 0.5 | 0.5 |
| ZnO | 2.5 | 2.5 | 2.5 | - |
| Sulfur | 0.8 | 0.8 | 0.8 | 0.8 |
| Accelerator | 2.4 | 2.4 | 2.4 | 2.4 |
| Accelerator | 1.5 | 1.5 | 1.5 | 1.5 |

**Table 2.** Silica content in the different samples. phr is parts per 100 g of rubber.

| Sample | Total phr | Spec Gravity | Silica phr | Silica Density (g/cm$^3$) | Silica Volume Fraction ($\phi$) |
|---|---|---|---|---|---|
| C | 217.5 | 1.14 | 60 | 2.2 | 0.143 |
| B | 185.1 | 1.07 | 30 | 2.2 | 0.079 |
| A | 152.7 | 0.98 | 0 | 2.2 | 0 |
| E | 147.2 | 0.97 | 0 | 2.2 | 0 |

## 2.2. Small Angle X-Ray Scattering (SAXS)

SAXS experiments were performed at ID02 [26], ESRF, (Grenoble, France) at detector distances 31, 8, and 2 m and at a wavelength 1 Å. The usable scattering vector $q$, defined as $q = \frac{4\pi}{\lambda} \sin \frac{\theta}{2}$, extended from $5 \times 10^{-4} < q < 0.4$ Å$^{-1}$. $\theta$ is the scattering angle. The $q$-range differs along $x$ and $y$ direction due to the asymmetric beam stop. 2D data were binned to $1920 \times 1920$ channels with $0.09 \times 0.09$ mm$^2$ width. They were corrected for empty beam, dark current, and radially averaged (where possible) using standard procedures of the beamline. Absolute scattering intensities in [cm$^{-1}$] units were obtained by calibration with the scattering of water at room temperature. A background, which was assigned to the left wing of the amorphous halo, was subtracted linearly as only data up to roughly $q = 0.1$ Å$^{-1}$ were used. For reference sake, sample E was re-measured and calibrated independently vs. a glassy carbon standard at the University of Erlangen. Intensities were found to match in excellent agreement.

## 2.3. Small Angle Neutron Scattering (SANS)

SANS experiments were conducted at the KWS-2 diffractometer of MLZ, (Garching, Germany) using three detector distances of respectively 20, 8, and 2 m at a wavelength of 10 Å. The scattering vector range spanned between $1 \times 10^{-3} < q < 0.4$ Å$^{-1}$. 2D-detector data were obtained in $142 \times 142$ channels of $8 \times 8$ mm$^2$ width. The wavelength distribution $\frac{\Delta \lambda}{\lambda}$ was 10%. For the absolute scaling to [cm$^{-1}$] the incoherent scattering level of a secondary standard Plexiglass was used. The two-dimensional data in $142 \times 142$ channels were corrected pixelwise for empty beam scattering, detector sensitivity, and background noise using B$_4$C as beam blocker and subsequently radially averaged. In the case of anisotropic scattering patterns, sectors with total opening angles of 10° along the main axes of the anisotropic patterns were applied to the scattering patterns. The incoherent background arising from the hydrogenous polymer fraction and the other additives was determined from the high $q$-range and was found to be in good agreement with the estimated incoherent scattering. Likewise, as for SAXS, sample E was re-measured and put to an absolute scale at the KWS-3 diffractometer of MLZ using the direct beam method. Moreover, here, the calibrated intensities of KWS-2 and KWS-3 matched ideally.

## 3. Results

The scattering model approach used in this work is developed for the non-filled samples with and without ZnO and further extended to the case of filled systems. Due to the strong scattering of the ZnO catalyst, this component cannot be neglected in the analysis despite its irrelevance in the reinforcing mechanism and mechanical performance of the rubber. It leads, therefore, to treatments that necessarily involve its structure. Although the matrix is ideally expected to contribute as a simple background to the scattering of filled rubbers, it is known that inhomogeneities due to the crosslinks possibly lead to an additional contribution [13,14]. This contribution is normally observed in the lowest $q$ region as a characteristic power law. The pure vulcanized SBR/oil sample has often been modeled then by means of a classical Ornstein-Zernike approach in analogy to the case of a semi-dilute polymer solution or swollen gel assuming that the processing oil acts as a contrast agent for the meshes of the network [25]. In the presence of the curing activator, on the other hand, at low $q$ ZnO nanocrystals or domains could become visible due to their strong and different contrast with the embedding matrix and at high $q$ they might exhibit Bragg peaks. The overall dimensions of the ZnO crystals could be estimated in principle with an additional contribution taken into account by a Debye-Bueche law. In the literature, the simple incoherent superposition of Ornstein-Zernike and Debye-Bueche laws was applied on a regular basis with limited success [25,27–29]. However, the estimation of the ZnO contribution to the total scattering could be more complex due to its interaction with other components of the mixture as sulfur and accelerators, which can be absorbed on the ZnO particles surface. These interactions could lead to a different average contrast between the ZnO and the matrix and possibly to a non-negligible deviation from the incoherent sum of two contrast-weighted partial structure factors.

The coherent scattering intensity of a general three-component mixture $I(q)$ is theoretically written as

$$I(q) = \sum_{i,j=1}^{n=3} \rho_i \rho_j S_{ij}(q) \qquad (1)$$

where the pre-factors of the partial structure factors $S_{ij}(q)$ are their scattering length densities $\rho_i, \rho_j$ (here abbreviated as SLD). The total scattering function is then defined as a linear combination of all partial structure factors resulting from all the intra ($S_{ii}$) and inter-component ($S_{ij}$) correlation functions. The approach is first applied to the determination of different contributions in non-filled samples, which can be approximated to two- and three-components systems. In the following section, we discuss first the case of the unfilled rubber and generalize the same concept then to the reinforced case.

### 3.1. The Unfilled Case

In the following section, the polymer-based approach will be applied to the description of the scattering function of an unfilled rubber. Sample E consists of the rubber mixture but was vulcanized without ZnO. The SBR and oil constitute then a classical two-component system. On the contrary, Sample A is already a three-component but two-phase mixture containing SBR polymer, processing extender oil, and ZnO. The other curing agents can be neglected in scattering evaluations. Taking sample A as the working example in further treatment and making use of the incompressibility condition, one obtains for the intensity [25]:

$$I(q) = (\rho_{pol} - \rho_{oil})^2 S_{pol} + (\rho_{ZnO} - \rho_{oil})^2 S_{ZnO} + 2(\rho_{pol} - \rho_{oil})(\rho_{ZnO} - \rho_{oil}) S_{pol-ZnO} \qquad (2)$$

Where $\rho_{pol}$, $\rho_{oil}$ and $\rho_{ZnO}$ are the SLD for the matrix, oil, and ZnO, respectively. Here, the oil is considered as the background component (or solvent in the polymer-based approach). $S_{pol}$, $S_{ZnO}$ are the partial structure factors respectively for the polymer and the ZnO while $S_{pol-ZnO}$ takes into account the interaction between these two components. The q-dependence of the structure factors was omitted for readability. The intensity from a background of oil does not appear explicitly therefore in the q-dependent scattering functions. It can be easily seen that this general three-component treatment for sample A can be simplified to obtain the case of sample E. If no ZnO is present in the compound, then the second and third term of the equation vanish, and the intensity for the resulting two-component system is written in the usual well-known form as $I(q) = (\rho_{pol} - \rho_{oil})^2 S_{pol}$. For sample E, the incompressibility condition leads to $\varnothing_{pol} + \varnothing_{oil} = 1$. $\varnothing$ corresponds to the volume fraction of the components. The scattering contrast remains, therefore, a delicate fitting parameter. For sample A, within the approximation that only the main components contribute to the scattering function, the following condition is valid: $\varnothing_{pol} + \varnothing_{oil} + \varnothing_{zn} = 1$. The small amount of activator allows neglecting interactions between neighboring ZnO nanocrystals. Therefore, the scattering contribution of the activator could ideally be described in terms of a particle form factor. Furthermore, the low concentration, as well as the non-specific interaction of the inorganic ZnO particles with the oil or SBR, are very good reasons to neglect also the cross-term $S_{pol-ZnO}$ in the scattering equation of the three-component system [12,25,29,30] which is then reduced to:

$$I(q) = (\rho_{pol} - \rho_{oil})^2 S_{pol} + (\rho_{ZnO} - \rho_{oil})^2 S_{ZnO} \qquad (3)$$

This approximation applies to sample A containing less than 1% of ZnO (2.5phr). However, the contrast factors in this solid cross-linked or vulcanized mixture differ from those of the system in which ZnO would be considered as one component in an effective matrix of SBR and oil due to interaction with the curing package. The subtraction of $S_{ZnO}$ from the scattering function is, therefore, not straightforward and requires an estimation of the contrast factors involved. Moreover, the size of the particular ZnO component is a pre-requisite for adequate correction. Different scenarios could determine the effect of the ZnO on the scattering function of the filled rubber. In the case of large ZnO particles, the Guinier regime of this component is expected at too small scattering vectors q. The only contribution would, therefore, be its surface scattering with its approximately observed characteristic $q^{-4}$ slope in addition to the Ornstein-Zernike behavior expected for the rubbery matrix. This particular $q^{-4}$ at a low scattering vector occurs in the case of smooth particles that have a well-defined flat and strong interface with the surrounding matrix. In the case of rough or fractal surfaces, however, the detected power law would differ from −4. The power law exponent X depends on the surface fractal dimension $D_s$ by $X = (6 - D_s)$. On the other hand, if the mixing and milling process results in better dispersed, relatively small ZnO particles, their form factor would interfere with the scattering contribution resulting from the fillers and corrections that require the estimation of contrast factors become necessary. Unlike a SVD method based on a number of contrast parameters that exceeds the number of unknown partial structure factors [12–14,23,27,29,30], in our particular case, it is impossible to induce additional contrast factors into a cross-linked rubber without a previous labeling of components by, e.g., deuteration and inevitably leading to intrinsically different samples [25]. For this reason, the decomposition of the measured signal intensity has been performed using a linear combination of experimental data obtained by X-ray and neutron scattering measurements applied to the same sample. Due to the different contrast factors in the two experiments, the partial structure factors for fillers and ZnO can be extracted for each q in analogy to the SVD approach. The resulting set of two linear equations with two unknowns, i.e., the partial structure factors can be solved exactly. On the other hand, the merging of SANS and SAXS data into one evaluation may be affected by the different q-resolution of the two methods. Especially in the case of SANS, resolution and polydispersity effects are fully correlated. In the lowest q range, this does not affect the data. Resolution- and non-resolution corrected data differ only in the order of 2%–3% and do not influence the results. As such, the determination of the scattering functions strongly depends on the quality of the absolute calibrations as well as on the correct evaluation of the SLD parameters for all the components taken into account within this approximation. Therefore, absolute care in the absolute normalization of the intensities is important.

In our presented approach, the partial form factors result from the exact solution of a system of two linear equations. For sample A the expression of $S_{pol}$ and $S_{ZnO}$ is given by:

$$S_{pol} = \frac{(\rho_{ZnO} - \rho_{oil})_X^2 I_{AN} - (\rho_{ZnO} - \rho_{oil})_N^2 I_{AX}}{(\rho_{pol} - \rho_{oil})_N^2 (\rho_{ZnO} - \rho_{oil})_X^2 - (\rho_{pol} - \rho_{oil})_X^2 (\rho_{ZnO} - \rho_{oil})_N^2} \tag{4}$$

$$S_{ZnO} = \frac{(\rho_{pol} - \rho_{oil})_N^2 I_{AX} - (\rho_{pol} - \rho_{oil})_X^2 I_{AN}}{(\rho_{pol} - \rho_{oil})_N^2 (\rho_{ZnO} - \rho_{oil})_X^2 - (\rho_{pol} - \rho_{oil})_X^2 (\rho_{ZnO} - \rho_{oil})_N^2} \tag{5}$$

The indices N and X stand for neutron and X-ray, respectively. $I_{AN}$ and $I_{AX}$ correspond to the experimentally measured absolute intensities of sample A in both N and X case. They are corrected for instrumental backgrounds of any source (see Experimental). It is obvious that the contrasts as weighting factors are of extreme importance in further evaluation. Therefore, SLDs of the components, as well as the consistency of absolute calibration of intensities, have to be verified with the greatest care. We remind that the structure factor is defined as $S(q) = \phi\, V\, P(q)$ with V the volume of the scattering entity and $P(q)$ its q-dependent form factor [25]. The weighed subtractions in Equations (7) and (8) were obtained in the common SANS $q$- range after a cubic spline was fitted to the SAXS data.

*3.2. Contrast Considerations*

As discussed above, deriving accurate partial structure factors of single components that give rise to the experimental scattering intensity requires accurate estimations of the contrasts of the particular components in the mixtures. Contrast is generally defined as the square of the difference in scattering length densities (SLD) between two scattering entities. Whereas for neutrons the scattering length $b$ is a nuclear property, for X-rays the equivalent parameter is a function of the electronic number of the component. The SLD is then the sum of the scattering lengths per volume and is defined for neutrons and X-ray respectively as: $SLDN = \frac{\sum_{i=1}^{n} b_i}{V_{molecule}}$, $SLDX = \frac{\sum_{i=1}^{n} e_i}{V_{molecule}} r_e$ where $n$ is the number of atoms of type $i$ and $e$ is the number of electrons. $r_e$ is the radius of the electron, $2.8 \times 10^{-13}$ cm.

The calculated parameters for the relevant components using the available chemical information are summarized in Table 3.

**Table 3.** Mass densities and scattering length densities for the most significant components of the mixture (amount > 2.5 phr). Only for SBR, the experimental value for the SLD in the case of X-rays was optimized against experimental data (see text).

| Component | Mass density g/cm$^3$ | SLD neutron cm$^{-2}$ | SLD X-ray cm$^{-2}$ | Remarks |
|---|---|---|---|---|
| Oil | 0.94 | $1.036 \times 10^{10}$ | $8.70 \times 10^{10}$ | C$_{7.4}$H$_{8.4}$ |
| SBR | 0.95 | $0.910 \times 10^{10}$ | $8.46 \times 10^{10}$ opt. | C$_{5.6}$H$_{6.8}$ 40% styrene |
| Silica | 2.10 | $3.320 \times 10^{10}$ | $17.9 \times 10^{10}$ | Solvay 165 m$^2$/g |
| ZnO | 5.60 | $4.760 \times 10^{10}$ | $43.3 \times 10^{10}$ | BET 5m$^2$/g |
| Coupling agent Silane (Si 266) | 1.03 | $0.149 \times 10^{10}$ | $9.54 \times 10^{10}$ | C$_{18}$H$_{42}$O$_6$S$_2$Si$_2$ |
| Filled rubber samples | 0.95 | | | SBR/oil/silane: $v/v$ fractions |
| Effective Matrix (B) | | $0.929 \times 10^{10}$ | $8.546 \times 10^{10}$ | 0.71/0.27/0.02 |
| Effective Matrix (C) | | $0.905 \times 10^{10}$ | $8.576 \times 10^{10}$ | 0.69/0.26/0.05. |

The contrast factors between the significant components were then estimated on the basis of the calculated SLD and are reported in Table 4. Sample E, considered as an ideal two-component system was used to verify the estimated contrast factors with the experimental scattering intensities. Whereas the SLD of SBR and the corresponding contrast of SBR-to-oil in sample E is in good agreement for the SANS case, a clear discrepancy was observed in SAXS, which led to higher-than-expected intensities. Doubts about the absolute calibrated intensity levels can be excluded, however, in view of the elaborate cross-references at different large-scale facilities. The reason for the reversed intensity order is then to be sought in the assumptions about the components.

**Table 4.** Contrasts for SANS and SAXS analysis.

| Neutron Contrasts [cm$^{-4}$] | Oil | SBR | Silica | ZnO |
|---|---|---|---|---|
| Oil | - | | | |
| SBR | $1.58 \times 10^{18}$ | - | | |
| Silica | $5.22 \times 10^{20}$ | $5.81 \times 10^{20}$ | - | |
| ZnO | $13.87 \times 10^{20}$ | $14.82 \times 10^{20}$ | $2.07 \times 10^{20}$ | - |
| Filled rubber samples | - | - | B: $5.72 \times 10^{20}$ C: $5.83 \times 10^{20}$ | B: $14.67 \times 10^{20}$ C: $14.86 \times 10^{20}$ |
| **X-ray contrasts [cm$^{-4}$]** | **Oil** | **SBR** | **Silica** | **ZnO** |
| Oil | - | | | |
| SBR | $5.76 \times 10^{18}$ | - | | |
| Silica | $84.64 \times 10^{20}$ | $89.1 \times 10^{20}$ | - | |
| ZnO | $119.7 \times 10^{21}$ | $121.4 \times 10^{21}$ | $6.45 \times 10^{22}$ | - |
| Filled rubber samples | - | - | B: $87.50 \times 10^{20}$ C: $86.94 \times 10^{20}$ | B: $120.8 \times 10^{21}$ C: $120.6 \times 10^{21}$ |

Deviations can also be partly due to uncertainties in the thickness of each sample, the statistical and systematic errors of the data themselves, and the quality and intrinsic accuracy of the absolute calibrations. Inaccurate estimation of the SLD of SBR could be also due to the neglect of the components present in phr amounts <2.5 phr. On the other hand, the SLD of the oil in both neutron and X-ray data, calculated from the element analysis can be accurately determined.

To comply with the aforementioned higher SAXS intensity, the SLD of the SBR in the X-ray case should then be $8.46 \times 10^{10}$ cm$^{-4}$, i.e., lower by 4%–5% lower than the theoretical value. It is worth noting that the SANS method is insensitive, in terms of contrast, to the alkane-like compounds of the matrix. The coherent SLD contribution of a $CH_2$ section is about 0 for SANS ($b_C$ = 6.65 fm, $b_H$ = −3.74 fm). This is not the case for SAXS where each atom always adds positively to the electron density. Thus, the optimized value for SBR indicated in Table 4 is assumed to affect only SAXS intensities. The silane component, used as a coupling agent in the filled samples, is in the same way involved in the uncertainty of the SLD determination due to its reaction with both the silica and the polymer matrix. In our estimation, we assume that it is distributed evenly in the matrix.

### 3.3. The Filled Case

The previous relations and considerations applied to the case of unfilled silica rubbers. In the more interesting case of silica-filled systems, the former three-component system now evolves to at least a four-component mixture. A further complication, besides the addition of silica, in the decomposition of the intensities of samples B and C (30 and 60 phr silica, respectively) stems from the presence of the silane component used as a coupling agent. For this reason, the estimation of the contrast factors is based on the estimation of an effective SLD of the matrix including the silane weighted volume fraction. We then obtain again workable equations like before. As in the case of the non-filled rubbers, we neglected all other ingredients with volume fraction below 2.5 phr as they contribute mainly to the incoherent background. The scattering intensity for samples B and C can then be expressed in terms of the partial contributions as:

$$I(q) = (\rho_{sil} - \rho_{eff})^2 S_{sil} + (\rho_{ZnO} - \rho_{eff})^2 S_{ZnO} + 2(\rho_{sil} - \rho_{eff})(\rho_{ZnO} - \rho_{eff}) S_{sil-ZnO} \qquad (6)$$

Note the similarity with Sample A, whereas now the effective matrix replaces the oil. Again, the cross term appearing in Equation (6) can be neglected due to the low fraction of ZnO, leading to negligible correlations with silica. The partial form factors can then be expressed using the same approach as before:

$$S_{sil} = \frac{(\rho_{ZnO} - \rho_{eff})_X^2 I_N - (\rho_{ZnO} - \rho_{eff})_N^2 I_X}{(\rho_{sil} - \rho_{eff})_N^2 (\rho_{ZnO} - \rho_{eff})_X^2 - (\rho_{sil} - \rho_{eff})_X^2 (\rho_{ZnO} - \rho_{eff})_N^2} \qquad (7)$$

$$S_{ZnO} = \frac{(\rho_{sil} - \rho_{eff})_N^2 I_X - (\rho_{sil} - \rho_{eff})_X^2 I_N}{(\rho_{sil} - \rho_{eff})_N^2 (\rho_{ZnO} - \rho_{eff})_X^2 - (\rho_{sil} - \rho_{eff})_X^2 (\rho_{ZnO} - \rho_{eff})_N^2} \qquad (8)$$

where $\rho_{sil}$ and $\rho_{eff}$ are respectively the SLD for silica and effective matrix including the silane component. In this way, the relevant structure factors of both contributing components can be extracted. A similar approach is lacking up-to-now in the literature. For the first time, microscopic details of the mixing process and/or vulcanization process in a reactive mixture could be accessible.

While interparticle interactions can be excluded for ZnO as well as interactions between the ZnO activator and silica filler particles, the same assumption cannot necessarily be applied to the silica particles. $S_{sil}$ as the cluster form factor generally consists of the product of a particle form factor with an intra-cluster and even an inter-cluster structure factor taking into consideration the interaction between them as we recently showed [31]. However, the full description of the hierarchical structure of the fillers is out of the scope of this work. Here, the decomposition of the different contributions in a mixed-filler system and the evaluation of the ZnO effect is in the foreground.

In order to obtain a common $q$- range and identical step size between the data points for X-ray and neutron results and allow the weighed subtractions in Equations (7) and (8), a cubic spline is applied to the SAXS data.

## 4. Results and Discussion

First, the unfilled samples are analyzed. Figure 1 shows the comparison between absolutely calibrated total SANS and SAXS results obtained for samples E and A, before the subtraction of the incoherent background. For sample A, the contribution of 2.5 phr ZnO is immediately evidenced by a ~ 10-fold increased intensity between A and E and an associated different $q$-dependence in the q-range $10^{-3} < q < 10^{-2}$ $\text{Å}^{-1}$ and a crystalline Bragg peak around $0.15$ $\text{Å}^{-1}$.

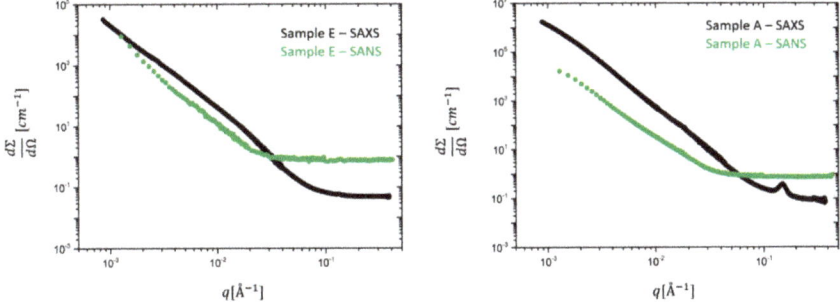

**Figure 1.** Absolutely calibrated SAXS (black) and SANS (green) data for samples E (left) and A (right). The data are not background-corrected.

In Figure 2, we compare extracted $S_{pol}$ and $S_{ZnO}$ with the SAXS intensities of sample A and E. The contribution of the ZnO form factor to the SANS curves results to be irrelevant as we show later. The partial form factors $S_{pol}$ and $S_{ZnO}$, respectively in red and orange, were shifted by an arbitrary factor along the $y$ axis for the sake of comparison with the experimental data.

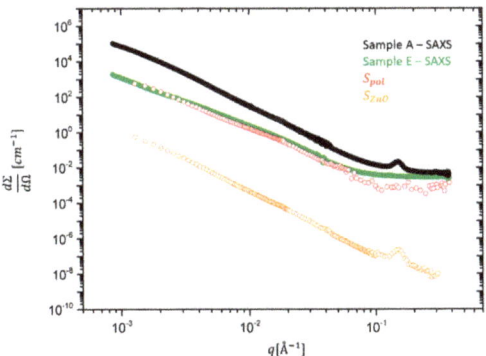

**Figure 2.** Partial structure factors $S_{pol}$ (red) and $S_{ZnO}$ (orange) as derived from sample A in comparison with absolutely calibrated SAXS data for samples A (black) and E (green). $S_{pol}$ was arbitrarily shifted to be compared with sample E.

The figure shows that the scaled $S_{pol}$ as derived from sample A is perfectly congruent with the intensity of sample E, leading to the conclusion that—maybe unexpectedly—the vulcanization with and without ZnO would not lead to differences in the matrix structure detectable by scattering in length scales smaller than ~3000 Å. On the other hand, $S_{ZnO}$ shows a visible power law contribution for $q < 10^{-1}$, indicative for additional scattering to $S_{pol}$.

The same approach defined for the extraction of the single contributions from the total scattering function can be transferred to the case of the filled samples.

Going to filled rubbers, Figure 3 shows a comparison between the SAXS and SANS averaged scattering intensities obtained for samples B and C (respectively, 30 and 60 phr silica). Besides the different absolute intensity and $q$-range due to instrumental factors, SANS and SAXS curves show different low-to-intermediate $q < 0.01$ Å$^{-1}$ behavior. In addition, it can be seen that for $q > 0.01$ Å$^{-1}$ the silica particle structure dominates in both SAXS and SANS.

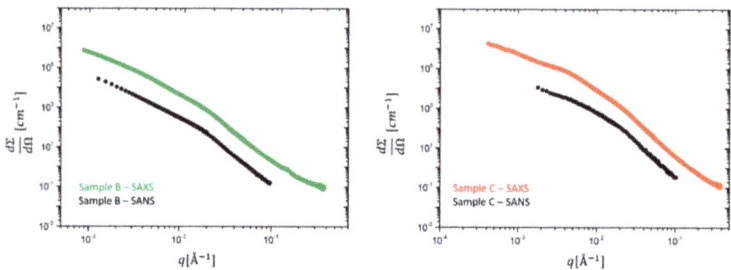

**Figure 3.** Experimental SANS and SAXS data in absolute intensity units cm$^{-1}$ for the ZnO-containing filled rubbers, sample B (left) and sample C(right). The number of data points of the ultrahigh-resolution SAXS data was about 4500 and was reduced to a SANS-compatible lower resolution of about 350 data points by means of a cubic spline fitting procedure.

The power law exponent in this scale range differs from −4 and stands for a deviation from the ideal spherical particle behavior. The value of approximately −3.7 observed here was also found in the literature for the same type of silica particles. Since this slope is not affected by particle polydispersity it indicates a rough silica surface. For the SANS data of sample C a closer inspection of the lower $q < 0.01$ Å$^{-1}$ reveals a power law region which is the signature of silica aggregates with a mass fractal dimension of ~2.2 and an onset of a Guinier plateau is seen. In sample B a Guinier region cannot be detected due to the obvious formation of larger clusters. This finding corroborates the finding that the cluster size decreases when the filling degree is increased. In addition, some orientation seems to be visible in the higher filling degree sample (C), and therefore, the data must be analyzed along two perpendicular directions. We have selected only the vertical direction as it allows obtaining lower $q$-values due to the horizontal alignment of the beam stop stick in SAXS.

Besides the different extension in the $q$-range due to instrumental factors, the SAXS data show in both cases a shallow peak or shoulder around $q \sim 4\ 10^{-3}$ Å$^{-1}$, which cannot be seen in the SANS results. This is attributed to ZnO, which has a much stronger contribution in the X-ray intensity than in the neutron data, as already observed for the unfilled rubbers. Interestingly, sample C presents a more pronounced peak with respect to sample B, probably due to the lower contribution of the smaller clusters to the scattering function. In Figure 4, the partial contributions $S_{ZnO}$ and $S_{Sil}$ are multiplied by the contrast factors, as shown in Equation (6) and the experimental SAXS intensities are re-constructed. ZnO contributes to 10% to the total intensity for sample B (30 phr) but amounts to ~50% for C. The scattering at high q is fully determined by silica and consolidates a rather particle-like morphology for ZnO, its form factor dropping with $q^{-4}$.

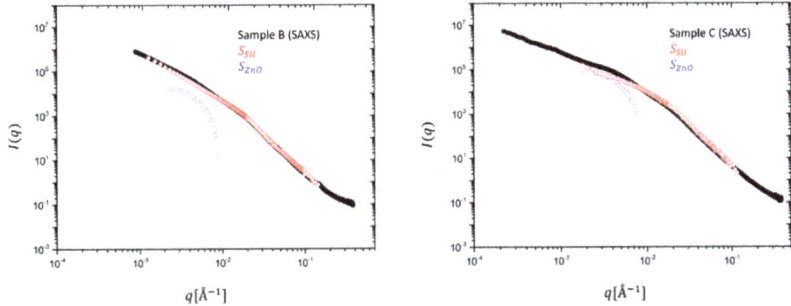

**Figure 4.** Experimental SAXS data in absolute intensity units cm$^{-1}$ for samples B (30 phr) (left) and C (60 phr) (right). Partial form factors for silica ($S_{pol}$, red) and ZnO ($S_{ZnO}$, blue) were multiplied by the contrast factors to re-construct the intensity.

In addition and as anticipated, the power law exponent of approximately −2, found in the ultra-low $q$ region of the SAXS data could be attributed to the connectivity of clusters into a network when the volume fraction is increased above their percolation threshold [2–4].

Except for these observations, however, the main result and highlight of this first decomposition of SAXS intensities in partial structure factors is the important extraction of the scattering contribution of minute amounts of ZnO and some microscopic details from a clear $q$-dependence. A comparison with Figure 5 showing the same decomposition for SANS data, highlights that due the different contrast, the contribution of ZnO to the total scattering intensity is about two orders of magnitude lower than the one of the silica. However, even the catalytic content of ZnO impedes the analysis in the full $q$ range since the intensity of the extracted partial structure factor drops very fast to below the sensitivity limit of the SAXS or SANS machine. Both SANS curves highlight that the main contribution comes almost entirely from silica. The catalyst has an almost negligible effect on the scattering profile. In SAXS, however, both partial intensities contribute to almost the same amount.

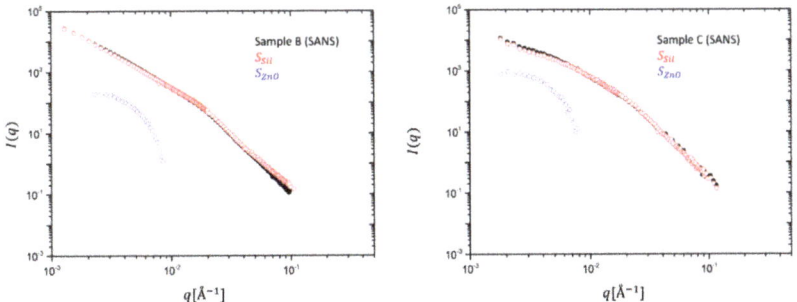

**Figure 5.** Experimental SANS data in absolute intensity units cm$^{-1}$ for samples B (30 phr) (left) and C (60 phr) (right). Partial form factors for silica ($S_{pol}$, red) and ZnO ($S_{ZnO}$, blue).

Thus, this analysis allowed us to evaluate the total scattering intensities as the sum of contributions of 2 components, similar to the case of a dual filler system. The structure of both in the mixture, as well as their mutual influence, can thereby be captured.

Let's first discuss ZnO. The stronger effect of the ZnO contribution in sample C over B is tentatively correlated to a simultaneous decrease of the silica contribution of smaller clusters.

The structure of ZnO, i.e., their $q$-dependence in the filled rubbers is, however, very different in comparison to that, extracted from the unfilled sample, as shown in Figure 6. In the case of sample A, the full scattering range could be accessed, and a typical power-law-like behavior as for silica is found.

The slope at high $q$ is close to $-3.8$ and indicates some deviation from smooth spherical behavior. Interestingly, the ZnO partial structure factor extracted from sample A does not exhibit any Guinier-like behavior at the lowest $q$, as shown by samples B and C. This indicates that either large ZnO clusters are formed in the case of the unfilled rubber in comparison to the silica-filled ones, which prevent the appearance of the Guinier region in the accessible $q$-range, or huge micron-sized strongly polydisperse and rough particles are still present, the size of which is larger than $R > 1/q_{min} = 1000$ Å $= 0.1$ $\mu$.

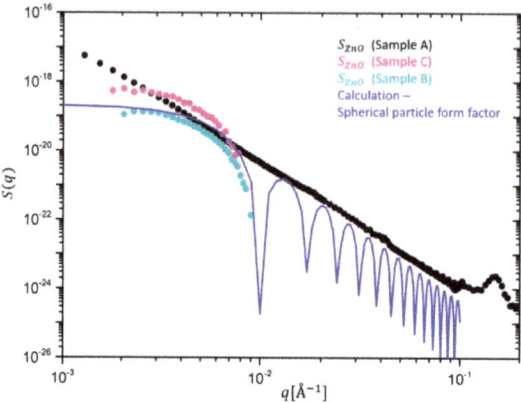

**Figure 6.** Partial structure factors for ZnO extracted from sample A (black), B (cyan), and C (magenta), and theoretical calculation of a spherical particle form factor (blue line).

The situation is entirely different if ZnO is mixed into the rubber together with a much larger amount of silica. Figure 6 clearly shows that after the mixing and vulcanization rather monodisperse single ZnO particles are dispersed, although the high $q$ region cannot be detected precisely anymore. We confirm this by comparing with a calculation of a monodisperse particle form factor (blue line in Figure 6). The computed intensity of a single ZnO particle follows in reasonable agreement the calculated spherical form factor, i.e., the Guinier plateau and the characteristic steep drop of the order of about 100 at q $\sim$ 0.01 Å$^{-1}$, i.e., $qR \sim 4.5$. The silica nanofillers, therefore, seem to affect the size of the ZnO nanocrystals and/or their aggregation. This could be explained considering the different shear forces imposed on the sample during the milling process. Such forces typically increase due to the presence of silica fillers in the rubber, which increase the melt viscosity by orders of magnitude and lead to a break-down of ZnO aggregates or crystals.

In order to compare the size of silica aggregates and ZnO species in the filled samples, the low $q$ scattering for both silica and ZnO partial structure factors is presented in a Guinier-type representation in Figures 7 and 8. The slopes are then related to $-R_g^2/3$.

Next, silica is discussed. For the silica-extracted form factor, shown in Figure 7, an average radius of gyration $R_g$ can be determined only for sample C because sample B does not exhibit a Guinier-like behavior or is more polydisperse. A cluster size of about 280 Å is obtained from the linear fit of the Guinier plot of $S_{sil}(q)$ and an increase is expected with a lower volume fraction. However, it has to be taken into account that the extraction of the silica cluster size from the Guinier-like description consists also of an approximation, as sample C does show a deviation from the simple Guinier behavior at low $q$. This extrapolation neither takes into account the polydispersity nor the interactions between the clusters. A more precise estimation of the silica cluster sizes is out of the scope of this work. A fit to a hierarchical scattering model is treated in a recently published work [31].

For samples B and C the extrapolation of the ZnO size, reported in Figure 8 yields almost comparable sizes of about 450–480 Å. It could not be applied to the ZnO form factor extracted from sample A as its low $q$-dependence clearly deviates from the Guinier behavior. Note that the maximum detectable size in the available q-range would approximately be ~1000 Å. Thus, the size of ZnO does not differ considerably from the silica aggregates in the case of the filled samples. Therefore, we infer that in the presence of silica, the processing conditions probably have a comparable effect on the de-agglomeration or destruction of the ZnO granulates. Besides this, the values found for the ZnO radius are comparable with the ones reported in the literature [22]. In other works, the size determination by means of other techniques for ZnO dispersed particles ranged from 300 to 500 Å as the result of the balance of fracture of aggregates and material resistance, i.e., intrinsic hardness in milling processes [32,33].

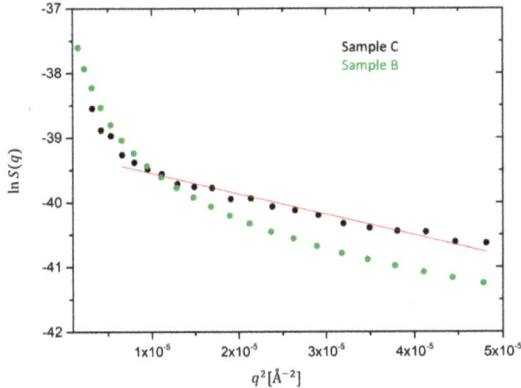

**Figure 7.** Guinier extrapolation for the silica cluster radius for sample C (black). The fit cannot be uniquely applied to sample B (green) due to the strong deviation from linearity.

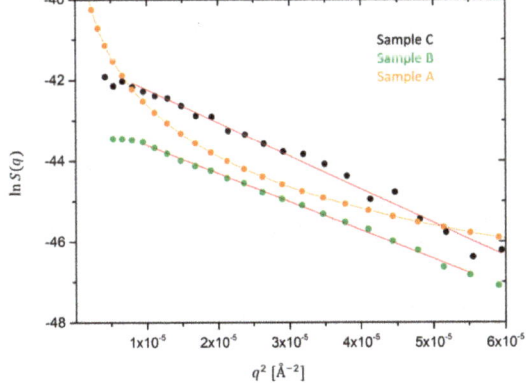

**Figure 8.** Guinier extrapolation of the ZnO radius for sample C (black) and B (green). Sample A (orange) clearly shows a deviation from the Guinier dependence.

## 5. Conclusions

We have shown that in industrially mixed rubber compounds a decomposition of the SAXS scattering intensity into its single contributing components, i.e., silica and ZnO is possible. We showed that SANS is hardly affected as it was postulated. The structure factor, the size and morphology of the silica aggregates and of the ZnO nanocrystals as an additional element of analysis can now be obtained. Latter strongly depends on the presence of silica during the mixing process. We observe that ZnO in silica-free samples forms aggregates or is of the micron-size and is characterized by a fractal surface. It appears to be structured on the same length scale as the silica filler. Therefore, a dependence of the size on the shear force exerted on the samples during the mixing process is suspected. Furthermore, the size of the ZnO catalyst is nearly constant for filled samples containing 30 and 60 phr of silica. As a consequence of this study, corrections for ZnO scattering based on computation with well-estimated contrast factors could be formulated. With estimates for the size of the resulting particle and assumptions for the average SLD of the effective matrix for the contrast, the contribution of $S_{ZnO}$ can be calculated theoretically for standard SAXS analyses. The proposed analysis is general and can be applied in the future to novel dual filler systems, which scatter both but differently strong in both SANS and SAXS.

**Author Contributions:** Conceptualization, M.S. (Mariapaola Staropoli) and W.P.; data curation, M.S. (Mariapaola Staropoli), A.R., M.S. (Michael Sztucki) and W.P.; formal analysis, M.S. (Mariapaola Staropoli) and W.P.; funding acquisition, S.W., D.L., B.D. and D.G.; investigation, D.G., M.S. (Michael Sztucki), A.R., M.S. (Mariapaola Staropoli) and W.P.; methodology, M.S. (Mariapaola Staropoli), W.P., D.G.; project administration, D.L., S.W., D.G. and B.D.; software, M.S. (Mariapaola Staropoli) and W.P.; supervision, D.L. and W.P.; visualization, M.S. (Mariapaola Staropoli); writing—original draft, M.S. (Mariapaola Staropoli) and W.P.; writing—review & editing, M.S. (Mariapaola Staropoli), D.G. and W.P. All authors have read and agreed to the published version of the manuscript.

**Funding:** The present project was supported by the National Research Fund Luxembourg (No (IPBG16/11514551/TireMat-Tech) whose financial support is acknowledged.

**Acknowledgments:** The authors thank T. Unruh and W. Gruber of the Friedrich-Alexander-Universität Erlangen-Nürnberg for the control measurement of the SAXS intensity of sample E and E. Scolan for scientific support. They thank V. Pipich of the MLZ for the corresponding service at KWS-3. The authors are grateful to the MLZ for the provision of neutron beam time at KWS-2 and KWS-3 in the framework of internal JCNS-1 access time. The authors thank Goodyear S.A. for the permission to publish the paper.

**Conflicts of Interest:** The authors declare no conflict of interest.

### References

1. Heinrich, G.; Straube, E.; Helmis, G. Rubber elasticity of polymer networks: Theories. In *Polymer Physics*; Springer: Berlin/Heidelberg, Germany, 1988; pp. 33–87.
2. Huber, G.; Vilgis, T.A.; Heinrich, G. Universal properties in the dynamical deformation of filled rubbers. *J. Phys. Condens. Matter* **1996**, *8*, L409. [CrossRef]
3. Medalia, A.I. Filler Aggregates and Their Effect on Reinforcement. *Rubber Chem. Technol.* **1974**, *47*, 411–433. [CrossRef]
4. Vilgis, T.A. Time scales in the reinforcement of elastomers. *Polymer* **2005**, *46*, 4223–4229. [CrossRef]
5. Baeza, G.P.; Genix, A.-C.; Degrandcourt, C.; Petitjean, L.; Gummel, J.; Couty, M.; Oberdisse, J. Multiscale Filler Structure in Simplified Industrial Nanocomposite Silica/SBR Systems Studied by SAXS and TEM. *Macromolecules* **2013**, *46*, 317–329. [CrossRef]
6. Belina, G.; Urban, V.; Straube, E.; Pyckhout-Hintzen, W.; Klüppel, M.; Heinrich, G. Microscopic deformation of filler particles in rubber under uniaxial deformation. *Macromol. Symp.* **2003**, *200*, 121–128. [CrossRef]
7. Ehrburger-Dolle, F.; Bley, F.; Geissler, E.; Livet, F.; Morfin, I.; Rochas, C. Filler networks in elastomers. *Macromol. Symp.* **2003**, *200*, 157–168. [CrossRef]
8. Jouault, N.; Vallat, P.; Dalmas, F.; Said, S.; Jestin, J.; Boué, F. Well-Dispersed Fractal Aggregates as Filler in Polymer-Silica Nanocomposites: Long-Range Effects in Rheology. *Macromolecules* **2009**, *42*, 2031–2040. [CrossRef]
9. Rharbi, Y.; Cabane, B.; Vacher, A.; Joanicot, M.; Boué, F. Modes of deformation in a soft/hard nanocomposite: A SANS study. *Europhys. Lett.* **1999**, *46*, 472–478. [CrossRef]
10. Schneider, G.J.; Göritz, D. Structural changes in precipitated silica induced by external forces. *J. Chem. Phys.* **2010**, *132*, 154903. [CrossRef]
11. Shinohara, Y.; Kishimoto, H.; Inoue, K.; Suzuki, Y.; Takeuchi, A.; Uesugi, K.; Yagi, N.; Muraoka, K.; Mizoguchi, T.; Amemiya, Y. Characterization of two-dimensional ultra-small-angle X-ray scattering apparatus for application to rubber filled with spherical silica under elongation. *J. Appl. Crystallogr.* **2007**, *40*, s397–s401. [CrossRef]
12. Endo, H.; Miyazaki, S.; Haraguchi, K.; Shibayama, M. Structure of Nanocomposite Hydrogel Investigated by Means of Contrast Variation Small-Angle Neutron Scattering. *Macromolecules* **2008**, *41*, 5406–5411. [CrossRef]

13. Ikeda, Y.; Higashitani, N.; Hijikata, K.; Kokubo, Y.; Morita, Y.; Shibayama, M.; Osaka, N.; Suzuki, T.; Endo, H.; Kohjiya, S. Vulcanization: New Focus on a Traditional Technology by Small-Angle Neutron Scattering. *Macromolecules* **2009**, *42*, 2741–2748. [CrossRef]
14. Takenaka, M.; Nishitsuji, S.; Amino, N.; Ishikawa, Y.; Yamaguchi, D.; Koizumi, S. Structure Analyses of Swollen Rubber-Filler Systems by Using Contrast Variation SANS. *Macromolecules* **2009**, *42*, 308–311. [CrossRef]
15. Heideman, G.; Datta, R.N.; Noordermeer, J.W.M.; van Baarle, B. Influence of zinc oxide during different stages of sulfur vulcanization. Elucidated by model compound studies. *J. Appl. Polym. Sci.* **2005**, *95*, 1388–1404. [CrossRef]
16. Kim, I.-J.; Kim, W.-S.; Lee, D.-H.; Kim, W.; Bae, J.-W. Effect of nano zinc oxide on the cure characteristics and mechanical properties of the silica-filled natural rubber/butadiene rubber compounds. *J. Appl. Polym. Sci.* **2010**, *117*, 1535–1543. [CrossRef]
17. Maghami, S.; Dierkes, W.K.; Noordermeer, J.W.M. Functionalized SBRs In Silica-Reinforced TireTread Compounds: Evidence for Interactions between Silica filler and Zinc Oxide. *Rubber Chem. Technol.* **2016**, *89*, 559–572. [CrossRef]
18. Panampilly, B.; Thomas, S. Nano ZnO as cure activator and reinforcing filler in natural rubber. *Polym. Eng. Sci.* **2013**, *53*, 1337–1346. [CrossRef]
19. Morfin, I.; Ehrburger-Dolle, F.; Grillo, I.; Livet, F.; Bley, F. ASAXS, SAXS and SANS investigations of vulcanized elastomers filled with carbon black. *J. Synchrotron Radiat.* **2006**, *13*, 445–452. [CrossRef]
20. Noda, Y.; Yamaguchi, D.; Hashimoto, T.; Shamoto, S.-i.; Koizumi, S.; Yuasa, T.; Tominaga, T.; Sone, T. Spin Contrast Variation Study of Fuel-efficient Tire Rubber. *Phys. Procedia* **2013**, *42*, 52–57. [CrossRef]
21. Otegui, J.; Miccio, L.A.; Arbe, A.; Schwartz, G.A.; Meyer, M.; Westermann, S. Determination of Filler Structure in Silica-filled SBR Compounds by means of SAXS and AFM. *Rubber Chem. Technol.* **2015**, *88*, 690–710. [CrossRef]
22. Yamaguchi, D.; Yuasa, T.; Sone, T.; Tominaga, T.; Noda, Y.; Koizumi, S.; Hashimoto, T. Hierarchically Self-Organized Dissipative Structures of Filler Particles in Poly(styrene-ran-butadiene) Rubbers. *Macromolecules* **2017**, *50*, 7739–7759. [CrossRef]
23. Noda, Y.; Koizumi, S.; Masui, T.; Mashita, R.; Kishimoto, H.; Yamaguchi, D.; Kumada, T.; Takata, S.-i.; Ohishi, K.; Suzuki, J.-i. Contrast variation by dynamic nuclear polarization and time-of-flight small-angle neutron scattering. I. Application to industrial multi-component nanocomposites. *J. Appl. Crystallogr.* **2016**, *49*, 2036–2045. [CrossRef] [PubMed]
24. Endo, H.; Mihailescu, M.; Monkenbusch, M.; Allgaier, J.; Gompper, G.; Richter, D.; Jakobs, B.; Sottmann, T.; Strey, R.; Grillo, I. Effect of amphiphilic block copolymers on the structure and phase behavior of oil–water-surfactant mixtures. *J. Chem. Phys.* **2001**, *115*, 580–600. [CrossRef]
25. Higgins, J.S.; Benoît, H. *Polymers and Neutron Scattering*; Clarendon Press: Oxford, UK, 1994.
26. Narayanan, T.; Sztucki, M.; Van Vaerenbergh, P.; Leonardon, J.; Gorini, J.; Claustre, L.; Sever, F.; Morse, J.; Boesecke, P. A multipurpose instrument for time-resolved ultra-small-angle and coherent X-ray scattering. *J. Appl. Crystallogr.* **2018**, *51*, 1511–1524. [CrossRef] [PubMed]
27. Endo, H. Study on multicomponent systems by means of contrast variation SANS. *Phys. B Condens. Matter* **2006**, *385–386*, 682–684. [CrossRef]
28. Suzuki, T.; Endo, H.; Osaka, N.; Shibayama, M. Dynamics and Microstructure Analysis of N-Isopropylacrylamide/Silica Hybrid Gels. *Langmuir* **2009**, *25*, 8824–8832. [CrossRef]
29. Suzuki, T.; Osaka, N.; Endo, H.; Shibayama, M.; Ikeda, Y.; Asai, H.; Higashitani, N.; Kokubo, Y.; Kohjiya, S. Nonuniformity in Cross-Linked Natural Rubber as Revealed by Contrast-Variation Small-Angle Neutron Scattering. *Macromolecules* **2010**, *43*, 1556–1563. [CrossRef]
30. Liu, D.; Chen, J.; Song, L.; Lu, A.; Wang, Y.; Sun, G. Parameterization of silica-filled silicone rubber morphology: A contrast variation SANS and TEM study. *Polymer* **2017**, *120*, 155–163. [CrossRef]
31. Staropoli, M.; Gerstner, D.; Sztucki, M.; Vehres, G.; Duez, B.; Westermann, S.; Lenoble, D.; Pyckhout-Hintzen, W. Hierarchical Scattering Function for Silica-Filled Rubbers under Deformation: Effect of the Initial Cluster Distribution. *Macromolecules* **2019**, *52*, 9735–9745. [CrossRef]

32. Conzatti, L.; Costa, G.; Castellano, M.; Turturro, A.; Negroni, F.M.; Gérard, J.-F. Morphology and Viscoelastic Behaviour of a Silica Filled Styrene/Butadiene Random Copolymer. *Macromol. Mater. Eng.* **2008**, *293*, 178–187. [CrossRef]
33. Salah, N.; Habib, S.S.; Khan, Z.H.; Memic, A.; Azam, A.; Alarfaj, E.; Zahed, N.; Al-Hamedi, S. High-energy ball milling technique for ZnO nanoparticles as antibacterial material. *Int. J. Nanomed.* **2011**, *6*, 863–869. [CrossRef] [PubMed]

© 2020 by the authors. Licensee MDPI, Basel, Switzerland. This article is an open access article distributed under the terms and conditions of the Creative Commons Attribution (CC BY) license (http://creativecommons.org/licenses/by/4.0/).

*Article*

# Efficient Chain Formation of Magnetic Particles in Elastomers with Limited Space

**Shota Akama** [1,2], **Yusuke Kobayashi** [1,2], **Mika Kawai** [1,2] **and Tetsu Mitsumata** [1,2,*]

1. Graduate School of Science and Technology, Niigata University, Niigata 950-2181, Japan; aaa.rock@icloud.com (S.A.); f19b131j@mail.cc.niigata-u.ac.jp (Y.K.); mikagoro@eng.niigata-u.ac.jp (M.K.)
2. ALCA, Japan Science and Technology Agency, Tokyo 102-0076, Japan
* Correspondence: tetsu@eng.niigata-u.ac.jp; Tel.: +81-(0)25-262-6884

Received: 27 December 2019; Accepted: 26 January 2020; Published: 1 February 2020

**Abstract:** The magnetic response of the storage modulus for bimodal magnetic elastomers containing magnetic particles with a diameter of 7.0 µm and plastic beads with a diameter of 200 µm were investigated by varying the volume fraction of plastic beads up to 0.60 while keeping the volume fraction of the magnetic particles at 0.10. The storage modulus at 0 mT for monomodal magnetic elastomers was $1.4 \times 10^4$ Pa, and it slightly increased with the volume fraction of plastic beads up to 0.6. The storage modulus at 500 mT for bimodal magnetic elastomers at volume fractions below 0.25 was constant, which was equal to that for the monomodal one (=$7.9 \times 10^4$ Pa). At volume fractions of 0.25–0.40, the storage modulus significantly increased with the volume fraction, showing a percolation behavior. At volume fractions of 0.40-0.60, the storage modulus was constant at $2.0 \times 10^5$ Pa, independently of the volume fraction. These results indicate that the enhanced increase in the storage modulus was caused by the chain formation of the magnetic particles in vacancies made of plastic beads.

**Keywords:** soft material; stimuli-responsive gel; magnetic elastomer; percolation

## 1. Introduction

Magnetic elastomers are a type of stimuli-responsive soft material [1–5] and its physical properties alter in response to magnetic fields. The magnetic response for a magnetic elastomer is in general drastic; therefore, the material attracts considerable attention as actuators in the next generation of materials [6–8]. Magnetic elastomers consist of polymeric matrices, such as polyurethane, and magnetic particles with nano or micron sizes in diameter. When a magnetic field is applied to a magnetic elastomer, the elasticity increases due to the chain structure formation (restructuring) of magnetic particles, which is called the magnetorheological (MR) effect. So far, we have investigated the MR effect for polyurethane-based magnetic elastomers and found that bimodal magnetic elastomers with magnetic and nonmagnetic particles exhibit a significant MR effect compared with monomodal magnetic elastomers [9–13].

In our past studies, zinc oxide with a diameter of 10.6 µm or aluminum hydroxide with a diameter of 1.4 µm was used as nonmagnetic particles [9,11]. The increase in the storage modulus for bimodal magnetic elastomers containing zinc oxide particles of 12 vol.% was 500 kPa, which was achieved by applying a magnetic field of 500 mT, which is 4.2 times that used for the monomodal one (120 kPa) [9]. The increase in the storage modulus for bimodal magnetic elastomers containing aluminum hydroxide particles of 6.6 vol.% was 3.27 MPa, which is 4.3 times that used for the monomodal one (754 kPa) [11]. Figure 1 shows the schematic illustrations representing the mechanism for the MR effect for bimodal magnetic elastomers in our previous study and the scenario for that in the present study. In the previous study, the enhanced magnetorheology mentioned above is caused by bridging the discontinuous chains of magnetic particles with nonmagnetic particles. However, this mechanism is not efficient since a considerable amount of nonmagnetic particles is needed to raise the elasticity of magnetic elastomers.

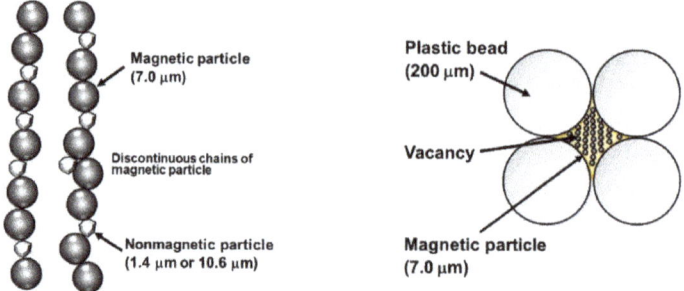

**Figure 1.** Schematic illustrations representing the mechanism for the magnetorheological (MR) effect for bimodal magnetic elastomers in our previous study and the scenario in the present study.

In our present study, we proposed a new concept for the enhanced magnetorheology of bimodal magnetic elastomers. A large bead of poly(methyl-2-methylpropenoate) (PMMA) with a diameter of 200 μm was used as a non-magnetic particle. The diameter of the bead was large, and the interaction with the matrix of polyurethane was weak compared with nonmagnetic particles in past studies. Beads with a large diameter made vacancies in the magnetic elastomer, and magnetic particles were localized in the vacancies made of the plastic beads. The density of the plastic was low compared with inorganic compounds; therefore, the beads were not precipitated during the synthesis; this contributed to the random dispersion of vacancies in the elastomer. Bimodal magnetic elastomers with plastic beads also have advantages of transparency and weight saving. The density of zinc oxide and aluminum hydroxide is 5.8 and 2.4 g/cm$^3$, respectively, meanwhile, the density of the plastic bead was 1.2 g/cm$^3$. Magnetic elastomers with both a lightweight and huge magnetic response could be obtained. Bimodal magnetic elastomers with plastic beads have a possibility to transmit visible light, although the light may reflect at the interface between the matrix and plastic beads. This should be helpful for biologists or medical scientists who investigate the effect of substrate elasticity on the cell behavior using magnetic elastomers [14].

Here, we prepared bimodal magnetic elastomers with various volume fractions of plastic beads while keeping the volume fraction of magnetic particles constant, and discuss the addition effect of plastic beads on the magnetic response, the transmissibility of visible light, and weight saving.

## 2. Experimental Procedures

### 2.1. Synthesis of Magnetic Elastomers

Polypropylene glycols (P2000, G3000B, Adeka Co., Tokyo, Japan) with molecular weights of $M_w$ = 2000 and 3000 were used for the matrix of magnetic elastomers. Tolyrene diisocyanate (Wako Pure Chemical Industries. Ltd., Osaka, Japan) and dioctyl phthalate (DOP, Wako Pure Chemical Industries. Ltd., Osaka, Japan) were used as a crosslinker and plasticizer, respectively. Carbonyl iron with a median diameter of 7.0 μm (CS Grade BASF SE., Ludwigshafen am Rhein, Germany) was used for magnetic particles. The saturation magnetization of carbonyl iron particles was 190 emu/g measured using a SQUID magnetometer (MPMS, Quantum Design Inc., San Diego, CA, USA). Plastic beads made of polymethylmethacrylate (PMMA) (Techpolymer, MBX-200, Sekisui Plastics Co., Ltd., Tokyo, Japan) with a mean diameter of 200 μm were used as nonmagnetic particles. Magnetic elastomers were synthesized using a prepolymer method. Polypropylene glycols are crosslinked with tolyrene diisocyanate. The concentration of the crosslinker remarkably affects not only the off-field modulus, but also the magnetorheological response. The molar ratio of –NCO to –OH group for the prepolymer was constant at 2.01 (=[NCO]/[OH]). Magnetic particles and plastic beads were mixed with prepolymer, linear polymer, plasticizer, and catalysis. The mixed liquid was poured into a silicon mold and cured for 30 min at 100 °C. DOP has a good chemical affinity with polypropylene glycols, which is a good solvent for crosslinked PPG elastomers. Similar to the crosslinker concentration, the concentration of the plasticizer strongly affects both the off-field modulus and the magnetorheological response. The weight concentration of DOP to the matrix without magnetic particles was fixed at 60 wt.%. The densities of the magnetic particles and plastic beads were 7.57 and 1.20 g/cm$^3$, respectively. The volume fraction of magnetic particles was kept at 0.1; meanwhile, for plastic beads, it was varied from 0.10 to 0.60.

## 2.2. Dynamic Viscoelastic Measurement

Dynamic viscoelastic measurements were carried out for magnetic elastomers using a rheometer (MCR301, Anton Paar Pty. Ltd., Graz, Austria) at 20 °C. The strain was varied from $10^{-5}$ to 1 and the frequency was kept at 1 Hz. The sample was a disk that was 20 mm in diameter and 1.5 mm in thickness. The normal force initially applied to the magnetic elastomer was approximately 0.3 N.

## 2.3. SEM Observation

The shape of the magnetic and nonmagnetic particles in the powder state and the particle morphology for monomodal and bimodal elastomers were observed using scanning electron microscopy (SEM, JCM-6000 Neoscope JEOL Ltd. Tokyo, Japan) with an accelerating voltage of 15 kV without a Au coating. SEM photographs for magnetic particles and plastic beads (nonmagnetic particles) are presented in Figure 2.

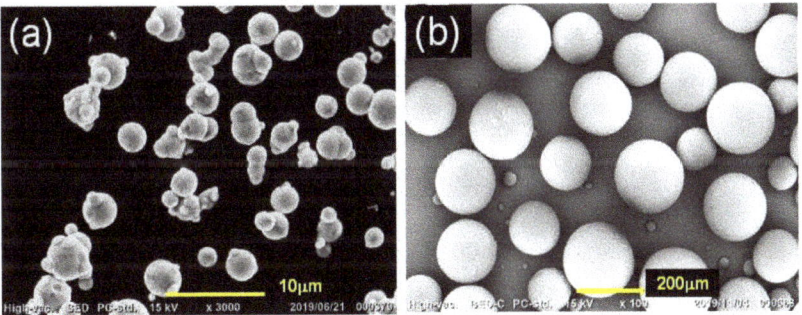

**Figure 2.** SEM photographs for (**a**) magnetic particles and (**b**) plastic beads (nonmagnetic particles).

## 3. Results and Discussion

Figure 3a exhibits the magnetic field response of the storage modulus at a strain of $10^{-4}$ (in the linear viscoelastic regime) for monomodal and bimodal magnetic elastomers containing plastic beads. A magnetic field of 500 mT was applied to the magnetic elastomers every 60 s. It was observed for all magnetic elastomers that the storage modulus was altered synchronously with the magnetic field, and completely recovered to the original modulus after removing the field. The storage modulus for bimodal magnetic elastomers at $\phi_{PB} < 0.3$ rapidly increased due to the application of the magnetic field. This result coincided with our previous study showing that the alignment time improved by adding aluminum hydroxide nonmagnetic particles [11]. Meanwhile, at $\phi_{PB} > 0.35$, the storage modulus gradually increased with time and it was not saturated within 60 s. It is considered that only magnetic particles move and form a chain structure at $\phi_{PB} < 0.3$. At $\phi_{PB} > 0.35$, large plastic beads moved, accompanying the movement of the magnetic particles, resulting in long relaxation time.

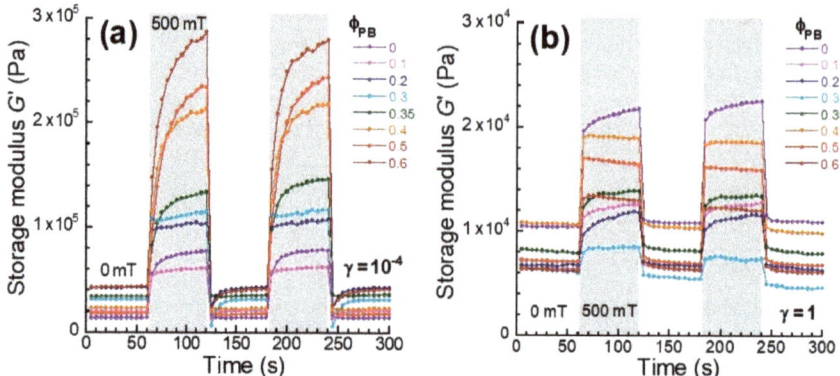

**Figure 3.** Magnetic field response of the storage modulus at strains of (**a**) $10^{-4}$ and (**b**) 1 for monomodal and bimodal magnetic elastomers containing plastic beads with various volume fractions.

Figure 3b demonstrates the magnetic-field response of the storage modulus at a strain of 1 (in the non-linear viscoelastic regime) for monomodal and bimodal magnetic elastomers containing plastic beads. Similar to the results at low strain, all magnetic elastomers responded to the magnetic field at high strain. However, the time development of the storage modulus was opposite at high strain. A gradual increase of the storage modulus with time was observed for monomodal and bimodal magnetic elastomers with $\phi_{PB} < 0.3$, and a rapid increase of the storage modulus with time was observed for bimodal magnetic elastomers with $\phi_{PB} > 0.35$.

Figure 4a depicts the storage modulus at $10^{-4}$ for monomodal and bimodal magnetic elastomers as a function of the volume fraction of plastic beads. The storage modulus at 0 mT was almost constant, although the volume fraction of the plastic beads increased. In general, when the volume fraction of fillers was high, the storage modulus for the composite elastomers was higher than that for the matrix. The low storage modulus seen at $\phi_{PB} = 0.60$ indicated that the plastic beads were randomly dispersed in the elastomer without direct contact between the beads. Of course, the interfacial effect between the matrix and plastic beads with a large diameter should be negligibly small compared with the small particles being several microns in size. Actually, when a plastic bead with a diameter of 8 μm was used, the storage modulus at 0 mT for the bimodal magnetic elastomer at $\phi_{PB} = 0.60$ was $3.6 \times 10^5$ Pa, which was far higher than that for the elastomer with large particles. In addition, the change in the storage modulus was very small and negative. This strongly indicated that the total interfacial area was dominant for the storage modulus of magnetic elastomers, i.e., the storage modulus could be raised by increasing the interfacial area, even though the interfacial interaction between the matrix and plastic beads was weak. The storage modulus at 500 mT for bimodal magnetic elastomers was equal to that of the monomodal one, and it was almost independent of the volume fraction at $\phi_{PB} < 0.25$. This behavior was unusual, and it indicated that bridging did not occur between discontinuous chains of magnetic particles via plastic beads, even though the volume fraction of plastic beads was considerably high. At $0.25 < \phi_{PB} < 0.40$, the storage modulus significantly increased with the volume fraction of the plastic beads. It was considered that the magnetic particles bridged the gap between the plastic beads. An interesting thing was that the significant increase in the storage modulus was only seen at 500 mT and did not appear at 0 mT. In addition, the influence of the matrix elasticity on the magnetorheological effect was investigated. The storage modulus for polyurethane elastomers with ($\phi_{PB}=0.40$) and without plastic beads was $(3.8 \pm 0.35) \times 10^4$ Pa and $(3.3 \pm 0.34) \times 10^4$ Pa, respectively.

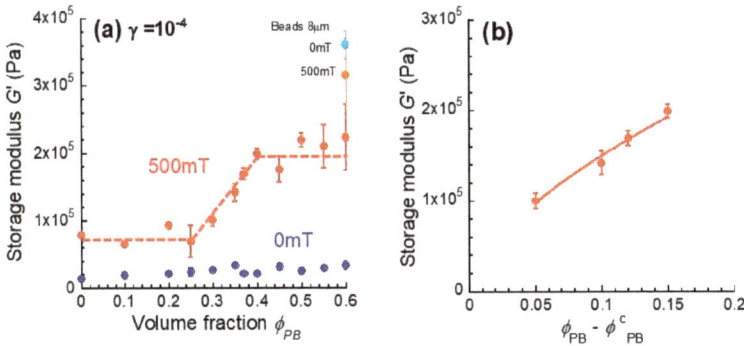

**Figure 4.** (a) Storage modulus at a strain of $10^{-4}$ at 0 and 500 mT as a function of the volume fraction of plastic beads and (b) storage modulus as a function of $\phi_{PB} - \phi^c_{PB}$ for monomodal and bimodal magnetic elastomers.

Therefore, an important factor for the stress transfer between plastic beads is the chain structure of magnetic particles, not the elasticity of the matrix. The onset volume fraction for the percolation is considered to be $0.25 < \phi_{PB} < 0.30$, which is close to the value of the cite percolation for a body-centered cubic structure (=0.248) [15,16]. Figure 4b shows the relation between the storage modulus and the volume fraction of plastic beads ($\phi_{PB} - \phi^c_{PB}$) for bimodal magnetic elastomers. The storage modulus at $\phi_{PB} > 0.25$ obeyed a power low of $G_{500} \sim (\phi_{PB} - \phi^c_{PB})^{0.64}$, although the critical exponent was relatively high compared with that of a 3D lattice ($\approx 0.4$) [15,16]. Accordingly, the plastic beads were packed in the structure of the body-centered cubic structure, and magnetic particles were filled in the vacancies between the beads. At $\phi_{PB} > 0.4$, the storage modulus was not raised, although the volume fraction of plastic beads was increased. This indicated that the percolation possibility reached a maximum at $\phi_{PB} \approx 0.5$, which was 74% of the maximum packing ratio for the lattice of a body-centered cubic structure. It might be that the effective paths contributing to the storage modulus were not increased at $\phi_{PB} > 0.4$ due to the increase in the number of branches of the percolated paths. Figure 5 exhibits the schematic illustrations representing the mechanism of the elasticity increase by the magnetic field for bimodal magnetic elastomers with plastic beads with a large diameter. The distance between the nearest neighbor beads was calculated to be 31 µm. Magnetic particles are forced to be localized in the vacancy and form a chain structure by the magnetic field, for example, a qualitative representation of the microstructure [17] and computed tomography images [18]. In our previous study, the storage modulus for bimodal magnetic elastomers containing small nonmagnetic particles (1.4 or 10.6 µm in diameter) gradually increased with the volume fraction of nonmagnetic particles, and a clear percolation threshold was not observed. This strongly indicated that the chains of magnetic particles were gradually connected via nonmagnetic particles with increases in the volume fraction of nonmagnetic particles [9–11]. The mechanism presented here was completely different from that observed in the past. Figure 5 also shows a realization of percolation behavior for bimodal magnetic elastomers with plastic beads. Each white circle represents a plastic bead, and the black part represents the magnetic elastomer, i.e., polyurethane and magnetic particles. The percolation of the stress between the plastic beads occurred via chains of magnetic particles at $\phi_{PB} = 0.25$. The percolation path increased with the volume fraction of plastic beads at $0.25 < \phi_{PB} < 0.40$. At $0.40 < \phi_{PB} < 0.60$, both the percolation path and the branch of the paths increased with the volume fraction.

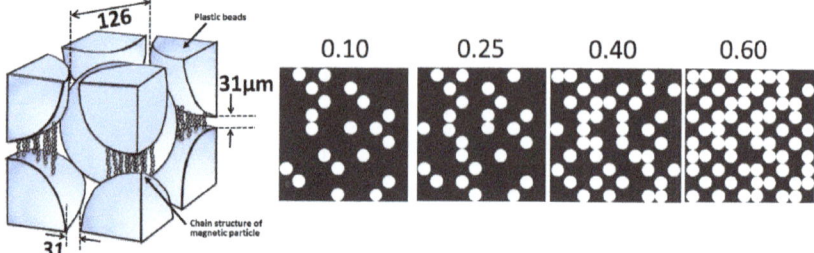

**Figure 5.** Schematic illustrations representing (**left**) the mechanism of the MR effect and (**right**) a realization of percolation behavior for bimodal magnetic elastomers with large plastic beads.

Figure 6a,e indicates the SEM photographs for monomodal magnetic elastomers taken at different magnifications. Figure 6e demonstrates the fact that no clear aggregations of magnetic particles were found in the polyurethane matrix. However, the photo showed that there were two regions with different concentrations of magnetic particles, i.e., densely and sparsely dispersed parts. This result coincided with the previous results showing that carbonyl iron particles form secondary particles within the polyurethane matrix when ultrasonication is not carried out [19]. Figure 6c,g displays the SEM photographs for bimodal magnetic elastomers with plastic beads above the percolation threshold ($\phi_{PB} = 0.4$). It was clear that magnetic particles were localized in the vacancy made by plastic beads, meaning the success of our scenario given that magnetic particles were gathered in a limited space. The hole made by removing the plastic bead showed a very smooth surface, suggesting that the interaction between the plastic beads and the matrix was weak (see Figure 6f). Figure 6d,h shows the SEM photographs for bimodal magnetic elastomers containing plastic beads with the maximum volume fraction ($\phi_{PB} = 0.60$). It was clearly observed that magnetic particles were distributed in the vacancy.

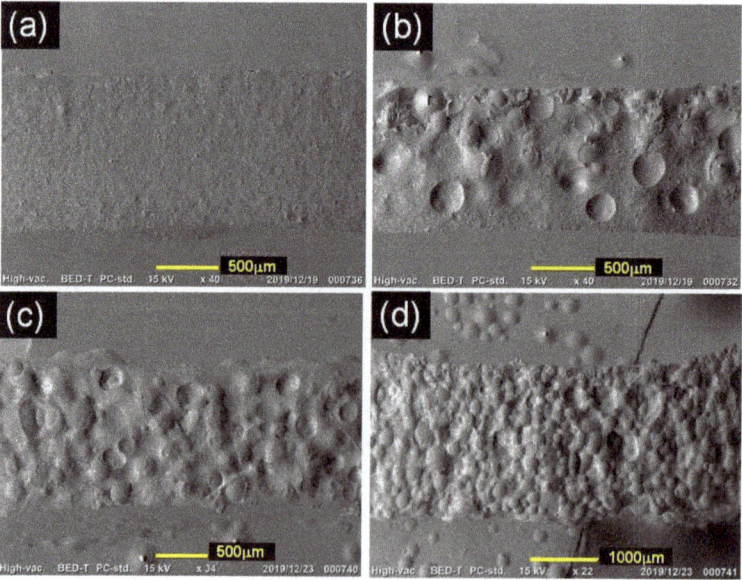

**Figure 6.** *Cont.*

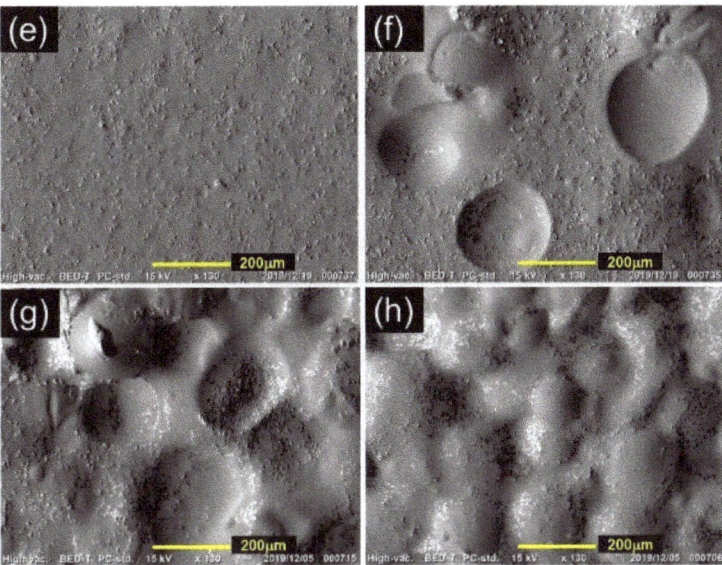

**Figure 6.** SEM photographs for (**a,e**) monomodal magnetic elastomers, (**b–d,f–h**) bimodal magnetic elastomers with various volume fractions of plastic beads (**b,f**) 0.25, (**c,g**) 0.40, and (**d,h**) 0.60.

Figure 7 demonstrates the photographs representing the transparency and the density for bimodal magnetic elastomers with plastic beads obtained in this study. The transparency for bimodal magnetic elastomers with plastic beads ($\phi_{PB}$~0.60) was compared with that for monomodal ones. The increase in the storage modulus for the bimodal magnetic elastomer (=(8.00 ± 0.53) × $10^4$ Pa) was equal to that for monomodal one (=(7.1 ± 0.26) × $10^4$ Pa). In this experiment, the volume fraction of the magnetic particles was kept at 0.07 because the difference in the transparency for these elastomers was not clear to see. The thickness of the monomodal and bimodal magnetic elastomers was approximately 180 and 290 µm, respectively. It was found that the paint, the mountain with snow, could only be seen through the bimodal magnetic elastomer. The density for the bimodal magnetic elastomers with plastic beads ($\phi_{PB} \approx 0.4, 0.6$) was also compared with that for a monomodal one. The increase in the storage modulus was (2.7 ± 0.99) × $10^5$ Pa for monomodal, (1.8 ± 0.01) × $10^5$ Pa for bimodal with $\phi_{PB} \approx 0.4$ and (1.9 ± 4.2) × $10^5$ Pa for bimodal with $\phi_{PB} \approx 0.6$. The density was 2.11 g/cm$^3$ for monomodal, 1.37 g/cm$^3$ for bimodal with $\phi_{PB} \approx 0.4$ and 1.15 g/cm$^3$ for bimodal with $\phi_{PB} \approx 0.6$. Weight saving of −59% was achieved by creating the limited space in the magnetic elastomer. It was the case that the particle distribution or the local rearrangement of magnetic particles [20,21] was considered and designed for an efficient MR effect in the vacancy.

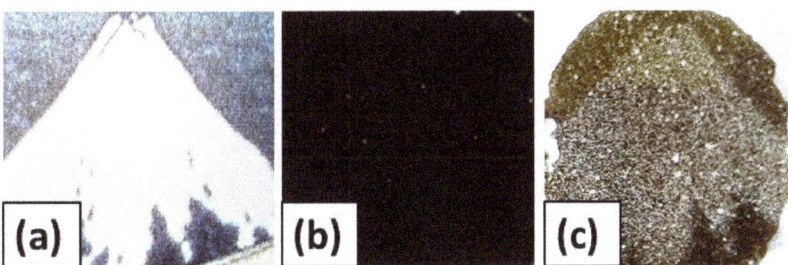

**Figure 7.** *Cont.*

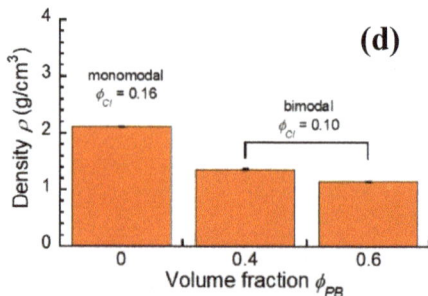

**Figure 7.** Photographs representing the transparency for monomodal and bimodal magnetic elastomers: (**a**) original paint, and (**b**) monomodal, and (**c**) bimodal magnetic elastomers were put on the original paint. (**d**) The density for monomodal and bimodal magnetic elastomers with plastic beads of $\phi_{PB} \approx 0.4$ and 0.6.

## 4. Conclusions

The magnetorheological response for bimodal magnetic elastomers with large plastic beads was investigated. It was found that the increment in the storage modulus for the bimodal magnetic elastomers was higher than twice that of the monomodal one. It could be considered that the enhanced magnetorheology was caused by an efficient chain formation of magnetic particles in a vacancy made by the plastic beads that packed into the body-centered cubic structure. In addition, it was found that the bimodal magnetic elastomer could partially transmit natural light, although the clarity was very low. At approximately half the weight of conventional elastomers, the bimodal magnetic elastomers showed the same level of elasticity changes. We believe that the magnetorheological features presented here are useful not only for applications but also for designing materials for new magnetorheological elastomers with a high efficiency and multiple functions.

**Author Contributions:** The idea for this work came from T.M. M.K. conceived and designed all the experiments; S.A. performed the sample synthesis, rheological experiment, and SEM observations for magnetic elastomers and wrote the paper; and Y.K. carried out the rheological experiment and drew the graphs and graphic abstract. Presentation of the results and the draft structure were discussed between all authors. All authors have read and agreed to the published version of the manuscript.

**Funding:** This research was partially supported by a grant from the Cooperative Research Program of Cooperative Research Program of the Network Joint Research Center for Materials and Devices (No. 20191257).

**Conflicts of Interest:** The authors declare no conflict of interest.

## References

1. Kose, O.; Boott, C.E.; Hamad, W.Y.; MacLachlan, M.J. Stimuli-Responsive Anisotropic Materials Based on Unidirectional Organization of Cellulose Nanocrystals in an Elastomers. *Macromolecules* **2019**, *52*, 5317–5324. [CrossRef]
2. Du, J.; Zhang, Z.L.; Liu, D.B.; Ren, T.B.; Wan, D.C.; Pu, H.T. Triple-stimuli responsive shape memory effect of novel polyolefin elastomer/lauric acid/carbon black. *Compos. Sci. Tech.* **2019**, *169*, 45–51. [CrossRef]
3. Kaiser, S.; Radl, S.V.; Manhart, J.; Ayalur-Karunakaran, S.; Griesser, T.; Moser, A.; Ganser, C.; Teichert, C.; Kern, W.; Schlogl, S. Switching "on" and "off" the adhesion in stimuli-responsive elastomers. *Soft Mater* **2018**, *14*, 2547–2559. [CrossRef] [PubMed]
4. Amaral, A.J.R.; Pasparakis, G. Stimuli responsive self-healing polymers: Gels, elastomers and membranes. *Polym. Chem.* **2017**, *8*, 6464–6484. [CrossRef]
5. Wang, M.; Sayed, S.M.; Guo, L.X.; Lin, B.P.; Zhang, X.Q.; Sun, Y.; Yang, H. Multi-Stimuli Responsive Carbon Nanotube Incorporated Polysiloxane Azobenzene Liquid Crystalline Elastomer Composites. *Macromolcules* **2016**, *49*, 663–671. [CrossRef]

6. Qi, S.; Fu, J.; Xie, Y.P.; Li, Y.P.; Gan, R.Y.; Yu, M. Versatile magnetorheological plastomer with 3D printability, switchable mechanics, shape memory, and self-healing capacity. *Compos. Sci. Tech.* **2019**, *183*, 107817. [CrossRef]
7. Kim, J.G.; Park, J.E.; Won, S.; Jeon, J.; Wie, J.J. Contactless Manipulation of Soft Robots. *Materials* **2019**, *12*, 19. [CrossRef] [PubMed]
8. Ditter, D.; Bluemler, P.; Klockner, B.; Hilgert, J.; Zentel, R. Microfluidic Synthesis of Liquid Crystalline Elastomer Particle Transport Systems which Can Be Remote-Controlled. *Adv. Funct. Mater.* **2019**, *29*, 29. [CrossRef]
9. Mitsumata, T.; Ohori, S.; Chiba, N.; Kawai, M. Enhancement of magnetoelastic behavior of bimodal magnetic elastomers by stress transfer via nonmagnetic particles. *Soft Matter* **2013**, *9*, 10108–10116. [CrossRef]
10. Nagashima, K.; Kanauchi, S.; Kawai, M.; Mitsumata, T.; Tamesue, S.; Yamauchi, T. Nonmagnetic particles enhance magnetoelastic response of magnetic elastomers. *J. Appl. Phys.* **2015**, *118*, 024903. [CrossRef]
11. Nanpo, J.; Nagashima, K.; Umehara, Y.; Kawai, M.; Mitsumata, T. Magnetic-Field Sensitivity of Storage Modulus for Bimodal Magnetic Elastomers. *J. Phys. Chem. B* **2016**, *120*, 12993–13000. [CrossRef] [PubMed]
12. Umehara, Y.; Yamanaga, Y.; Akama, S.; Kato, S.; Kamoshita, S.; Kawai, M.; Mitsumata, T. Railway Actuator Made of Magnetic Elastomers and Driven by a Magnetic Field. *Polymers* **2018**, *10*, 1351. [CrossRef] [PubMed]
13. Kobayashi, Y.; Akama, S.; Ohori, S.; Kawai, M.; Mitsumata, T. Magnetic Elastomers with Smart Variable Elasticity Mimetic to Sea Cucumber. *Biomimetics* **2019**, *4*, 68. [CrossRef] [PubMed]
14. Mayer, M.; Rabindranath, R.; Börner, J.; Hörner, E.; Bentz, A.; Salgado, J.; Han, H.; Böse, H.; Probst, J.; Shamonin, M.; et al. Ultra-soft PDMS-based magnetoactive elastomers as dynamic cell culture substrata. *PloS ONE* **2013**, *8*, e76196. [CrossRef] [PubMed]
15. Stauffer, D. *Introduction to Percolation Theory*; Taylor & Fran-cis: London, UK, 1985.
16. Grimmett, G. *Percolation*, 2nd ed.; Springer: Cambridge, UK, 1999.
17. Snarskii, A.A.; Zorinets, D.; Shamonin, M.; Kalita, V.M. Theoretical method for calculation of effective properties of composite materials with reconfigurable microstructure: Electric and magnetic phenomena. *Physica A* **2019**, *535*, 122467. [CrossRef]
18. Watanabe, M.; Takeda, Y.; Maruyama, R.; Ikeda, J.; Kawai, M.; Mitsumata, T. Chain Structure in a Cross-Linked Polyurethane Magnetic Elastomer Under a Magnetic Field. *Int. J. Mol. Sci.* **2019**, *20*, 2879. [CrossRef] [PubMed]
19. Watanabe, M.; Ikeda, J.; Takeda, Y.; Kawai, M.; Mitsumata, T. Effect of Sonication Time on Magnetorheological Effect for Monomodal Magnetic Elastomers. *Gels* **2018**, *4*, 49. [CrossRef] [PubMed]
20. Ivaneyko, D.; Toshchevikov, V.; Saphiannikova, M.; Heinrich, G. Effects of particle distribution on mechanical properties of magneto-sensitive elastomers in a homogeneous magnetic field. *Condens. Matter Phys.* **2012**, *15*, 33601. [CrossRef]
21. Romeis, D.; Toshchevikov, V.; Saphiannikova, M. Effects of local rearrangement of magnetic particles on deformation in magneto-sensitive elastomers. *Soft Matter* **2019**, *15*, 3552. [CrossRef] [PubMed]

© 2020 by the authors. Licensee MDPI, Basel, Switzerland. This article is an open access article distributed under the terms and conditions of the Creative Commons Attribution (CC BY) license (http://creativecommons.org/licenses/by/4.0/).

*Article*

# Cationic Copolymerization of Isobutylene with 4-Vinylbenzenecyclobutylene: Characteristics and Mechanisms

Zhifei Chen [1,2], Shuxin Li [1,*], Yuwei Shang [1], Shan Huang [1,2], Kangda Wu [1], Wenli Guo [1] and Yibo Wu [1,*]

[1] Beijing Key Lab of Special Elastomeric Composite Materials, Department of Materials Science and Engineering, Beijing Institute of Petrochemical Technology, Beijing 102617, China; chenzhifei1615@163.com (Z.C.); shangyuwei@bipt.edu.cn (Y.S.); h18811044437@163.com (S.H.); biptwkd@163.com (K.W.); gwenli@bipt.edu.cn (W.G.)

[2] College of Material Science and Engineering, Beijing University of Chemical Technology, Beijing 100029, China

* Correspondence: lishuxin@bipt.edu.cn (S.L.); wuyibo@bipt.edu.cn (Y.W.)

Received: 7 November 2019; Accepted: 3 January 2020; Published: 13 January 2020

**Abstract:** A random copolymer of isobutylene (IB) and 4-vinylbenzenecyclobutylene (4-VBCB) was synthesized by cationic polymerization at −80 °C using 2-chloro-2,4,4-trimethylpentane (TMPCl) as initiator. The laws of copolymerization were investigated by changing the feed quantities of 4-VBCB. The molecular weight of the copolymer decreased, and its molecular weight distribution (MWD) increased with increasing 4-VBCB content. We proposed a possible copolymerization mechanism behind the increase in the chain transfer reaction to 4-VBCB with increasing of feed quantities of 4-VBCB. The thermal properties of the copolymers were studied by solid-phase heating and crosslinking. After crosslinking, the decomposition and glass transition temperatures ($T_g$) of the copolymer increased, the network structure that formed did not break when reheated, and the mechanical properties remarkably improved.

**Keywords:** cationic polymerization; 4-vinylbenzocyclobutene; random copolymer; crosslink

## 1. Introduction

Over the years, isobutylene (IB) has played a crucial role in cationic polymerization due to its special structure. The living polymerization of IB has been realized, and researchers have also proposed living-polymerization mechanisms [1–11]. Copolymers with different structures and functions are obtained by the copolymerization of IB and other monomers [12–16]. Obtaining a functionalized copolymer by introducing a specific monomer during IB polymerization is an important research direction. Benzocyclobutene (BCB) and its derivatives are attracting considerable attention because of their special thermal properties. As shown in Scheme 1, the BCB's strained four-membered ring opens when heated to >180 °C to form a high-activity intermediate, *o*-quinodimethane. This intermediate can react with dienophiles through the Diels–Alder reaction, thereby forming a six-membered ring. In the absence of dienophile monomers, the intermediate reacts with itself to form an eight-membered ring or a homopolymer of BCB [17–20].

Similar to styrene, 4-VBCB is cationically polymerizable. The electronegativity of the aromatic ring adjacent to the vinyl group of 4-VBCB stabilizes the vinyl secondary carbon to form a stable carbocation. Therefore, 4-VBCB can be introduced into cationic copolymerization with IB. The vulcanization of the copolymer of IB and 4-VBCB has significant advantages over the vulcanization of other IB-based copolymers, such as IB–isoprene rubber and brominated IB-*co*-p-methylstyrene [14]. The crosslinking of 4-VBCB involves only heat and does not need the addition of a vulcanizing agent and an accelerator, such as sulfur, zinc oxide, xanthates, and quinoid chemical reagents, which are harmful to the human body and difficult to remove from the rubber matrix completely [21]. Therefore, the copolymer of IB and 4-VBCB does not require the removal of small-molecule chemicals after crosslinking, and can be autoclaved without considering the volatilization of toxic substances, which is a crucial characteristic of biomedical materials.

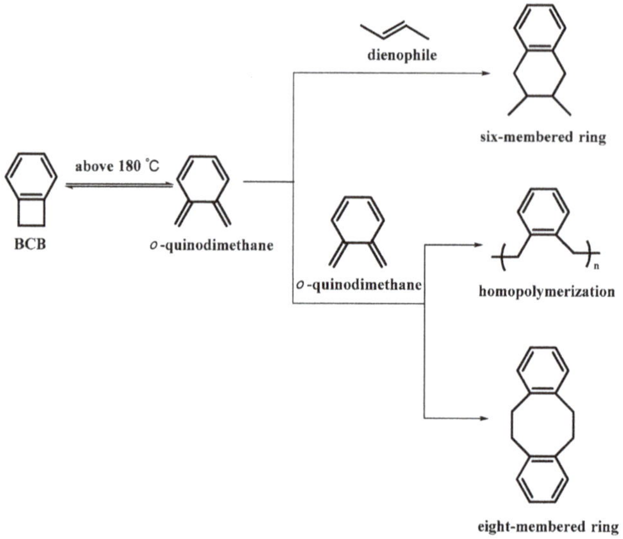

**Scheme 1.** Thermal properties of benzocyclobutene (BCB) ring.

Many studies have been conducted on the excellent properties of BCB-functionalized copolymers. The thermal and mechanical properties of BCB functional polymers and copolymers are drastically improved, and the dielectric constant is reduced after the thermal crosslinking reaction of the BCB ring. Therefore, the development of microelectronic materials exhibits great potential. Yang et al. synthesized oligomer poly(DVS-BCB-co-POSS) through the Heck reaction using 1,1,3,3-tetramethyl-1,3-divinyldisiloxane (DVS), BCB, and octavinyl-T8-silsesquioxane (vinyl-POSS). This oligomer has good thermal stability and low dielectric constants [22]. So et al. synthesized a copolymer of styrene and 4-VBCB through free radical polymerization and found that BCB dramatically increases the $T_g$ of polystyene after ring opening and crosslinking [23]. Huang et al. synthesized a copolymer containing BCB and silacyclobutane side groups with low dielectric constants (approximately 2.30 MHz to 10 MHz) through atom transfer radical copolymerization [24]. The thermal crosslinking of BCB can also be applied in the preparation of nanoparticles. Sakellariou heated a poly(4-VBCB-b-butadiene) diblock copolymer in decane to induce the crosslinking of poly(4-VBCB) segments in the micelle core and obtain surface-functionalized nanospheres [25]. Harth prepared single-molecule nanoparticles by regulating the collapse of linear polymer chains [26]. BCB-functionalized polymers and copolymers are synthesized through free-radical [27,28] and anion polymerization [25,29]. Few works have investigated the cationic polymerization of 4-VBCB, except for the research of Sheriff et al. on the application of poly(styrene-coblock-4-VBCB)-polyisobutylene-poly(styrene-coblock-4-VBCB) in prosthetic heart valves [30]. Moreover, the mechanism underlying the cationic copolymerization of IB with 4-VBCB is unclear.

In reference to these results, we described the cationic random polymerization of IB and 4-VBCB with 2-chloro-2,4,4-trimethylpentane (TMPCl) as the initiator and $TiCl_4$ as the coinitiator.

Copolymerization laws were studied by changing the feed ratio of 4-VBCB. We proposed a possible polymerization mechanism using Gaussian 09 software for calculations. We also studied thermal and mechanical properties through the solid-phase crosslinking of poly(IB-co-4-VBCB).

## 2. Materials and Methods

### 2.1. Materials

IB (purity: 99.9%, Beijing Yanshan Petrochemical Co., Ltd., Beijing, China) and chloromethane ($CH_3Cl$, purity: 99.9%, Beijing Yanshan Petrochemical Co., Ltd.) were dried in the gaseous state via passage through in-line gas-purifier columns packed with $CaSO_4$ (purity: 99%, Beijing Chemical Reagents Company, Beijing, China)/drierite. They were condensed in a cold bath in a glove box prior to polymerization. N-hexane (purity: 99.5%, Beijing Chemical Reagents Company) was refluxed with sodium for several hours and distilled before use under nitrogen ($N_2$) atmosphere. Dichloromethane ($CH_2Cl_2$, purity: 99.5%, Beijing Chemical Reagents Company) was dried over $CaH_2$ (purity: 99.5%, Beijing Chemical Reagents Company) for one week and then rectified by reflux under $N_2$ atmosphere. TMPCl was prepared by passing dry HCl into a 2,4,4-trimethyl-1-pentene (purity: 98.0%, Tokyo Chemical Industry Co., Ltd., Tokyo, Japan)/$CH_2Cl_2$ 50/50($v/v$) mixture at 0 °C for 5 h; subsequently, the mixture was washed with $NaHCO_3$ (purity: 99%, Beijing Chemical Reagents Company) until neutral, $MgSO_4$ (purity: 99%, Beijing Chemical Reagents Company) was added to remove water, the mixture was filtered, and the filtrate was purified by using a rotary evaporator [31]. Titanium tetrachloride ($TiCl_4$, purity: 99.0%, Energy Chemical, Shanghai, China) was purified from $P_4O_{10}$ (purity: 99.0%, Beijing Chemical Reagents Company) through distillation. 4-VBCB (purity: 97.5%, Bide Pharmatech Ltd., Shanghai, China), 2,6-di-tert-butylpyridine (DTBP, purity: 99.80%, Bide Pharmatech Ltd.), isopropanol (A.R. Beijing Chemical Reagents Company), and methanol (A.R. Beijing Chemical Reagents Company) were used as received.

### 2.2. Polymerization

The cationic copolymerization of IB and 4-VBCB was conducted in a glove box under dry $N_2$ atmosphere. In a typical reaction, 15 mL of cyclohexane, 10 mL of $CH_3Cl$, $3.54 \times 10^{-5}$ mol TMPCl, $8.84 \times 10^{-5}$ mol DTBP, and $1.13 \times 10^{-3}$ mol $TiCl_4$ were added sequentially to a screw-cap vial that was drained of water at −80 °C. The mixture was stirred evenly and held for 30 min to form TMPCl and $TiCl_4$ complex. A mixture of 2 mL of IB (1.42 g, 0.025 mol) and $2.53 \times 10^{-3}$ mol 4-VBCB dissolved in cyclohexane was added to the reaction mixture. Polymerization was continued for 2 h, and excessive methanol was added to terminate the reaction. Polymerization was quenched with 10 mL of prechilled ethanol. Polymer products were dried to a constant weight in a vacuum oven at 40 °C overnight. Monomer conversion was determined gravimetrically.

### 2.3. Measurements

Gel-permeation chromatography (GPC) was used to determine the molecular weights and molecular weight distributions ($M_w/M_n$) of the polymer. The unit was equipped with a Waters e2695 Separations Module, a Waters 2414 RI Detector, a Waters 2489 UV Detector, and four Waters styragel columns (E 2695, Waters, Milford, MA, USA) connected in the series of 500, $10^3$, $10^4$, and $10^5$ at 30 °C. Tetrahydrofuran was used as the mobile phase at a flow rate of 1 mL·min$^{-1}$ at room temperature. PSS WinGPC software (Waters) was used to acquire and analyze chromatograms. The molecular weights of polymers were calculated relative to those of linear polystyrene standards. The microstructure of the polymer was analyzed by $^1H$ NMR and $^1C$ NMR using a nuclear magnetic resonance spectrometer (AVANCE AV400, Bruker, Switzerland). Tetramethylsilane was the internal standard and $CDCl_3$ was the solvent. The polymers were subjected to thermogravimetric analysis using a thermogravimetric analyzer (TGA Q500, TA Instruments, New castle, DE, USA) in $N_2$ atmosphere at 10 °C·min$^{-1}$. The test temperature was 25 °C to 550 °C, and the heating rate was 10 °C·min$^{-1}$.

The glass transition temperature was analyzed using a differential thermal analyzer (DSC Q2000, TA Instruments) in a $N_2$ atmosphere of 50 mL·min$^{-1}$ gas flow. The test temperature was −90 °C to 150 °C, the temperature rise rate was 10 °C·min$^{-1}$. The changes of crosslinked structure were characterized using a Fourier transform infrared spectrometer (Nicolet6700, Thermo Fisher Scientific, Waltham, MA, USA). The mechanical properties of the copolymer were tested by a microcomputer controlled electronic universal testing machine (Instron3366, Instron, Shanghai, China). The stretch rate was 40 mm·min$^{-1}$.

## 3. Results and Discussion

### 3.1. Random Copolymers of IB and 4-VBCB

#### 3.1.1. Copolymerization of IB and 4-VBCB

Scheme 2 outlines the synthesis route of the poly(IB-co-4-VBCB) random copolymer. We first added IB to the initiation system to avoid the self-polymerization of 4-VBCB due to its high reactivity ratio. After the initiation of IB, a mixture of 4-VBCB and IB formulated at a certain ratio was gradually added to the reaction. Polymerization was continued for 2 h. Finally, excessive prechilled methanol was used to terminate the reaction.

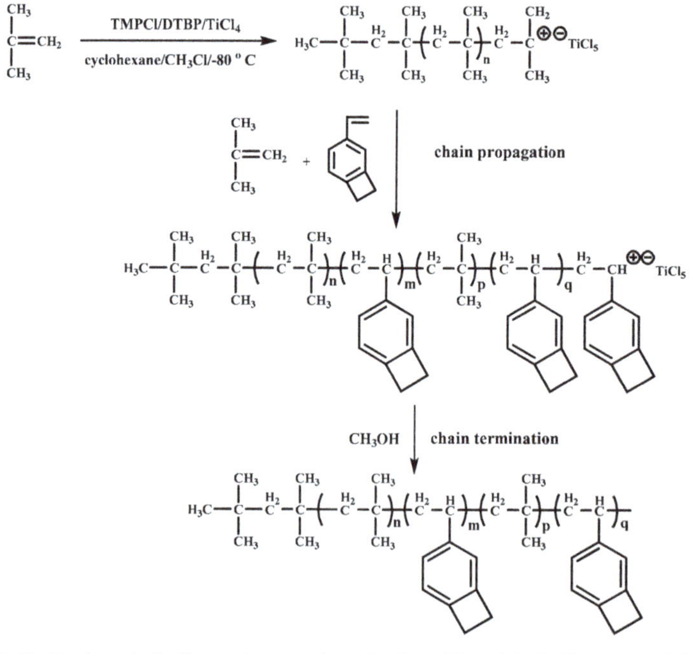

**Scheme 2.** Synthesis route for the random copolymerization of IB and 4-vinylbenzenecyclobutylene (4-VBCB) initiated by the 2-chloro-2,4,4-trimethylpentane (TMPCl)/2,6-di-tert-butylpyridine (DTBP)/TiCl$_4$ system.

The structures of the copolymer were characterized by $^1$H NMR. Figure 1a shows the $^1$H NMR spectrum of polyisobutylene (PIB). The characteristic resonance values at δ = 1.1 ppm (peak a) and δ = 1.4 ppm (peak b) were assigned to methyl and methylene, respectively. Figure 1b shows the $^1$H NMR spectrum of 4-VBCB. Aromatic ring proton peaks were located at δ: 6.6–7.24 ppm (peak c), vinyl proton peaks were located at δ = 5.1 ppm (peak a) and δ = 5.7ppm (peak b), and two methylene protons peak of BCB four-member ring were found at δ = 3.1 ppm (peak d). Figure 1c shows the $^1$H NMR spectrum of the poly(IB-co-4VBCB) random copolymer, and the characteristic peaks of each proton in poly(IB-co-4VBCB) random copolymer were consistent with the structures. Comparing Figure 1c with Figure 1a,b reveals that the $^1$H NMR spectrum of the random copolymers lacked vinyl protons at $\delta$ = 5.1 ppm and $\delta$ = 5.7 ppm, and protons associated with the P(IB) and P(4-VBCB) segment backbone and BCB side groups were present. The poly(IB-co-4VBCB) random copolymer was successfully synthesized.

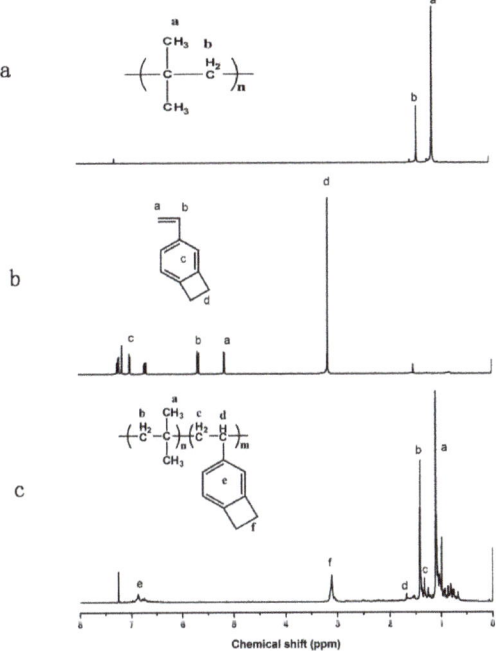

**Figure 1.** $^1$H NMR spectra of polyisobutylene (PIB) (**a**), 4-VBCB (**b**), poly(IB-co-4-VBCB) (**c**).

3.1.2. Reactivity Ratio of IB and 4-VBCB

The reactivity of monomers directly determines the copolymer composition. Thus, we first studied and calculated the reactivity ratio of IB and 4-VBCB using the Yezreielv–Brokhina–Roskin method [32]. We obtained a series of copolymers with different monomer compositions by changing the feed ratio of 4-VBCB to IB. The polymerization lasted for only 20 min to ensure that conversion rates were less than 15%. Then, the content of two monomers in the copolymer was calculated on the basis of the integrated area of the corresponding peak in the $^1$H NMR spectra (Table 1). The calculation method is as follows:

$$F_{IB} = \frac{A_{1.1}}{6} / \left( \frac{A_{1.1}}{6} + \frac{A_{3.1}}{4} \right)$$

$$F_{4-VBCB} = \frac{A_{3.1}}{4} / \left( \frac{A_{1.1}}{6} + \frac{A_{3.1}}{4} \right)$$

where $A_{1.1}$ is the resonance area of the two –CH$_3$ on isobutene, and $A_{3.1}$ is the resonance area of the four-membered ring on 4-VBCB.

Table 1. IB and 4-VBCB contents in the copolymer calculated by $^1$H NMR.

| Run | Conversion (%) | Monomer Feed | | Compositions of Copolymer | |
| --- | --- | --- | --- | --- | --- |
| | | IB (%) | 4-VBCB (%) | IB (%) | 4-VBCB (%) |
| 1 | 14.05 | 90.91 | 9.09 | 84.33 | 15.67 |
| 2 | 12.88 | 95.24 | 4.76 | 88.28 | 11.72 |
| 3 | 12.15 | 97.56 | 2.44 | 94.25 | 5.75 |
| 4 | 14.42 | 98.36 | 1.64 | 96.97 | 3.03 |
| 5 | 14.23 | 98.77 | 1.23 | 97.47 | 2.53 |

In accordance with the Yezreielv–Brokhina–Roskin method, we calculated monomer reactivity with the following equation:

$$\left(\frac{x}{y^{1/2}}\right)r_1 - \left(\frac{y^{1/2}}{x}\right)r_2 + \left(\frac{1}{y^{1/2}} - y^{1/2}\right) = 0$$

where x is the starting molar ratio of IB to 4-VBCB; y is the molar ratio of IB to 4-VBCB in the copolymer; and $r_1$ and $r_2$ are the reactivity ratio of IB and 4-VBCB, respectively, which are calculated as follows:

$$r_1 = \left[\sum_{i=1}^{n}\frac{y_i}{x_i^2} \times \sum_{i=1}^{n}x_i\left(1-\frac{1}{y_i}\right) + n\sum_{i=1}^{n}\frac{y_i}{x_i}\left(\frac{1}{y_i}-1\right)\right] / \left(\sum_{i=1}^{n}\frac{x_i}{y_i^2} \times \sum_{i=1}^{n}\frac{y_i}{x_i^2} - n^2\right),$$

$$r_2 = \left[\sum_{i=1}^{n}\frac{x_i^2}{y_i} \times \sum_{i=1}^{n}\frac{y_i}{x_i}\left(\frac{1}{y_i}-1\right) + n\sum_{i=1}^{n}x_i\left(1-\frac{1}{y_i}\right)\right] / \left(\sum_{i=1}^{n}\frac{x_i}{y_i^2} \times \sum_{i=1}^{n}\frac{y_i}{x_i^2} - n^2\right).$$

The reactivity ratios of IB and 4-VBCB are $r_1 = 0.47$ and $r_2 = 2.08$, respectively. Although 4-VBCB tended to self-polymerize, the two monomers tended to form random copolymers when its concentration was considerably lower than IB.

### 3.1.3. Characteristics and Mechanisms

The controlled polymerization of IB initiated by TMPCl/TiCl$_4$/DTBP system was achieved (Figures S1–S3). To verify the living characteristics of the random copolymerization of 4-VBCB and IB, we conducted a series of experiments where only the 4-VBCB feed ratio was changed, and other conditions remained unchanged. The details are shown in Table 2. Figure 2 shows the GPC traces of PIB segments and random copolymers obtained with different 4-VBCB feeding quantities. As the feed quantities of 4-VBCB were increased, the peak position of the random copolymers moved toward low molecular weights, and the widths of the peak gradually increased. Figure 3 shows the average molecular weight $M_n$ and the MWD of the PIB and the random copolymers obtained from various experiments as a function of the feeding quantity of 4-VBCB. The $M_n$ value gradually decreased with the increase in 4-VBCB feed quantity, whereas MWD exhibited an increasing trend.

Table 2. Poly(IB-co-4VBCB) with different feed quantities of 4-VBCB initiated by the TMPCl/TiCl$_4$/DTBP system, reacting in cyclohexane/CH$_3$Cl 60/40 (v/v) at −80 °C, [IB] = 1.01 M, [TMPCl] = 1.41 × 10$^{-3}$ M, [DTBP] = 3.53 × 10$^{-3}$ M, and [TiCl$_4$] = 4.52 × 10$^{-2}$ M.

| Run | [4-VBCB]/[TMPCl] | $M_n$ (g·mol$^{-1}$) | $M_w/M_n$ | Yield (%) |
| --- | --- | --- | --- | --- |
| A | 0 | 26,100 | 1.23 | 97.80 |
| B | 4.45 | 15,800 | 1.35 | 98.00 |
| C | 8.91 | 14,300 | 1.48 | 93.96 |
| D | 17.86 | 13,100 | 1.41 | 96.67 |
| E | 35.71 | 9920 | 1.52 | 97.30 |
| F | 71.43 | 8860 | 1.54 | 96.95 |

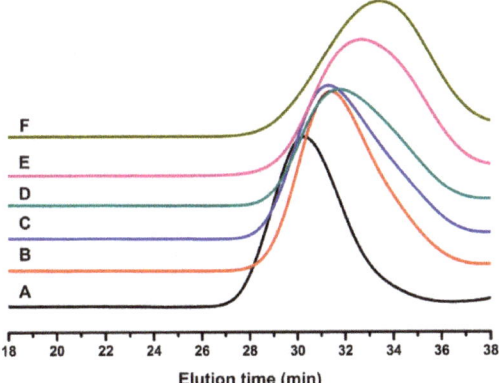

**Figure 2.** Gel-permeation chromatography (GPC) traces of poly(IB-*co*-4-VBCB) initiated by the TMPCl/DTBP/TiCl$_4$ system in cyclohexane/CH$_3$Cl (60/40 v/v) at −80 °C, [IB] = 1.01 M, [TMPCl] = 1.41 × 10$^{-3}$ M, [DTBP] = 3.53 × 10$^{-3}$ M, and [TiCl$_4$] = 4.52 × 10$^{-2}$ M (A, B, C, D, E, and F correspond to the terms in Table 1).

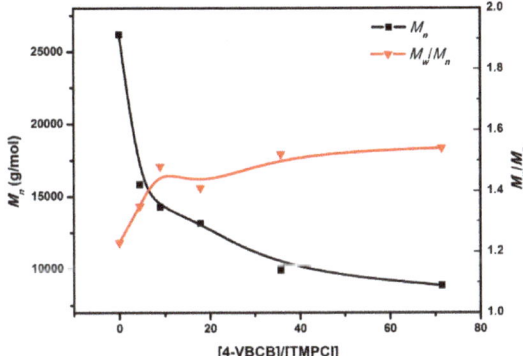

**Figure 3.** $Mn$ and $Mw/Mn$ values of poly(IB-*co*-4-VBCB) in different feed ratios of 4-VBCB to TMPCl initiated by TMPCl/DTBP/TiCl$_4$ system in cyclohexane/CH$_3$Cl (60/40 v/v) at −80 °C, [IB] = 1.01 M, [TMPCl] = 1.41 × 10$^{-3}$ M, [DTBP] = 3.53 × 10$^{-3}$ M, and [TiCl$_4$] = 4.52 × 10$^{-2}$ M.

The most likely reason for this result was that the chain transfer reaction to 4-VBCB increased rapidly with increasing feeding quantities of 4-VBCB. The active center of the copolymer chain was more likely to transfer to 4-VBCB with the simultaneous existence of two monomers. The copolymers of IB and 4-VBCB were assumed to have only four chain end groups, α-olefin, β-olefin, *tert*-Cl [33], and indane-BCB. Figure 4 shows the $^1$H NMR spectra of the four end groups. The fractional molar amount of each chain end was calculated separately as follows:

$$F_{\alpha\text{-olefin}} = \frac{(A_{4.64} + A_{4.85})/2}{(A_{4.64} + A_{4.85})/2 + A_{1.68}/6 + A_{1.7}/6 + (A_{2.51} + A_{2.83})/2}$$

$$F_{\beta\text{-olefin}} = \frac{A_{1.7}/6}{(A_{4.64} + A_{4.85})/2 + A_{1.68}/6 + A_{1.7}/6 + (A_{2.51} + A_{2.83})/2}$$

$$F_{tert\text{-Cl}} = \frac{A_{1.68}/6}{(A_{4.64} + A_{4.85})/2 + A_{1.68}/6 + A_{1.7}/6 + (A_{2.51} + A_{2.83})/2}$$

$$F_{indane-BCB} = \frac{(A_{2.51} + A_{2.83})/2}{(A_{4.64} + A_{4.85})/2 + A_{1.68}/6 + A_{1.7}/6 + (A_{2.51} + A_{2.83})/2}$$

where $A_{4.64}$ and $A_{4.85}$ are the resonance areas of $\alpha$-olefin, $A_{1.68}$ is the resonance area of the two methyl groups attached to $\beta$-olefin, and $A_{1.7}$ is the area of two methyl groups resonance attached to the *tert*-Cl carbon. $A_{2.51}$ and $A_{2.83}$ are related to the methylene protons of the indane structure produced by chain-transfer to 4-VBCB. The results are shown in Table 3.

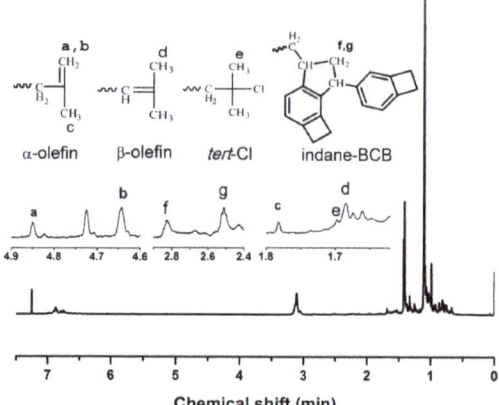

**Figure 4.** $^1$H NMR spectra of the copolymer end group.

**Table 3.** End group composition of different 4-VBCB feed quantities.

| [4-VBCB]/[IB] | 4-VBCB Components in Copolymer (mol %) | $M_n$ (g·mol$^{-1}$) | $M_w/M_n$ | End-Group Composition (mol %) | | | |
|---|---|---|---|---|---|---|---|
| | | | | $\alpha$-olefin | $\beta$-olefin | *tert*-Cl | Indane-BCB |
| 0 | 0 | 26,100 | 1.23 | 63.49 | 33.31 | 3.20 | 0 |
| 1/10 | 0.61 | 15,800 | 1.35 | 36.52 | 27.64 | 12.49 | 23.35 |
| 1/20 | 1.32 | 14,300 | 1.48 | 28.60 | 21.17 | 9.96 | 40.28 |
| 1/40 | 2.68 | 13,100 | 1.41 | 20.76 | 17.26 | 9.13 | 52.85 |
| 1/60 | 5.25 | 9920 | 1.52 | 14.89 | 16.35 | 8.94 | 59.81 |
| 1/80 | 10.64 | 8860 | 1.54 | 13.21 | 16.72 | 7.71 | 62.36 |

The corresponding reaction mechanism of IB and 4-VBCB cationic copolymerization initiated by TMPCl/DTBP/TiCl$_4$ was proposed on the basis of the results for the end groups in the above copolymerization reaction (Scheme 3). In the initiation reaction, TiCl$_4$ extracted –Cl from TMPCl to induce the generation of carbon cationic and formed the [TiCl$_5$]$^-$ counterion. With the help of the proton trapping agent DTBP, electrophilic carbocation preferentially attacked the monomers IB and 4-VBCB to initiate polymerization. Various chain transitions and chain terminations occurred frequently during polymerization. Chain transfer broke the chain by transferring the active center to the monomers IB and 4-VBCB. The broken polymer chain generated two groups at the end, namely, $\alpha$-olefin and $\beta$-olefin, when the reaction chain was transferred to IB and the indane-BCB end group when the reaction chain was transferred to 4-VBCB. Then, the new carbocation active center restarted the reaction to produce a new polymer chain. Moreover, the growing carbocation combined with the negative part of the counter ion to terminate the chain and produce the *tert*-Cl end group.

### 3.2. Thermal Properties of Poly(IB-co-4-VBCB) Random Copolymer

Figure 5 shows the DSC curve of the poly(IB-*co*-4-VBCB) containing 6 mol % of 4-VBCB. It shows the heat change of the copolymer with temperature raised from −90 °C to 350 °C, retained for 10 min at 260 °C for crosslinking, cooled to −90 °C, and reheated to 350 °C. $T_g$ was 25.76 °C during the first heating process. The BCB ring began to open when the copolymer was heated to 225 °C, and the maximum BCB ring-opening temperature was approximately 257 °C. $T_g$ was 38.56 °C during the second heating process after BCB ring opening, which was higher than the first $T_g$. This finding could be attributed the crosslinking of the BCB ring. The new C–C bond formed by crosslinking was stable and did not open even when the copolymer was heated to 350 °C.

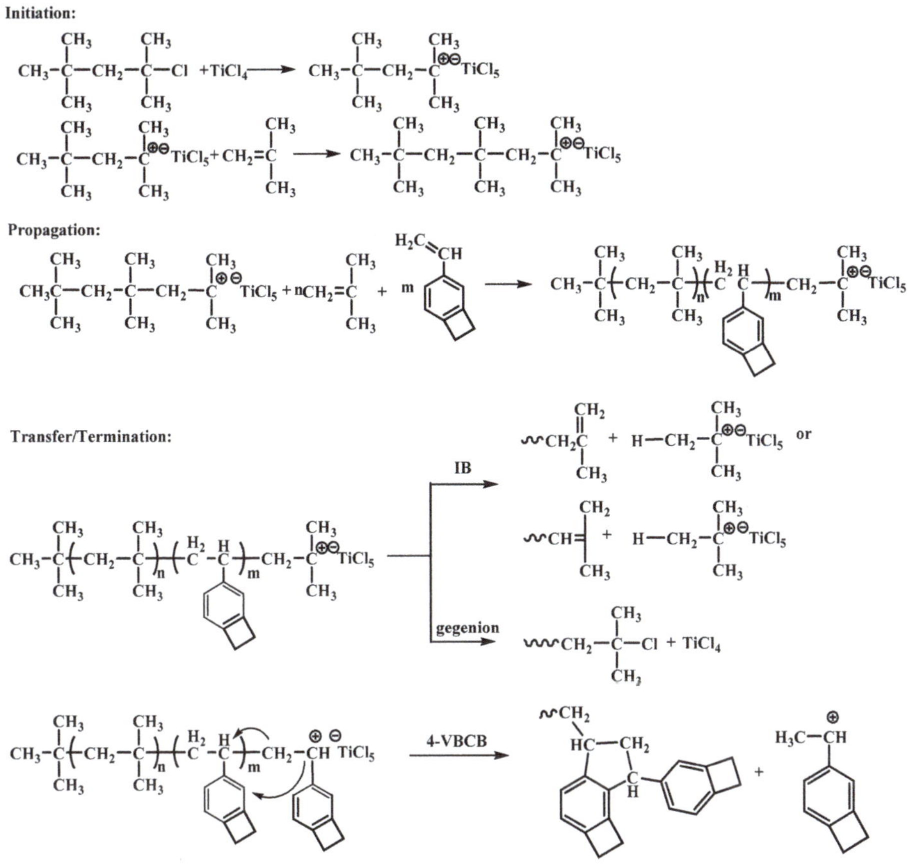

**Scheme 3.** Proposed mechanism for IB and 4-VBCB cationic copolymerization initiated by TMPCl/DTBP/TiCl$_4$.

We studied the $^1$H NMR spectrum of the copolymer to further determine the crosslinking structure of the copolymer. As shown in Figure 6, the absence of a signal peak (peak a) at 3.11 ppm on the $^1$H NMR spectrum of the crosslinked copolymer, indicated that all of the BCB olefin rings were open. A crosslinked eight-membered ring proton peak can be found at 3.18 ppm (peak b). The new signal peak at 5.15 ppm (peak c) may be ascribed to the structure in which no crosslinking occurred after the opening of the BCB ring. In the $^{13}$C NMR spectra (Figure 7), the characteristic signals of the PIB segment appeared at 31.4, 38.3, and 59.7 ppm, and the characteristic signals of the P4VBCB segment appeared at approximately 29.7 ppm. After crosslinking, we found that the characteristic signals (peak f) of the BCB four-member ring disappeared in the $^{13}$C NMR spectra, and a new signal peak (peak d), which was attributed to the crosslinking structure appeared. The same evidence can be observed from the FTIR spectrum of the copolymer (Figure 8). The bands at 2949, 1471, 1388, 1365, and 1230 cm$^{-1}$ were attributed to the PIB segment. The bands at 823 and 710.05 cm$^{-1}$ disappeared after crosslinking, whereas a new absorption peak appeared at 1072 cm$^{-1}$, thereby corresponding to the opening of the four-membered ring of BCB and the formation of the eight-membered ring structure by crosslinking.

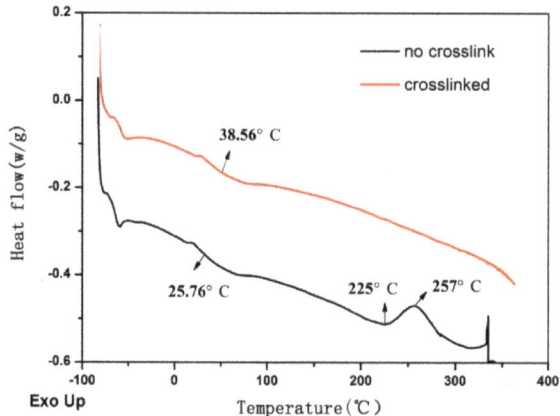

**Figure 5.** DSC curve of poly(IB-*co*-4-VBCB) containing 6 mol% 4-VBCB.

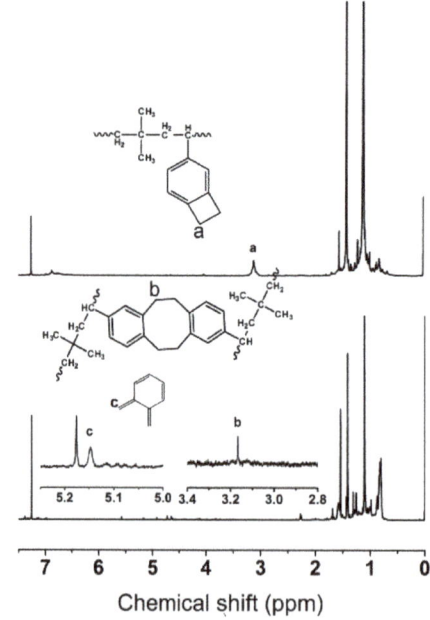

**Figure 6.** $^1$H NMR spectra of the copolymer before thermal crosslinking (**top**) and after thermal crosslinking (**bottom**).

The thermal stability of the copolymer containing 6 mol % 4-VBCB is shown in Figure 9. The epitaxial starting temperatures of the uncrosslinked and crosslinked copolymers were 390 °C and 406 °C, respectively. This result was due to the increased thermal stability caused by the crosslinking of the copolymer.

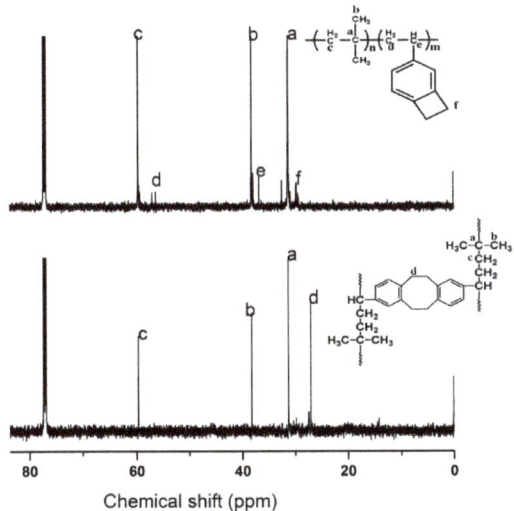

**Figure 7.** $^{13}$C NMR spectra of the copolymer before thermal crosslinking (**top**) and after thermal crosslinking (**bottom**).

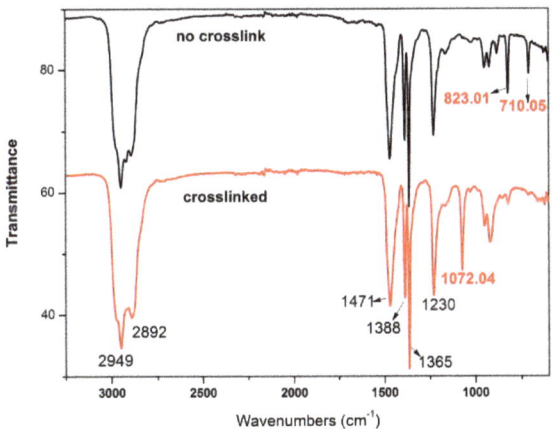

**Figure 8.** FTIR spectra of the uncrosslinked copolymer (**top**) and crosslinked copolymer (**bottom**).

### 3.3. Mechanical Properties of Poly(IB-co-4-VBCB) Random Copolymer.

We studied the mechanical properties of copolymers after thermal crosslinking. The copolymer with a molecular weight of approximately 40,000 was placed in a dumbbell mold. Then, the copolymer was heated to 250 °C at 4 MPa for 10 min. During heating, the copolymer became viscous at 250 °C. The 4-VBCB's four-membered ring opened and reacted with itself to form solid C–C bond. The stress-strain behavior of the module is illustrated in Table 4 and Figure 10. As the components of 4-VBCB in the copolymer increased, the elongation at break of the crosslinked copolymer increased slightly, and the maximum tensile stress did not change considerably. The mechanical properties of the copolymer were significantly improved compared with its viscous and gummy form before crosslinking.

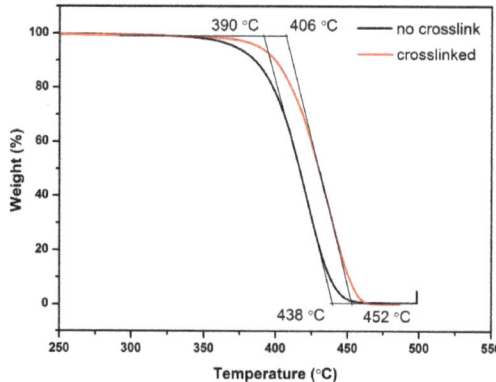

**Figure 9.** TGA of poly(IB-*co*-4-VBCB) containing 6 mol % 4-VBCB.

**Table 4.** Stress-strain behavior of poly(IB-*co*-4-VBCB) after thermal crosslinking.

| Num | 4-VBCB Components in Copolymer (mol %) | Maximum Tensile Stress (MPa) | Elongation at Break (%) |
|-----|---------------------------------------|------------------------------|-------------------------|
| A | 0.72 | 0.37 | 114 |
| B | 1.47 | 0.43 | 117 |
| C | 3.29 | 0.47 | 120 |
| D | 6.86 | 0.47 | 122 |
| E | 11.67 | 0.53 | 130 |

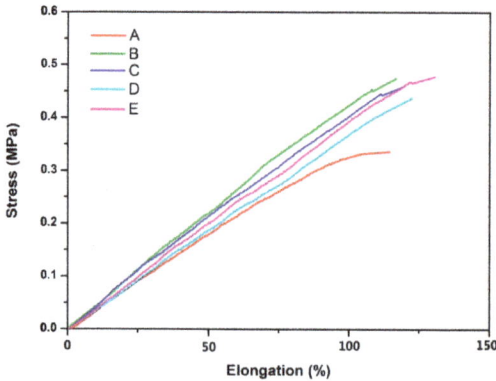

**Figure 10.** Stress-strain behavior of poly(IB-*co*-4-VBCB) after thermal crosslinking (A, B, C, D, and E correspond to the terms in Table 4).

## 4. Conclusions

A poly(IB-*co*-4-VBCB) random copolymer was successfully synthesized through cationic polymerization initiated by the TMPCl/DTBP/TiCl$_4$ system. The reactivity ratios of IB and 4-VBCB, which were r$_1$ = 0.47 and r$_2$ = 2.08, respectively, in the copolymerization reaction were calculated using the Yezreielv–Brokhina–Roskin method. The two monomers formed a random copolymer. The molecular weight and molecular weight distribution of the copolymer were characterized by GPC. The content of 4-VBCB in the copolymer also increased with the increase in the feed ratio of 4-VBCB to IB. However, the molecular weight of the copolymer decreased while the molecular weight distribution increased. The possible reaction mechanism was that the chain transfer reaction to 4-VBCB increased with the increase in the feed ratio of 4-VBCB to IB, thereby breaking the polymer chains.

The thermal properties of the copolymer were determined by DSC and TGA. When the copolymer was heated to approximately 220 °C, the BCB ring began to open, and the newly formed intermediates crosslinked with each other to form a network structure between copolymer chains. The crosslinked structure did not break even when the copolymer was reheated to the temperature at which the BCB ring can be opened. At the same time, the glass transition and decomposition temperatures of the copolymer increased after crosslinking due to its thermal crosslinking property. The state of the copolymer changed from viscous to fixed with a certain mechanical strength after crosslinking.

**Supplementary Materials:** The following are available online at http://www.mdpi.com/2073-4360/12/1/201/s1, Figure S1: GPC traces of PIB initiated by TMPCl/DTBP/TiCl$_4$ system in cyclohexane/CH$_3$Cl 60/40 (v/v) at −80 °C, [TMPCl] = 2.87 × 10$^{-3}$ M, [DTBP] = 7.18 × 10$^{-3}$ M, [TiCl$_4$] = 0.09 M, and [IB] = 1.01 M.; Figure S2: Mn and Mw/Mn as a function of conversion for the polymerization of IB initiated by TMPCl/DTBP/TiCl$_4$ system in cyclohexane/CH3Cl 60/40 (v/v) at −80 °C, [TMPCl] = 2.87 × 10$^{-3}$ M, [DTBP] = 7.18 × 10$^{-3}$ M, [TiCl$_4$] = 0.09 M and [IB] = 1.01M.; and Figure S3: First–order plot for the polymerization of IB initiated by TMPCl/DTBP/TiCl$_4$ system in cyclohexane/CH$_3$Cl 60/40 (v/v) at −80 °C, [TMPCl] = 2.87×10$^{-3}$ M, [DTBP] = 7.18×10$^{-3}$ M, [TiCl$_4$] = 0.09 M and [IB] = 1.01 M.

**Author Contributions:** Z.C. and Y.W. conceived and designed the experiments; Z.C., S.H., and K.W. performed the experiments; Z.C. and Y.W. analyzed the data; W.G., S.L., Y.S., and Y.W. contributed resources and supervision; Z.C. and Y.W. wrote the paper. All authors have read and agreed to the published version of the manuscript.

**Funding:** This research was funded by the National Natural Science Foundation of China (No.51573020), Beijing Natural Science Foundation (No. 2172022), Scientific Research Project of Beijing Educational Committee (KM201810017008), Project of Petrochina (No. KYWK18002), and URT Program (No. 2018J00029).

**Conflicts of Interest:** The authors declare no conflict of interest.

### References

1. Faust, R.; Kennedy, J.P. Living carbocationic polymerization. IV. Living polymerization of isobutylene. *Polym. Sci. Part A Polym. Chem* **1987**, *25*, 1847–1869. [CrossRef]
2. Faust, R.; Kennedy, J.P. Living carbocationic polymerization. *Polym. Bull.* **1986**, *15*, 317–323. [CrossRef]
3. Roth, M.; Patz, M.; Freter, H.; Mayr, H. Living oligomerization of isobutylene using di- and triisobutylene hydrochlorides as Initiators. *Macromolecules* **1997**, *30*, 722–725. [CrossRef]
4. Storey, R.F.; Chisholm, B.J.; Curry, C.L. Carbocation rearrangement in controlled/living isobutylene polymerization. *Macromolecules* **1995**, *28*, 4055–4061. [CrossRef]
5. Storey, R.F.; Choate, K.R., Jr. Kinetic investigation of the living cationic polymerization of isobutylene using a t-Bu-m-DCC/TiCl$_4$/2,4-DMP initiating system. *Macromolecules* **1997**, *30*, 4799–4806. [CrossRef]
6. Tawada, M.; Faust, R. Living cationic polymerization of isobutylene with mixtures of titanium tetrachloride/titanium tetrabromide. *Macromolecules* **2005**, *38*, 4989–4995. [CrossRef]
7. Puskas, J.E.; Brister, L.B.; Michel, A.J.; Lanzendorfer, M.G.; Jamieson, D.; Pattern, W.G. Novel substituted epoxide initiators for the carbocationic polymerization of isobutylene. *J. Polym. Sci. Part A Polym. Chem.* **2000**, *38*, 444–452. [CrossRef]
8. Bahadur, M.; Shaffer, T.D.; Ashbaugh, J.R. Dimethylaluminum chloride catalyzed living isobutylene polymerization. *Macromolecules* **2000**, *33*, 9548–9552. [CrossRef]
9. Wu, Y.; Ren, P.; Guo, W.; Li, S.; Yang, X.; Shang, Y. Living cationic sequential block copolymerization: Synthesis and characterization of poly(4-(2-hydroxyethyl)styrene-b-isobutylene-b-4-(2-hydroxyethyl)styrene) triblock copolymers. *Polym. J.* **2010**, *42*, 268–272. [CrossRef]
10. Wu, Y.; Guo, W.; Li, S.; Gong, H. Synthesis of poly(isobutylene-b-alpha-methylstyrene) copolymers by living cationic polymerization. *Acta Polym. Sin.* **2008**, *8*, 574–580. [CrossRef]
11. Wu, Y.; Guo, W.; Li, S.; Gong, H. Kinetic studies on homopolymerization of alpha-methylstyrene and sequential block copolymerization of isobutylene with alpha-methylstyrene by living/controlled cationic polymerization. *Polym. Korea* **2008**, *32*, 366–371.
12. Zhang, X.; Guo, W.; Wu, Y.; Shang, Y.; Li, S.; Xiong, W. Synthesis of random copolymer of isobutylene with p-methylstyrene by cationic polymerization in ionic liquids. *E-Polymers* **2018**, *18*, 423–431. [CrossRef]
13. Yang, S.; Fan, Z.; Zhang, F.; Li, S.; Wu, Y. Functionalized Copolymers of Isobutylene with Vinyl Phenol: Synthesis, Characterization, and Property. *Chin. J. Polym. Sci.* **2019**, *37*, 919–929. [CrossRef]

14. Xie, Y.; Chang, J.; Wu, Y.; Yang, D.; Wang, H.; Zhang, T.; Li, S.; Guo, W. Synthesis and properties of bromide-functionalized poly(isobutylene-*co*-p-methylstyrene) random copolymer. *Polym. Int.* **2017**, *66*, 468–476. [CrossRef]
15. Kennedy, J.P.; Midha, S.; Tsunogae, Y. Tsunogae Living carbocationic polymerization. 56. Polyisobutylene-containing block polymers by sequential monomer addition. 8. Synthesis, characterization, and physical properties of poly(indene-b-isobutylene-b-indene) thermoplastic elastomers. *Macromolecules* **1993**, *26*, 429–435. [CrossRef]
16. Li, J.; Wu, K.; Huang, S.; Zhang, J.; Zhao, M.; Gong, G.; Guo, W.; Wu, Y. Synthesis and Properties of Hydroxytelechelic Polyisobutylenes by End Capping with tert-Butyl-dimethyl-(4-methyl-pent-4-enyloxy)-silane. *Chin. J. Polym. Sci.* **2019**, *37*, 936–942. [CrossRef]
17. Hahn, S.F.; Martin, S.J.; McKelvy, M.L.; Patrick, D.W. Thermal polymerization of bis(benzocyclobutene) monomers containing. alpha.,.beta.-disubstituted ethenes. *Macromolecules* **1993**, *26*, 3870–3877. [CrossRef]
18. Kirchhoff, R.A.; Bruza, K.J. Benzocyclobutenes in polymer synthesis. *Prog. Polym. Sci.* **1993**, *18*, 85–185. [CrossRef]
19. John, M.W.; William, C.P.; Robert, A.D. Benzocyclobutenes as styrene monomer scavengers and molecular weight "stabilizers" in atactic and syndiotactic polystyrenes. *J. Appl. Polym. Sci.* **2000**, *78*, 2008–2015.
20. So, Y.H.; Foster, P.; Im, J.H.; Garrou, P.; Hetzner, J.; Stark, E. Divinylsiloxane-bisbenzocyclobutene-based polymer modified with polystyrene polybutadiene–polystyrene triblock copolymers. *J. Polym. Sci. Part A Polym. Chem.* **2006**, *44*, 1591–1599. [CrossRef]
21. Okay, O.; Erman, B.; Durmaz, S. Solution cross-linked poly(isobutylene) gels: Synthesis and swelling behavior. *Macromolecules* **2000**, *33*, 4822–4827. [CrossRef]
22. Yang, J.; Cheng, Y.; Xiao, F. Synthesis, thermal and mechanical properties of benzocyclobutene-functionalized siloxane thermosets with different geometric structures. *Eur. Polym. J.* **2012**, *48*, 751–760. [CrossRef]
23. So, Y.H.; Hahn, S.F.; Li, Y.; Reinhard, M.T. Styrene 4-vinylbenzocyclobutene copolymer for microelectronic applications. *J. Polym. Sci. Part A Polym. Chem.* **2008**, *46*, 2799–2806. [CrossRef]
24. Huang, Y.; Zhang, S.; Hu, H.; Wei, X.; Yu, H.; Yang, J. Photoactive polymers with benzocyclobutene/silacyclobutane dual crosslinked structure and low dielectric constant. *J. Polym. Sci. Part A Polym. Chem.* **2017**, *55*, 1920–1928. [CrossRef]
25. Sakellariou, G.; Avgeropoulos, A.; Hadjichristidis, H.; Mays, J.W.; Baskaran, D. Functionalized organic nanoparticles from core-crosslinked poly(4-vinylbenzocyclobutene-b-butadiene) diblock copolymer micelles. *Polymer* **2009**, *10*, 038. [CrossRef]
26. Harth, V.; Horn, B.V.; Lee, V.Y.; Germack, D.S.; Gonzales, C.P.; Miller, R.D.; Hawker, C.J. A facile approach to architecturally defined nanoparticles via intramolecular chain collapse. *J. Am. Chem. Soc.* **2002**, *124*, 8653–8660. [CrossRef]
27. Takeshi, E.; Toshio, K.; Toshikazu, T.; Keisuke, C. Synthesis of poly(4-vinylbenzocyclobutene) and its reaction with dienophiles. *J. Polym. Sci. Part A Polym. Chem.* **1995**, *33*, 707–715.
28. Xu, Y.; Zhu, F.; Xie, L.; Yang, J.; Zhang, L.; Xie, R. Synthesis and property studies of oligomer obtained from the reaction of 4-vinylbenzocyclobutene and styrene with divinyl tetramethyl disiloxane-bisbenzocyclobutene. *E-Polymers* **2010**, *10*, 117. [CrossRef]
29. Sakellariou, G.; Baskaran, D.; Hadjichristidis, N.; Mays, J.W. Well-defined poly(4-vinylbenzocyclobutene): #160 synthesis by living anionic polymerization and characterization. *Macromolecules* **2006**, *39*, 2525–3530.
30. Sheriff, J.; Thomas, E.C.; Phat, L.; Kothadia, R.; George, S.; Kato, Y.; Pinchuk, L.; Slepian, M.J.; Bluestein, D. Physical characterization and platelet interactions under shear flows of a novel thermoset polyisobutylene-based co-polymer. *ACS Appl. Mater. Interfaces* **2015**, *7*, 22058–22066. [CrossRef] [PubMed]
31. Hadjikyriacou, S.; Acar, M.; Faust, R. Living and controlled polymerization of isobutylene with alkylaluminum halides as coinitiators. *Macromolecules* **2004**, *37*, 7543–7547. [CrossRef]

32. Gronowski, A.; Wojtczak, Z. Copolymerizations of styrene with magnesium, calcium, strontium and barium acrylates in dimethyl sulfoxide. *Macromol. Chem. Phys.* **1989**, *190*, 2063–2069. [CrossRef]
33. Yan, P.F.; Guo, A.R.; Liu, Q. Living cationic polymerization of isobutylene coinitiated by FeCl$_3$ in the presence of isopropanol. *J. Polym. Sci. Part A Polym. Chem.* **2012**, *50*, 3383–3392. [CrossRef]

© 2020 by the authors. Licensee MDPI, Basel, Switzerland. This article is an open access article distributed under the terms and conditions of the Creative Commons Attribution (CC BY) license (http://creativecommons.org/licenses/by/4.0/).

*Article*

# A Nonequilibrium Model for Particle Networking/Jamming and Time-Dependent Dynamic Rheology of Filled Polymers

Christopher G. Robertson [1,*], Sankar Raman Vaikuntam [2] and Gert Heinrich [2,3]

1 Endurica LLC, Findlay, OH 45840, USA
2 Leibniz-Institut für Polymerforschung Dresden e.V., 01069 Dresden, Germany; sankarraman.vaikuntam@gmail.com (S.R.V.); gheinrich@ipfdd.de (G.H.)
3 Institut für Textilmaschinen und Textile Hochleistungswerkstofftechnik, Technische Universität Dresden, 01069 Dresden, Germany
* Correspondence: cgrobertson@endurica.com

Received: 19 November 2019; Accepted: 5 January 2020; Published: 10 January 2020

**Abstract:** We describe an approach for modeling the filler network formation kinetics of particle-reinforced rubbery polymers—commonly called filler flocculation—that was developed by employing parallels between deformation effects in jammed particle systems and the influence of temperature on glass-forming materials. Experimental dynamic viscosity results were obtained concerning the strain-induced particle network breakdown and subsequent time-dependent reformation behavior for uncross-linked elastomers reinforced with carbon black and silica nanoparticles. Using a relaxation time function that depends on both actual dynamic strain amplitude and fictive (structural) strain, the model effectively represented the experimental data for three different levels of dynamic strain down-jump with a single set of parameters. This fictive strain model for filler networking is analogous to the established Tool–Narayanaswamy–Moynihan model for structural relaxation (physical aging) of nonequilibrium glasses. Compared to carbon black, precipitated silica particles without silane surface modification exhibited a greater overall extent of filler networking and showed more self-limiting behavior in terms of network formation kinetics in filled ethylene-propylene-diene rubber (EPDM). The EPDM compounds with silica or carbon black filler were stable during the dynamic shearing and recovery experiments at 160 °C, whereas irreversible dynamic modulus increases were noted when the polymer matrix was styrene-butadiene rubber (SBR), presumably due to branching/cross-linking of SBR in the rheometer. Care must be taken when measuring and interpreting the time-dependent filler networking in unsaturated elastomers at high temperatures.

**Keywords:** polymer nanocomposites; filled rubber; particle network; filler flocculation; fictive strain; structural relaxation; Tool–Narayanaswamy–Moynihan model; jamming

## 1. Introduction

There is considerable academic and industrial interest in the rheology/viscoelasticity of polymer systems due the influence on both the processability in manufacturing and the final product performance. One important example is in the field of automobile tire technology where the dynamic mechanical behavior of the tread compound is closely connected to—and predictive of—the fuel economy, traction, and handling/cornering performance characteristics of a tire [1,2]. For general background information about polymer viscoelasticity, several excellent books on this technical area are suggested [3–6].

Temperature and frequency are the main experimental variables in dynamic mechanical characterization of polymer materials. Strain amplitude effects are also of critical importance in polymer composites reinforced by particles. First documented by Dillon, Prettyman, and Hall in 1944 [7], the Payne effect [8–10] is a well-known viscoelastic phenomenon in particle-filled elastomers that is characterized by a significant reduction in dynamic storage modulus and the appearance of a peak in loss tangent (tanδ) as the oscillatory strain amplitude is increased. This hysteretic softening occurs at small dynamic strains, with most of the storage modulus reduction taking place in the range from 0.1 to 10% strain. Unfilled elastomers do not typically exhibit strain-dependent viscoelastic response until much higher strain amplitudes (>50%). The key features of the Payne effect are described in a chapter by Heinrich and Klüppel [11] and in a brief review [12]. The tire industry is particularly interested in understanding the Payne effect and developing materials technologies to reduce its magnitude because of the impact of this viscoelastic behavior on fuel economy. The global oil consumption each day for ground vehicles is over 50 million barrels of oil [13]. About 10% of the fuel used by an automobile is consumed to overcome the rolling resistance of tires, and at least half of that rolling resistance is from the Payne effect of the rubber compounds within the tires. Therefore, more than 2.5 million barrels of oil are wasted each day around the world due to the Payne effect of tire compounds.

The presence of a strain-sensitive filler network composed of percolated particle–particle contacts is responsible for the majority of the Payne effect [11,14–17], although polymer dynamics at the polymer–filler interfaces also have a secondary contribution [15,18–20]. Once the shearing of an uncross-linked-filled elastomer compound is ceased, the filler network strengthens with time which leads to a larger Payne effect. This is commonly referred to as filler flocculation, and it has been widely studied across the past two decades [21–36]. Once a filler network is formed, it can be broken by deforming the material above about 20% strain, and the filler network will reform with time after the strain is removed or reduced. This is schematically illustrated in Figure 1 for nano-structured fillers such as precipitated silica and carbon black made up of fused aggregates of nanometer-scale primary particles. Only very small movements of aggregates are needed to break the connectivity of the filler network, and reformation of the network takes place across correspondingly small distances. Therefore, the term flocculation is somewhat of a misnomer for this filler network build-up phenomenon.

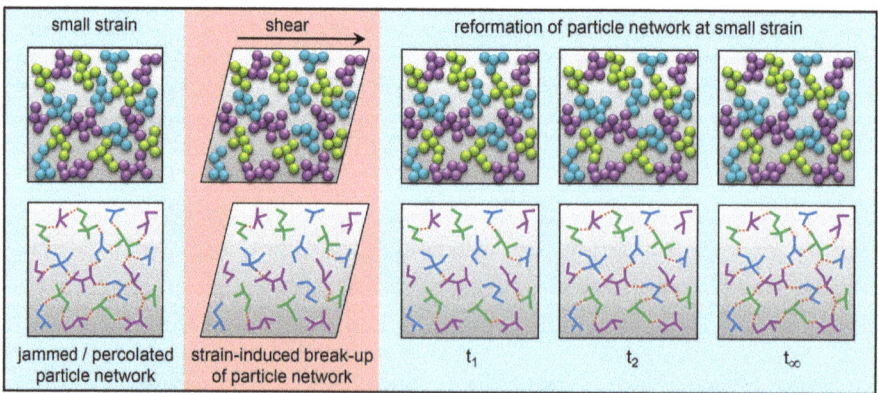

**Figure 1.** Illustration of the effect of shear strain on the particle network in a filled elastomer and the subsequent time-dependent reformation process. The different colored aggregates in the upper diagrams represent fused primary particles, and these aggregates are the smallest dispersible units of nano-structured fillers such as carbon black and precipitated silica. The lower diagrams illustrate the connectivity of the filler network, with solid lines showing fused connections of primary particles within aggregates and dashed lines showing aggregate–aggregate (filler–filler) contacts.

The physics of particle systems have similarities to glassy behavior, but with deformation (stress, strain, vibration) driving the response instead of temperature as the key parameter. The noted similarities between temperature effects in glass-forming materials and influence of deformation on particle systems led to the creation of jamming phase diagrams [37,38]. These ideas have been extended to particle-filled rubber [39,40]. Phenomenological modeling of time-dependent properties during relaxation in the nonequilibrium glassy state is well established, which provides the opportunity to borrow those concepts for modeling the filler networking/jamming process in filled polymers by substituting deformation for temperature. This present study will show dynamic rheological results that reveal filler networking/flocculation in model-filled rubber formulations, and glassy modeling approaches will be adapted to fit the time-dependent data and give new insights into the behavior. Given the known importance of relative surface energies for polymer and particles to the filler flocculation process [33–35,41,42], our investigation included two different elastomers, ethylene-propylene-diene rubber (EPDM) and styrene-butadiene rubber (SBR), that are reinforced with two different fillers: precipitated silica and carbon black (CB).

## 2. Materials and Methods

The polymers included in this investigation were EPDM (Buna EP G 3440) and solution SBR (VSL 2525 0 m) from Arlanxeo Deutschland GmbH (Dormagen, Germany). The precipitated silica (Ultrasil VN3 grade) from Evonik Industries AG (Essen, Germany) and carbon black (N115 grade) from Orion Engineered Carbons GmbH (Cologne, Germany) have nitrogen surface areas of 180 and 137 $m^2/g$, respectively, as reported by the suppliers. The antioxidant used was rubber grade 6PPD (N-(1,3-dimethylbutyl)-N'-phenyl-p-phenylenediamine).

The simple model rubber formulations in Table 1 were mixed using a Haake Rheomix 600P (Thermo Fisher Scientific GmbH in Karlsruhe, Germany). This mixer has a chamber volume of 80 cc, which was filled 70% with compound during mixing. Using a starting temperature of 110 °C and rotor speed of 60 rpm, the compounds were mixed for 8 min to a final temperature of 140 to 150 °C. The compounds were then milled for 2 min using a two-roll mill (Polymix-110L; Servitec Maschinenservice GmbH in Wustermark, Germany) at 50 °C with a friction ratio of 1:1.2. Densities for the raw materials were used to determine values of filler volume fraction ($\varphi$), which was nearly constant at 0.15 to 0.17 for the four compounds (Table 1).

Table 1. Model formulations for particle-filled rubber compounds (phr).

| Compound: | SBR-CB | SBR-Silica | EPDM-CB | EPDM-Silica |
|---|---|---|---|---|
| SBR | 100 | 100 | | |
| EPDM | | | 100 | 100 |
| Carbon Black (N115) | 40 | | 40 | |
| Precipitated Silica | | 40 | | 40 |
| Antioxidant (6PPD) | 2 | 2 | 2 | 2 |
| Total (phr): | 142 | 142 | 142 | 142 |
| Filler Volume Fraction, $\varphi$: | 0.17 | 0.15 | 0.16 | 0.15 |

Oscillatory shear rheology measurements were conducted using a Rubber Process Analyzer (RPA) made by Scarabaeus GmbH (Wetzlar, Germany; model SIS-V50) that uses a serrated biconical die geometry. All RPA testing was performed at a temperature of 160 °C and frequency (f) of 1.67 Hz (angular frequency ($\omega$) = 10.5 rad/s). The dynamic strain amplitude ($\gamma$) was varied in a time sequence: (1) 5 min at $\gamma$ = 0.25 to break up the filler network; (2) 120 min at the flocculation $\gamma$ value; and (3) 10 min each at sequentially increasing strains up to and including $\gamma$ = 0.25. A fresh rubber compound specimen was used for each RPA testing sequence, and the volume of compound used for testing was approximately 5 $cm^3$. We report results for shear storage modulus (G') and the magnitude of complex viscosity, $|\eta^*|$, which is commonly called dynamic viscosity and represented without the brackets, $\eta^*$.

## 3. Results and Discussion

This study considered simple SBR and EPDM formulations containing only polymer, filler, and an antioxidant (Table 1). The filler volume fraction was nearly constant ($\varphi = 0.15$ to $0.17$) for these materials which did not contain any cross-linking additives. The CB and silica filler grades used have specific surface areas of 137 m$^2$/g and 180 m$^2$/g. Although these are nano-structured fillers (see Figure 1), we can get an idea of the size of these filler materials by converting the specific surface areas to values of equivalent spherical particle diameter (d). The surface to volume ratio of a sphere is 6/d which leads to d = 17 nm for silica and d = 24 nm for CB using densities for silica and carbon black of 2.0 and 1.8 g/cm$^3$, respectively. The filled elastomers investigated here are clearly polymer nanocomposites.

Oscillatory shear rheology was used to study filler networking kinetics for CB-filled and silica-filled SBR and EPDM. It is common to use temperatures in the 150 to 170 °C range for studying filler flocculation in rubber, because this is the typical range where curing (vulcanization) of rubber compounds takes place commercially. Most of the filler networking occurs during the early stages of curing, before the polymer chains become cross-linked [30]. We employed a temperature of 160 °C for all of our measurements. The testing protocol involved: (1) breaking up the filler network at a strain amplitude ($\gamma$) of 0.25 (25%); (2) down-jump to the flocculation $\gamma$ where changes in rheological properties were monitored for 120 min; and (3) sequential break-up of the filler network at various increasing strains, each applied for 10 min, up to the final $\gamma$ of 0.25. So, the testing series started and ended at the same conditions. The storage modulus (G') results for this testing series are shown in Figure 2 for the CB-filled polymers using a flocculation $\gamma$ of 0.014 and subsequent filler network break-up $\gamma$ values of 0.05, 0.10, and 0.25, and the results for the silica filler are given in Figure 3 using the same conditions. For both types of filled EPDM, it is evident that the starting and ending G' at $\gamma = 0.25$ were essentially the same. When the polymer matrix was SBR, however, irreversible increases in G' were noted, presumably due to branching/cross-linking of SBR in the rheometer. Cross-linking of unsaturated elastomers without vulcanization agents at high temperatures is a known occurrence [43–45]. This grade of SBR has a structure with 75 wt.% butadiene (unsaturated; C=C double bond in every polymer repeat unit from butadiene), whereas the EPDM has a nearly fully saturated structure except for the sparse double bonds from the 4.1 wt.% ethylidene norbornene comonomer. For filler flocculation studies, we recommend a testing sequence like the one utilized here and/or a time sweep on the unfilled polymer to ensure that the material is stable, such that any kinetic model fitting—and parameter interpretation therefrom—can be considered valid and meaningful. The remainder of the experiments to be discussed and the phenomenological model fitting will only involve the stable EPDM-CB and EPDM-silica materials.

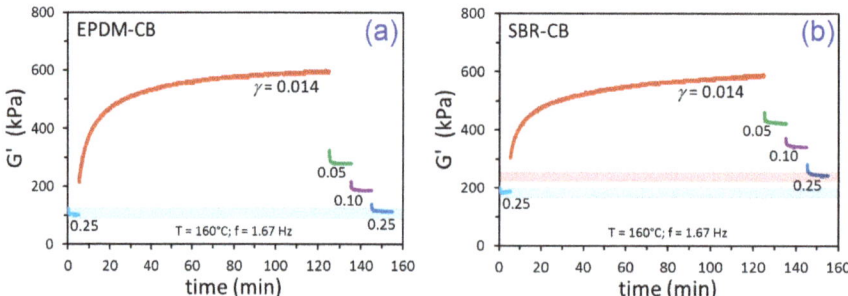

**Figure 2.** Time-dependent G' results for the indicated strain amplitude history for EPDM-CB (**a**) and SBR-CB (**b**). The blue bar in (**a**) represents the similar starting and ending values of G' at $\gamma = 0.25$. In (**b**), the blue bar is the starting G' at $\gamma = 0.25$ and the pink bar is the final G' at $\gamma = 0.25$, which is higher.

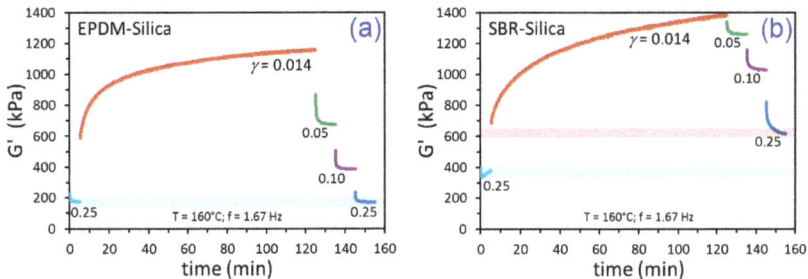

**Figure 3.** Time-dependent G′ results from the indicated strain amplitude history for EPDM-Silica (**a**) and SBR-Silica (**b**). The blue bar in (**a**) represents the similar starting and ending values of G′ at $\gamma$ = 0.25. In (**b**), the blue bar is the starting G′ at $\gamma$ = 0.25 and the pink bar is the final G′ at $\gamma$ = 0.25, which is higher.

For studying filler flocculation, the time dependence of G′ is often used, and examples from our testing were just presented in Figures 2 and 3. However, for our detailed analysis and model fitting, we use the magnitude of the complex viscosity, $|\eta^*|$, which is equal to $|G^*|/\omega$ where $|G^*|$ is the magnitude of the complex modulus and $\omega$ is the angular frequency ($\omega = 2\pi f$). We represent this simply as $\eta^*$ (without the brackets) and refer to this quantity as dynamic viscosity. We study $\eta^*$ for two reasons: (1) it is directly determined from the experimental stress and strain amplitudes without having to separate the dynamic shear response into storage modulus and loss modulus and accordingly prove the validity of linear-nonlinear dichotomy for our filled rubber compounds at the testing conditions [20,30,46]; and (2) viscosity is a typical property used to describe glass-forming materials, to which we will draw parallels later.

The build-up of dynamic viscosity after a down-jump from equilibrium at $\gamma$ = 0.25 was studied at three different networking/flocculation $\gamma$ values of 0.014, 0.03, and 0.05. The data in Figure 4 indicate significantly more extensive filler networking for silica relative to CB in the EPDM matrix. This is expected since the bare silica without silane surface modification is polar (surface covered with -OH groups) and consequently has much stronger filler–filler interactions as flocculation driving forces compared to carbon black. Another observation for both materials is that the degree of viscosity growth increased as the flocculation strain was reduced.

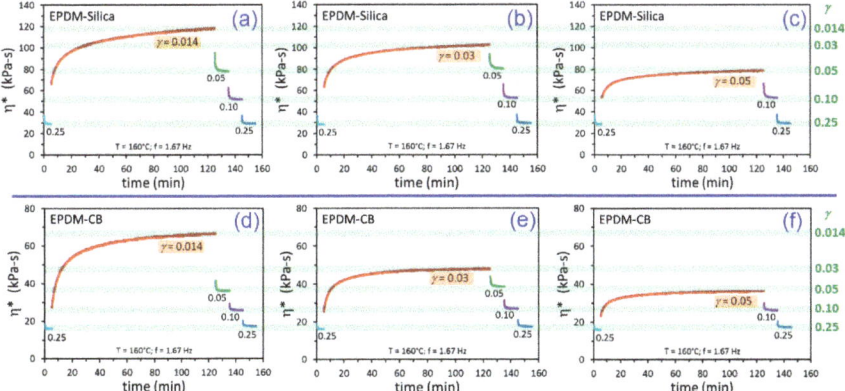

**Figure 4.** Time-dependent dynamic viscosity results for EPDM-Silica (**a**–**c**) and EPDM-CB (**d**–**f**) from testing using the indicated strain amplitude histories. The green bars represent the apparent equilibrium $\eta^*$ values at the various values of $\gamma$. Note the different y-axis scalings for EPDM-Silica (**a**–**c**) and EPDM-CB (**d**–**f**).

A key observation from Figure 4 is that for each polymer–filler system, there appears to be an equilibrium dynamic viscosity, $\eta^*_{eq}$, at each value of $\gamma$ that is independent of the prior strain history path. This is highlighted by the green bars in Figure 4, and the $\gamma$-dependence of $\eta^*_{eq}$ displays a power law behavior for both EPDM-CB and EPDM-Silica (Figure 5).

$$\eta^*_{eq} = c\, \gamma^{-\alpha} \tag{1}$$

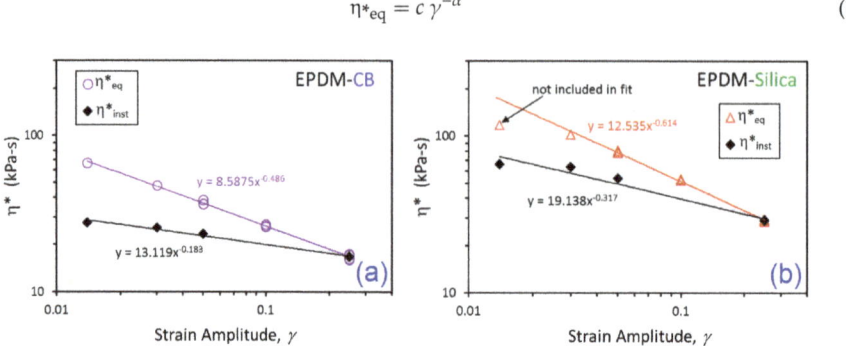

**Figure 5.** $\gamma$-dependent values of $\eta^*_{eq}$ and $\eta^*_{inst}$ (symbols) and power law fits (solid lines) for EPDM-CB (**a**) and EPDM-Silica (**b**).

The power law parameters are summarized in Table 2. Switching the independent and dependent variables produces an expression that will be used later to define the time dependence of fictive strain, $\gamma_f$, in terms of the dynamic viscosity evolution with time:

$$\gamma_f(t) = c^{1/\alpha}\, [\eta^*(t)]^{-1/\alpha} \tag{2}$$

**Table 2.** Power Law Parameters for $\gamma$-dependences of $\eta^*_{eq}$ and $\eta^*_{inst}$.

| Compound | $\eta^*_{eq}$ [Equations (1) and (2)] | | $\eta^*_{inst}$ [Equation (3)] | |
|---|---|---|---|---|
| | c (kPa-s) | $\alpha$ | b (kPa-s) | $\theta$ |
| EPDM-CB | 8.588 | 0.486 | 13.119 | 0.183 |
| EPDM-Silica | 12.535 | 0.614 | 19.138 | 0.317 |

Another noted aspect of the general behavior in Figure 4 is that there seems to be a discontinuity in $\eta^*$ when the strain is suddenly reduced from equilibrium at $\gamma = 0.25$ to the flocculation $\gamma$. The time between the last datapoint at $\gamma = 0.25$ and the first datapoint at the flocculation strain is 6 s, so it is possible that the noted viscosity jump in each case is a consequence of this time gap (i.e., missing data). Interestingly, however, this "instantaneous" dynamic viscosity, $\eta^*_{inst}$, also shows a power law dependence with respect to $\gamma$, with parameters reported in Table 2.

$$\eta^*_{inst} = b\, \gamma^{-\theta} \tag{3}$$

The apparent viscosity jumps may reveal some real physics of the flocculation process, namely that it is composed of an instantaneous part and a time-dependent part.

The nonequilibrium behavior of glasses is modeled using actual temperature ($T$) and fictive temperature ($T_f$) [47–50], and the analogous concept of fictive strain was recently introduced for filled rubber [32]. Given the similarities between the influence of deformation on jammed particle systems and temperature effects on glass-forming materials, we adapt the Tool–Narayanaswamy–Moynihan (TNM) model [47,48,51,52] that is used to represent the structural relaxation (physical aging) process in the nonequilibrium glassy state. Substituting $\gamma$ and $\gamma_f$ for $T$ and $T_f$ in the TNM approach leads to the fictive strain model:

$$\eta^*(t) = \eta*_0 + (\eta*_\infty - \eta*_0)\left\{1 - \exp\left[-\left(\int_0^t \frac{dt}{\tau(\gamma, \gamma_f(t))}\right)^\beta\right]\right\} \quad (4)$$

$$\tau = A \exp\left[\frac{a}{\gamma} + \frac{s}{\gamma_f(t)}\right] \quad (5)$$

It should be noted that the functional form of Equation (5) is different than a preliminary expression proposed earlier for the relaxation time, $\tau$ [32]. At first glance, the expression in Equation (4) looks like a typical stretched exponential growth function, with stretching exponent $\beta$. However, the unique part is the relaxation time function, which depends on actual strain through the parameter $a$ and fictive/structural strain through the parameter $s$ (Equation (5)). The fictive strain decreases toward the actual strain during flocculation, thereby imparting a time-dependent increasing nature to $\tau$ as filler networking progresses. The $\gamma_f(t)$ is assigned from the measured $\eta^*(t)$ using Equation (2), and a visual example of the connection between the time dependence of dynamic viscosity and fictive strain is presented in Figure 6. In fitting the experimental data, we set $\eta^*_\infty = \eta^*_{eq}$ from Equation (1). We also fix $\eta^*_0 = \eta^*_{inst}$, where $\eta^*_{inst}$ was assigned from the first datapoint after the strain down-jump when fitting experimental results and was determined using Equation (3) when predicting behavior outside the range of measured dynamic rheology results. The fitting used four varying parameters, $a$, $s$, $A$, and $\beta$, for the nonexponential (stretched exponential) version of the model. Only three parameters were allowed to vary for the exponential version of the model, because $\beta$ was fixed at a value of 1 for that case.

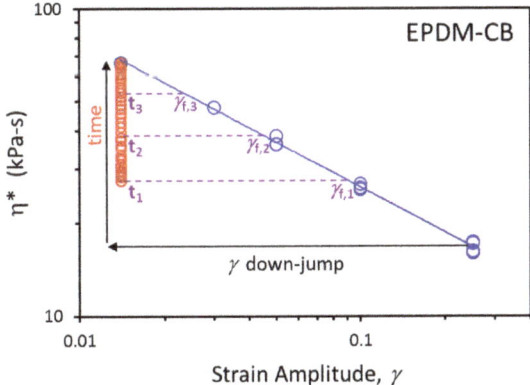

**Figure 6.** Illustration of fictive strain determination during time-dependent $\eta^*$ build-up (red symbols) at $\gamma = 0.014$ after strain down-jump from $\gamma = 0.25$ for EPDM-CB. The blue symbols are $\eta^*_{eq}$ vs. $\gamma$ along with the related power law fit (solid line).

The $\eta^*(t)$ responses from the three flocculation experiments at $\gamma = 0.014$, 0.03, and 0.05 were simultaneously fit for each material using exponential and nonexponential versions of the fictive strain model. The fits are shown in Figures 7 and 8, and the resulting model parameters are summarized in Table 3. The modeling approach was able to effectively represent the time-dependent dynamic viscosity responses for EPDM-CB and EPDM-Silica systems at three levels of strain down-jump using a single set of fitting parameters for each material. The stretched exponential model gave some modest improvement in fitting compared to the exponential model due to an additional fitting parameter (4 versus 3).

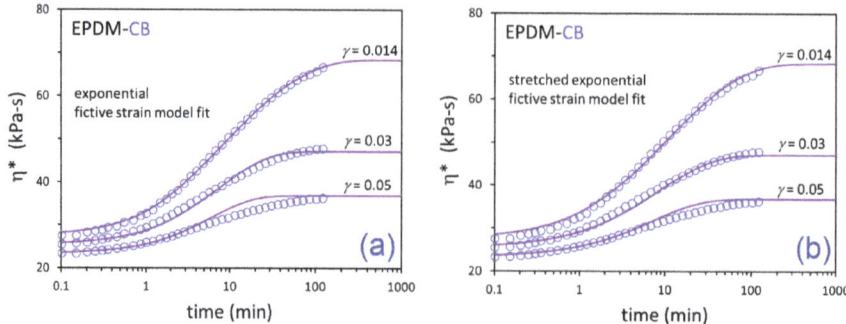

**Figure 7.** Time-dependent η* data (symbols) after down-jump from equilibrium at $\gamma = 0.25$ to the indicated $\gamma$ values and fictive strain model fits (solid lines) for EPDM-CB using the exponential (**a**) and stretched exponential (**b**) versions of the model.

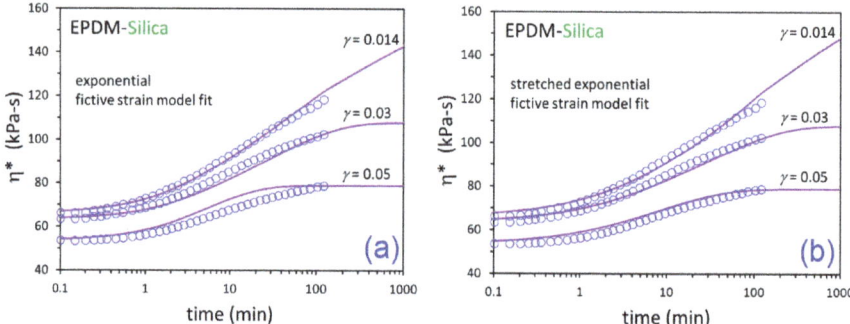

**Figure 8.** Time-dependent η* data (symbols) after down-jump from equilibrium at $\gamma = 0.25$ to the indicated $\gamma$ values and fictive strain model fits (solid lines) for EPDM-Silica using the exponential (**a**) and stretched exponential (**b**) versions of the model.

**Table 3.** Fictive strain model parameters.

| Compound | Model | A (min) | a | s | β |
|---|---|---|---|---|---|
| EPDM-CB | Exponential | 3.76 | $4.13 \times 10^{-4}$ | $4.14 \times 10^{-2}$ | 1.0 (fixed) |
| EPDM-Silica | | 0.765 | $9.01 \times 10^{-3}$ | $1.48 \times 10^{-1}$ | 1.0 (fixed) |
| EPDM-CB | Stretched Exponential | 5.06 | $2.47 \times 10^{-3}$ | $2.71 \times 10^{-2}$ | 0.835 |
| EPDM-Silica | | 1.61 | $2.20 \times 10^{-2}$ | $9.45 \times 10^{-2}$ | 0.723 |

The relaxation time using this fictive strain modeling approach is not a constant, nor are there two or more fixed relaxation times. Rather, τ increases with time as the particles become progressively networked/jammed, and the fictive strain is a metric for the evolving filler network structure. Therefore, our methodology describes particle networking/flocculation as a self-limiting process wherein the rate of networking slows down as the filler network builds up. The relaxation time functions from the fitting are compared in Figure 9, where it is evident that untreated silica has more self-limiting filler networking behavior than carbon black in these uncross-linked EPDM nanocomposites. Although the modeling approach does not explicitly include parameters related to filler–filler and polymer–filler interactions, the fitting results show clear differences between the nature of carbon black relative to the significantly more polar silica particles. For both compounds, $s \gg a$, so the relaxation time depends more on the fictive strain than the actual measurement strain.

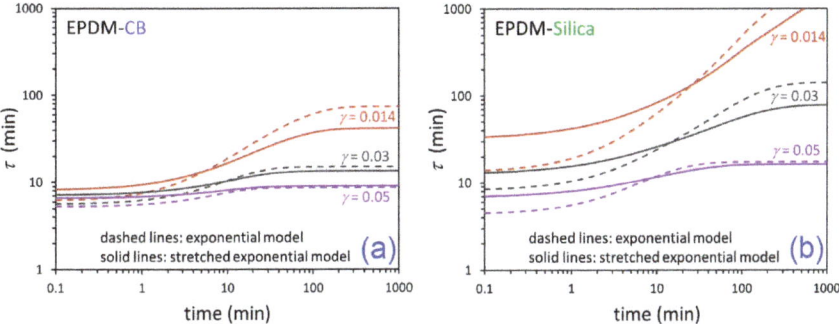

**Figure 9.** Time-dependent τ from fictive strain model fits for down-jump experiments from equilibrium at γ = 0.25 to the indicated γ values for EPDM-CB (**a**) and EPDM-Silica (**b**). The dashed lines are from the exponential model and the solid lines are from the stretched exponential model.

The parameters quantified from fitting the experimental data can be used to predict the filler flocculation behavior at γ and time conditions that are outside the ranges probed in the dynamic rheological measurements. The results are shown in Figure 10. This exercise showed that filler networking at strain amplitudes below 0.006 (0.6%) is predicted to not reach equilibrium for both EPDM-CB and EPDM-Silica until sometime after a flocculation time of 100,000 min (~70 days), which was the longest time considered in the model predictions. Extending the model to these long-time conditions gives such insights that are not otherwise experimentally possible. It would not be realistic to perform rheological experiments for months, and even mostly saturated EPDM would undergo branching/cross-linking during annealing at 160 °C for this duration.

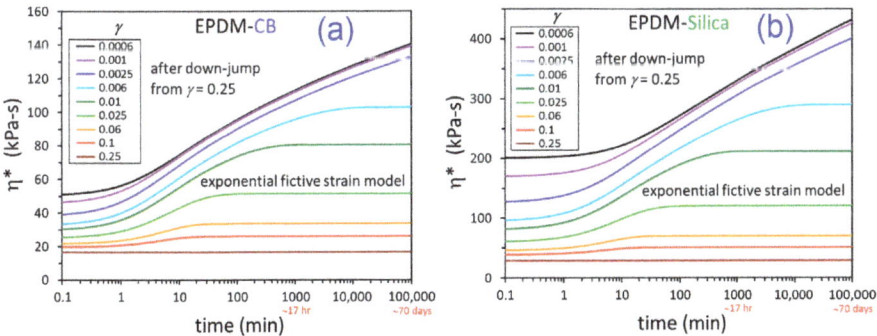

**Figure 10.** Predicted η* vs. time at the indicated strain amplitudes for EPDM-CB (**a**) and EPDM-Silica (**b**) using parameters from fitting the experimental data (Figures 8 and 9) to the exponential fictive strain model. Note the different y-axis scalings for (**a**,**b**).

The departure from equilibrium as γ is reduced can be viewed by replotting the results in Figure 10 as a function of strain amplitude for various flocculation times in Figure 11. This bears a striking resemblance to the behavior of amorphous polymers and small-molecule glass-formers as they are cooled into the glassy state [53]. Zhao, Simon, and McKenna [54] studied a fossilized amber resin glass and they demonstrated that the glass did not reach the extrapolated equilibrium liquid behavior, even after 20 million years. In Figure 12, we compare our γ-dependent predictions for EPDM-CB with their T-dependent measurements for the amber resin to further reinforce the known similarities between jammed particle systems and glass-forming materials. A critical jamming transition strain amplitude, $\gamma_j$, in particle-filled elastomers is analogous to the glass transition temperature, $T_g$.

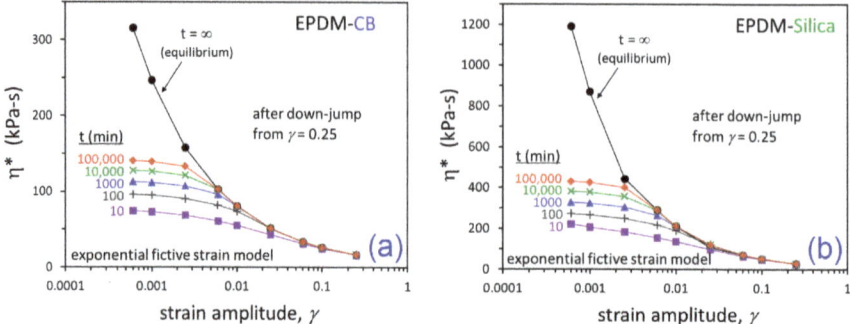

**Figure 11.** Predicted η* vs. γ at the indicated times for EPDM-CB (**a**) and EPDM-Silica (**b**) using parameters from fitting the experimental data (Figures 8 and 9). Note the different y-axis scalings for (**a**,**b**).

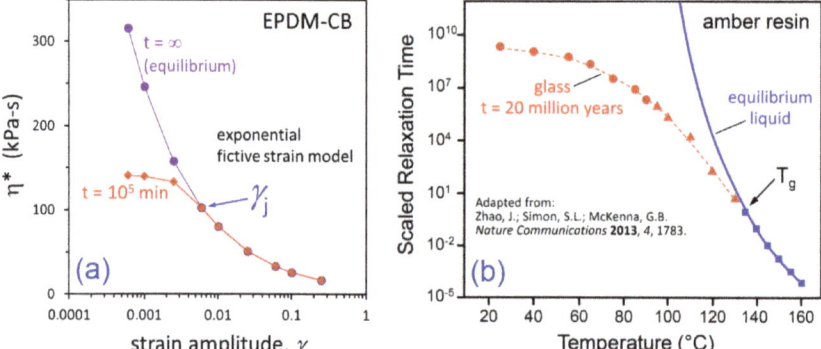

**Figure 12.** Comparison of departure from equilibrium as strain amplitude is decreased for EPDM-CB (**a**) with departure from equilibrium liquid behavior as temperature is decreased for 20-million-year-old amber resin glass from Zhao, Simon, and McKenna [54] (**b**).

## 4. Conclusions

The fictive strain model was able to capture the time-dependent dynamic viscosity responses for EPDM-CB and EPDM-Silica systems at three levels of strain down-jump using a single set of fitting parameters for each material. Fitting was improved by using a stretched exponential model versus an exponential model, at the expense of introducing an additional fitting parameter (4 versus 3). The utility of this glass-like phenomenological treatment of the time-dependent filler networking/flocculation process in particle-filled elastomers is that it provides new insights into the relative networking characteristics of different polymer nanocomposite systems. Compared to carbon black in filled EPDM compounds, the untreated precipitated silica exhibited more extensive filler networking/flocculation, and the filler network build-up displayed more self-limiting behavior. For both filler types, the relaxation time function was more dependent on structural (fictive) strain than actual strain amplitude, as determined from fitting parameter results indicating $s \gg a$.

For dynamic strains less than 0.006 (0.6%), the filler network is predicted to not reach equilibrium, even after 70 days at 160 °C for both filled EPDM materials. There is a tendency to think about the Payne effect in terms of dynamic mechanical properties changing as strain amplitude is increased. However, based on our fictive strain model predictions in Figure 11, we propose that the typical reverse-sigmoidal shape of the Payne effect—as observed for G' or η* when plotted versus log(γ)—is a consequence of the transition from equilibrium state to jammed nonequilibrium filler network as strain amplitude is decreased, similar to the departure of glass-forming liquids into the glassy state as temperature is reduced. A critical strain amplitude for jamming, $\gamma_j$, in particle-filled elastomers (polymer nanocomposites) is thus analogous to the glass transition temperature for glass-formers.

The EPDM compounds with silica or carbon black filler were stable during the 155-min-long dynamic rheology experiments at 160 °C, but irreversible dynamic storage modulus increases were noted when the polymer matrix was SBR, presumably due to branching/cross-linking of the unsaturated SBR in the rheometer. We recommend verifying the stability of a material when studying filler flocculation in order to have meaningful interpretation of any model parameters derived from fitting the time-dependent physical phenomenon without any contribution from underlying chemical changes in the polymer–particle system.

**Author Contributions:** Conceptualization, C.G.R.; methodology, C.G.R., S.R.V. and G.H.; formal analysis, C.G.R.; investigation, C.G.R. and S.R.V.; resources, G.H.; data curation, C.G.R and S.R.V.; writing—original draft preparation, C.G.R.; writing—review and editing, C.G.R., S.R.V., and G.H.; project administration, G.H. All authors have read and agreed to the published version of the manuscript.

**Funding:** This research received no external funding.

**Acknowledgments:** C.G.R. is grateful to Amit Das, Klaus Werner Stöckelhuber, René Jurk, and Aladdin Sallat in the Elastomer Group at IPF–Dresden for technical discussions, equipment assistance, and being great hosts during a visiting scientist experience in Germany during January 2016 when the experiments were performed for this paper.

**Conflicts of Interest:** The authors declare no conflicts of interest.

## References

1. Futamura, S. Effect of Material Properties on Tire Performance Characteristics—Part II, Tread Material. *Tire Sci. Technol.* **1990**, *18*, 2–12. [CrossRef]
2. Warasitthinon, N.; Robertson, C.G. Interpretation of the tanδ Peak Height for Particle-Filled Rubber and Polymer Nanocomposites with Relevance to Tire Tread Performance Balance. *Rubber Chem. Technol.* **2018**, *91*, 577–594. [CrossRef]
3. Ferry, J.D. *Viscoelastic Properties of Polymers*, 3rd ed.; John Wiley & Sons, Inc.: New York, NY, USA, 1980.
4. Roland, C.M. *Viscoelastic Behavior of Rubbery Materials*; Oxford University Press, Inc.: New York, NY, USA, 2011.
5. Shaw, M.T. *Introduction to Polymer Rheology*; John Wiley & Sons, Inc.: Hoboken, NJ, USA, 2012.
6. Osswald, T.A.; Rudolph, N. *Polymer Rheology: Fundamentals and Applications*; Carl Hanser Verlag: München, Germany, 2015.
7. Dillon, J.H.; Prettyman, I.B.; Hall, G.L. Hysteretic and Elastic Properties of Rubberlike Materials Under Dynamic Shear Stresses. *J. Appl. Phys.* **1944**, *15*, 309–323. [CrossRef]
8. Payne, A.R. The dynamic properties of carbon black-loaded natural rubber vulcanizates. Part I. *J. Appl. Polym. Sci.* **1962**, *6*, 57–63. [CrossRef]
9. Payne, A.R. Strainwork dependence of filler-loaded vulcanizates. *J. Appl. Polym. Sci.* **1964**, *8*, 2661–2686. [CrossRef]
10. Fletcher, W.P.; Gent, A.N. Nonlinearity in the Dynamic Properties of Vulcanized Rubber Compounds. *Trans. Inst. Rub. Ind.* **1953**, *29*, 266–280. [CrossRef]
11. Heinrich, G.; Klüppel, M. Recent advances in the theory of filler networking in elastomers. *Adv. Polm. Sci.* **2002**, *160*, 1–44.
12. Robertson, C.G. Dynamic Mechanical Properties. In *Encyclopedia of Polymeric Nanomaterials*; Kobayashi, S., Müllen, K., Eds.; Springer: Berlin/Heidelberg, Germany, 2015; pp. 647–654.
13. World Petrol Demand 'Likely to Peak by 2030 as Electric Car Sales Rise. Available online: https://www.theguardian.com/business/2017/oct/16/world-petrol-demand-peak-electric-car-wood-mackenzie-oil (accessed on 9 January 2020).
14. Smith, S.M.; Simmons, D.S. Poisson ratio mismatch drives low-strain reinforcement in elastomeric nanocomposites. *Soft Matter* **2019**, *15*, 656–670. [CrossRef]

15. Warasitthinon, N.; Genix, A.-C.; Sztucki, M.; Oberdisse, J.; Robertson, C.G. The Payne effect: Primarily polymer-related or filler-related phenomenon? *Rubber Chem. Technol.* **2019**, *92*, 599–611. [CrossRef]
16. Kraus, G. Mechanical Losses in Carbon-Black-Filled Rubbers. *Appl. Polym. Symp.* **1984**, *39*, 75–92.
17. Hentschke, R. The Payne effect revisited. *Express Polym. Lett.* **2017**, *11*, 278–292. [CrossRef]
18. Mujtaba, A.; Keller, M.; Ilisch, S.; Radusch, H.J.; Beiner, M.; Thurn-Albrecht, T.; Saalwächter, K. Detection of Surface-Immobilized Components and Their Role in Viscoelastic Reinforcement of Rubber–Silica Nanocomposites. *ACS Macro Lett.* **2014**, *3*, 481–485. [CrossRef]
19. Berriot, J.; Lequeux, F.; Monnerie, L.; Montes, H.; Long, D.; Sotta, P. Filler-Elastomer Interaction in Model Filled Rubbers, a H-1 NMR Study. *J. Non-Cryst. Solids* **2002**, *307*, 719–724. [CrossRef]
20. Xiong, W.; Wang, X. Linear-nonlinear dichotomy of rheological responses in particle-filled polymer melts. *J. Rheol.* **2018**, *62*, 171–181. [CrossRef]
21. Böhm, G.G.A.; Nguyen, M.N. Flocculation of carbon black in filled rubber compounds. I. Flocculation occurring in unvulcanized compounds during annealing at elevated temperatures. *J. Appl. Polym. Sci.* **1995**, *55*, 1041–1050. [CrossRef]
22. Lin, C.J.; Hergenrother, W.L.; Alexanian, E.; Böhm, G.G.A. On the Filler Flocculation in Silica-Filled Rubbers Part I. Quantifying and Tracking the Filler Flocculation and Polymer-Filler Interactions in the Unvulcanized Rubber Compounds. *Rubber Chem. Technol.* **2002**, *75*, 865–890. [CrossRef]
23. Schwartz, G.A.; Cerveny, S.; Marzocca, A.J.; Gerspacher, M.; Nikiel, L. Thermal aging of carbon black filled rubber compounds. I. Experimental evidence for bridging flocculation. *Polymer* **2003**, *44*, 7229–7240. [CrossRef]
24. Scurati, A.; Lin, C.J. The hysteresis temperature and strain dependences in filled rubbers. *Rubber Chem. Technol.* **2006**, *79*, 170–197. [CrossRef]
25. Meier, J.G.; Klüppel, M. Carbon black networking in elastomers monitored by dynamic mechanical and dielectric spectroscopy. *Macromol. Mater. Eng.* **2008**, *293*, 12–38. [CrossRef]
26. Mihara, S.; Datta, R.N.; Noordermeer, J.W.M. Flocculation in silica reinforced rubber compounds. *Rubber Chem. Technol.* **2009**, *82*, 524–540. [CrossRef]
27. Böhm, G.A.; Tomaszewski, W.; Cole, W.; Hogan, T. Furthering the understanding of the non linear response of filler reinforced elastomers. *Polymer* **2010**, *51*, 2057–2068. [CrossRef]
28. Richter, S.; Saphiannikova, M.; Stöckelhuber, K.W.; Heinrich, G. Jamming in filled polymer systems. *Macromol. Symp.* **2010**, *291–292*, 193–201. [CrossRef]
29. Robertson, C.G.; Lin, C.J.; Bogoslovov, R.B.; Rackaitis, M.; Sadhukhan, P.; Quinn, J.D.; Roland, C.M. Flocculation, reinforcement, and glass transition effects in silica-filled styrene-butadiene rubber. *Rubber Chem. Technol.* **2011**, *84*, 507–519. [CrossRef]
30. Randall, A.M.; Robertson, C.G. Linear-Nonlinear Dichotomy of the Rheological Response of Particle-Filled Polymers. *J. Appl. Polym. Sci.* **2014**, *131*, 40818. [CrossRef]
31. Tunnicliffe, L.B.; Kadlcak, J.; Morris, M.D.; Shi, Y.; Thomas, A.G.; Busfield, J.J.C. Flocculation and Viscoelastic Behaviour in Carbon Black-Filled Natural Rubber. *Macromol. Mater. Eng.* **2014**, *299*, 1474–1483. [CrossRef]
32. Robertson, C.G. Flocculation in elastomeric polymers containing nanoparticles: Jamming and the new concept of fictive dynamic strain. *Rubber Chem. Technol.* **2015**, *88*, 463–474. [CrossRef]
33. Stöckelhuber, K.W.; Wießner, S.; Das, A.; Heinrich, G. Filler flocculation in polymers–a simplified model derived from thermodynamics and game theory. *Soft Matter* **2017**, *13*, 3701–3709. [CrossRef]
34. Gundlach, N.; Hentschke, R.; Karimi-Varzaneh, H.A. Filler flocculation in elastomer blends-an approach based on measured surface tensions and monte carlo simulation. *Soft Mater.* **2019**, *17*, 283–296. [CrossRef]
35. Gundlach, N.; Hentschke, R. Modelling Filler Dispersion in Elastomers: Relating Filler Morphology to Interface Free Energies via SAXS and TEM Simulation Studies. *Polymers* **2018**, *10*, 446. [CrossRef]
36. Hayichelaeh, C.; Reuvekamp, L.A.E.M.; Dierkes, W.K.; Blume, A.; Noordermeer, J.W.M.; Sahakaro, K. Enhancing the Silanization Reaction of the Silica-Silane System by Different Amines in Model and Practical Silica-Filled Natural Rubber Compounds. *Polymers* **2018**, *10*, 584. [CrossRef]
37. Liu, A.J.; Nagel, S.R. Nonlinear dynamics: Jamming is not just cool any more. *Nature* **1998**, *396*, 21–22. [CrossRef]
38. Trappe, V.; Prasad, V.; Cipelletti, L.; Segre, P.N.; Weitz, D.A. Jamming phase diagram for attractive particles. *Nature* **2001**, *411*, 772–775. [CrossRef] [PubMed]

39. Wang, X.; Robertson, C.G. Strain-induced nonlinearity of filled rubbers. *Phys. Rev. E* **2005**, *72*. [CrossRef] [PubMed]
40. Robertson, C.G.; Wang, X. Isoenergetic jamming transition in particle-filled systems. *Phys. Rev. Lett.* **2005**, *95*. [CrossRef] [PubMed]
41. Stöckelhuber, K.W.; Svistkov, A.S.; Pelevin, A.G.; Heinrich, G. Impact of Filler Surface Modification on Large Scale Mechanics of Styrene Butadiene/Silica Rubber Composites. *Macromolecules* **2011**, *44*, 4366–4381. [CrossRef]
42. Stöckelhuber, K.W.; Das, A.; Jurk, R.; Heinrich, G. Contribution of physico-chemical properties of interfaces on dispersibility, adhesion and flocculation of filler particles in rubber. *Polymer* **2010**, *51*, 1954–1963. [CrossRef]
43. Frenkin, E.I.; Podolsky, Y.Y.; Vinogradov, G.V. Thermal Vulcanization of Rubbers at High Pressure. *Rubber Chem. Technol.* **1985**, *58*, 23–36. [CrossRef]
44. Bellander, M.; Stenberg, B.; Perrson, S. Crosslinking of polybutadiene rubber without any vulcanization agent. *Polym. Eng. Sci.* **1998**, *38*, 1254–1260. [CrossRef]
45. Robertson, C.G.; Wang, X. Nanoscale cooperative length of local segmental motion in polybutadiene. *Macromolecules* **2004**, *37*, 4266–4270. [CrossRef]
46. Robertson, C.G.; Wang, X. Spectral hole burning to probe the nature of unjamming (Payne effect) in particle-filled elastomers. *Europhys. Lett.* **2006**, *76*, 278–284. [CrossRef]
47. Tool, A.Q. Relation between inelastic deformability and thermal expansion of glass in its annealing range. *J. Am. Ceram. Soc.* **1946**, *29*, 240–253. [CrossRef]
48. Narayanaswamy, O.S. A Model of Structural Relaxation in Glass. *J. Am. Ceram. Soc.* **1971**, *54*, 491–498. [CrossRef]
49. McKenna, G.B. *Comprehensive Polymer Science, Polymer Properties*; Booth, C., Price, C., Eds.; Pergamon: Oxford, UK, 1989; Volume 2, pp. 311–362.
50. Simon, S.L.; Sobieski, J.W.; Plazek, D.J. Volume and enthalpy recovery of polystyrene. *Polymer* **2001**, *42*, 2555–2567. [CrossRef]
51. Moynihan, C.T.; Easteal, A.J.; DeBolt, M.A.; Tucker, J. Dependence of the Fictive Temperature of Glass on Cooling Rate. *J. Am. Ceram. Soc.* **1976**, *59*, 12–16. [CrossRef]
52. Hodge, I.M. Adam-Gibbs Formulation of Enthalpy Relaxation Near the Glass Transition. *J. Res. Natl. Inst. Stand. Technol.* **1997**, *102*, 195–205. [CrossRef]
53. Kobayashi, H.; Takahashi, H.; Hiki, Y. Viscosity measurement of organic glasses below and above glass transition temperature. *J. Non-Cryst. Solids* **2001**, *290*, 32–40. [CrossRef]
54. Zhao, J.; Simon, S.L.; McKenna, G.B. Using 20-million-year-old amber to test the super-Arrhenius behaviour of glass-forming systems. *Nat. Commun.* **2013**, *4*, 1783. [CrossRef]

© 2020 by the authors. Licensee MDPI, Basel, Switzerland. This article is an open access article distributed under the terms and conditions of the Creative Commons Attribution (CC BY) license (http://creativecommons.org/licenses/by/4.0/).

MDPI
St. Alban-Anlage 66
4052 Basel
Switzerland
Tel. +41 61 683 77 34
Fax +41 61 302 89 18
www.mdpi.com

*Polymers* Editorial Office
E-mail: polymers@mdpi.com
www.mdpi.com/journal/polymers

www.ingramcontent.com/pod-product-compliance
Lightning Source LLC
LaVergne TN
LVHW070359100526
838202LV00014B/1351